Robot Systems for Rail Transit Applications

Robot Systems for Rail Transit Applications

Hui Liu
School of Traffic and
Transportation Engineering,
Central South University,
Changsha, China

ELSEVIER

Elsevier
Radarweg 29, PO Box 211, 1000 AE Amsterdam, Netherlands
The Boulevard, Langford Lane, Kidlington, Oxford OX5 1GB, United Kingdom
50 Hampshire Street, 5th Floor, Cambridge, MA 02139, United States

Notices

Library of Congress Cataloging-in-Publication Data
A catalog record for this book is available from the Library of Congress

British Library Cataloguing-in-Publication Data
A catalogue record for this book is available from the British Library

ISBN: 978-0-12-822968-2

For information on all Elsevier publications visit our website at
https://www.elsevier.com/books-and-journals

Publisher: Matthew Deans
Acquisitions Editor: Glyn Jones
Editorial Project Manager: Naomi Robertson
Production Project Manager: Nirmala Arumugam
Cover Designer: Matthew Limbert

Typeset by TNQ Technologies

Contents

List of figures and tables

Preface

Rail transit is the lifeblood of many national economies and the backbone of transportation. Safe and efficient rail transit is based on highly reliable manufacturing and high-quality maintenance. Rail transit robots can replace manual repetitive tasks and improve the automation level of the rail transit system. As a result, work efficiency can be improved, accidental failures due to human negligence can be avoided, and the safety of the rail transit system can be improved.

Rail transit robots involve the intersection of automatic control, artificial intelligence, signal processing, pattern recognition, mechanical engineering, and transportation engineering. When applied to the rail transit system, the robot is faced with the key problems to be solved urgently. Therefore, rail transit robots are currently recognized as research hotspots among scientific problems. Based on research from the past 10 years, the author puts forward a framework of rail transit robot technology and completes the related work.

This book covers seven mainstream rail transit robots, including assembly robots, collaborative robots, automated guided vehicles, autonomous rail rapid transit, inspection robots, channel robots, and inspection unmanned aerial vehicles. The key problems of robots are described in detail, including positioning navigation, path planning, human–robot interaction, and power management, etc. For students and managers in related departments, this book can provide valuable information about rail transit robots. For researchers and doctoral students, this book can provide some ideas and encourage future research in rail transit robots.

This book contains eight chapters:

Chapter 1: Introduction

This chapter first outlines the rail transit robot. Then the chapter describes three basic issues of robotics, including navigation, human–robot interaction, and power control.

Chapter 2: Rail transit assembly robot systems

This chapter first introduces the development progress and key technologies of assembly robots, and then introduces the main components of assembly robots. After that, the dynamics models for the assembly robot systems are explained. Finally, the artificial neural network algorithm for the inverse dynamic optimization calculation of the robot arm is introduced.

Chapter 3: Rail transit collaborative robot systems

This chapter first gives the basic definition of a collaborative robot. Then the chapter summarizes the development history, application field and collaborative mode of collaborative robots. The components of the collaborative robot are summarized. Finally, the basic concepts of visual perception in human-robot collaboration are introduced.

Chapter 4: Automatic Guided Vehicles (AGVs) in the rail transit intelligent manufacturing environment

This chapter first introduces the development progress and types of the AGV in rail transit intelligent manufacturing environment. Then the main components of AGVs are introduced. After that, the key technologies and the applications of the AGV are introduced.

Chapter 5: Autonomous rail Rapid Transit (ART)

This chapter firstly introduces the hardware of ART. Then, the technologies of ART are introduced. Finally, the pedestrian detection algorithms of ART are introduced in detail.

Chapter 6: Rail transit inspection robots

This chapter first introduces the development history, function and main components of the rail transit inspection robot. Then, two key technologies to ensure that the inspection robots normally complete the inspection work, positioning methods, path planning methods, and hand—eye vision system are introduced in detail.

Chapter 7: Rail transit channel robot systems

This chapter firstly introduces the development history and main components of the rail transit channel robot, including the ground rail, dual-arm robot, infrared thermometer, laser sensor, etc. Then the TEDS intelligent sensing system is described in detail. Finally, fault diagnosis algorithms based on deep learning models are introduced.

Chapter 8: Rail transit inspection Unmanned Aerial Vehicle (UAV)

This chapter first introduces the development history of the UAV and its applications in various fields. Secondly, it introduces the basic structure of fixed-wing UAVs, unmanned helicopters and rotary-wing UAVs. Various sensors are applied to rail transit inspection. Then the UAV technologies are described in detail. Finally, the applications of the UAV in the detection of rail transit intruding detection are introduced.

Prof. Dr.-Ing. habil. Hui Liu

Changsha, China

November 2019

Acknowledgement

The studies in the book are supported by the National Natural Science Foundation of China, the National Key R&D Program of China, and the Innovation Drive of Central South University, China. The publication of the book is funded by the High-level Postgraduate Text Book Project of the Hunan Province of China. In the process of writing the book, Mr. Zhu Duan, Mr. Jiahao Huang, Mr. Kairong Jin, Mr. Yu Xia, Mr. Rui Yang, Ms. Shi Yin, Mr. Ye Li, Mr. Guangji Zheng, Ms. Jing Tan, Mr. Huipeng Shi, Mr. Haiping Wu, Mr. Chao Chen, Mr. Zhihao Long, and other team members have done a lot of model verification and other work. These team members as mentioned have the same contribution to this book.

Nomenclature list

#

2D	Two-Dimensional
3C	Computer, Communication, Consumer Electronic
3D	Three-Dimensional

A

ABB	Asea Brown Boveri
AC	Alternating Current
ACMS	Aircraft Condition Monitoring System
ACO	Ant Colony Optimization
ADC	Analog-to-Digital Converter
ADU	Automatic Drilling Unit
AGV	Automatic Guided Vehicle
AGVS	Automated Guided Vehicle System
AMR	Anisotropic Magnetoresistive
ANN	Artificial Neural Network
ANIFS	Adaptive Network-Based Fuzzy Inference System
AR	Augmented Reality
ARM	Advanced RISC Machine
AP	Access Point
APF	Artificial Potential Field
API	Application Programming Interface
ARMA	Autoregressive Moving Average
ART	Autonomous rail Rapid Transit
ASK	Amplitude Shift Keying
ATC	Automatic Train Control
ATO	Automatic Train Operation
ATP	Automatic Train Protection
ATS	Automatic Train Supervision
AUC	Area Under Curve

B

BFS	Best-First Search
BP	Back Propagation
BPNN	Back Propagation Neural Network
BRIEF	Binary Robust Independent Elementary Features

C

CAD	Computer-Aided Design
CCD	Charge Coupled Device
CCOT	Continuous Convolution Operators Tracker
CIMS	Computer Integrated Manufacturing System
CNC	Computerized Numerical Control
CNN	Convolutional Neural Network
CNR	China Northern Locomotive Rolling Stock Industry Group
CPU	Central Processing Unit
CRF	Conditional Random Fields
CSM	Correlation Scan Match
CSR	China Southern Locomotive Rolling Stock Industry Group
CW	Continuous Wave
CWT	Continue Wavelet Transform

D

DC	Direct Current
D-DCOP	Dynamic Distributed Constraint Optimization Problem
Dec-MDP	Decentralized Markov Decision Process
DEM	Digital Elevation Model
DFT	Discrete Fourier Transform
D-H	Denavit-Hartenberg
DLT	Deep Learning Tracker
DMPC	Distributed Model Predictive Control
DNFO	Dynamic Network Flow Optimization
DNN	Deep Neural Network
DOF	Degree of Freedom
DSP	Digital Signal Processor
DTW	Dynamic Time Warping

E

ECO	Efficient Convolution Operator
EKF	Extended Kalman Filter
EKF-SLAM	Extended Kalman Filter SLAM
ELM	Extreme Learning Machine
EMD	Empirical Mode Decomposition
EMU	Electric Multiple Units

F

FAS	Flexible Assembly System
FAST	Features from Accelerated Segment Test
Faster RCNN	Faster Regional Convolutional Neural Network
FDD	Frequency Division Duplexing
FFT	Fast Fourier Transform
FMS	Flexible Manufacturing System
FSK	Frequency Shift Keying
FTP	File Transfer Protocol

G

GA	Genetic Algorithm
GOA	Grade of Automation
GPRS	General Packet Radio Service
GPS	Global Positioning System

H

HDT	Hedged Deep Tracking
HDFS	Hadoop Distributed File System
HMM	Hidden Markov Model
HOG	Histogram of Oriented Gradient
HRI	Human-Robot Interaction
HSB	Hue-Saturation-Brightness

I

ID	Identity Document
IDIM-LS	Inverse Dynamic Identification Model and Linear Least Squares Technique
IFF	Identification Friend or Foe
IFR	International Federation of Robots
IGBT	Insulated Gate Bipolar Translator
IL	Imitation Learning
IMU	Inertial Measurement Unit
IMW	Intelligent Manufacturing Workshop
IMF	Intrinsic Mode Function
IoU	Intersection over Union
INS	Inertial Navigation System
ISM	Industrial Scientific Medical

J

JFET	Junction Field-Effect Transistor

K

KCF	Kernelized Correlation Filters
KF	Kalman Filter
KNN	K-Nearest Neighbors

L

LAN	Local Area Network
LBP	Local Binary Pattern
LCD	Liquid Crystal Display
LiDAR	Light Detection and Ranging
LRR	Long-Range Radar

M

MAE	Mean Absolute Error
MANET	Mobile Ad Hoc Network
mAP	mean Average Precision
MBTA	Massachusetts Bay Transit Authority
MDPs	Markov Decision Processes
MEEM	Multiple Experts using Entropy Minimization
MEMS	Micro-Electro-Mechanical Systems
MF	Morphological Filter
MIL	Multiple Instance Learning
MILP	Mixed Integer Linear Programming
MLP	Multilayer Perceptron
MSE	Mean Square Error
MTSP	Multiple Traveling Salesman Problem
MVB	Multifunction Vehicle Bus

N

NFS	Network File Systems
NMS	Nonmaximun Suppression
NOMA	Nonorthogonal Multiple Access
NP	Nondeterministic Polynomial

O

OFDM	Orthogonal Frequency Division Multiplexing
OGA−PSO	Optimum Genetic Algorithm−Particle Swarm Optimization algorithm
ORB	Oriented FAST and Rotated BRIEF

P

PG	Policy Gradient
PLC	Programmable Logic Controller
POS	Point of Sale
PRM	Probabilistic Road Map
PSK	Phase Shift Keying
PSO	Particle Swarm Optimization
PTZ	Pan-Tilt-Zoom
PUMA	Programmable Universal Machine for Assembly

Q

QPSO	Quantum Particle Swarm Optimization
QR	Quick Response

R

RANSAC	Random Sample Consensus
RBF	Radial Basis Function
RBPF	Rao-Blackwellized Particle Filter
RCC	Remote Center Compliance
RCNN	Regional Convolutional Neural Network
R–CNN	Regions with CNN features
RDD	Resilient Distributed Dataset
R–FCN	Region-based Fully Convolutional Networks
RFID	Radio Frequency Identification
RGB	Red-Green-Blue
RGB-D	Red-Green-Blue-Deep
RL	Reinforcement Learning
RNN	Recurrent Neural Network
ROC	Receiver Operating Characteristic curve
ROI	Region of Interest
ROS	Robot Operating System
RPN	Region Proposal Networks
RRT	Rapid-Exploration Random Tree
RSSI	Received Signal Strength Indication
RTP	Real-Time Protocols

S

SARSA	State Action Reward State Action
SCARA	Selective Compliant Assembly Robot Arm
SDA	Stacked Denoising Autoencoder
SEA	Series Elastic Actuator
SfM	Structure from Motion
SIA	Swarm Intelligence Algorithm
SIFT	Scale Invariant Feature Transform
SLAM	Simultaneous Localization and Mapping
SMT	Surface Mount Technology
SPP	Spatial Pyramid Pooling
SRDCF	Spatially Regularized Discriminative Correlation Filters
SRR	Short-Range Radar
SSD	Single Shot multibox Detector
SURF	Speeded Up Robust Features
SVM	Support Vector Machine
SMT	Surface Mount Technology

T

TCN	Train Communication Network
TCP	Transmission Control Protocol
TCSN	Train Control and Service Network
TDD	Time Division Duplexing
TEDS	Trouble of moving EMU Detection System
TOF	Time of Flight

U

UAV	Unmanned Aerial Vehicle
UDP	User Datagram Protocol
UHV	Ultrahigh Voltage
UNECE	United Nations Economic Commission for Europe
USB	Universal Serial Bus
UWB	Ultrawideband

V

VR	Virtual Reality
VRP	Vehicle Routing Problem
VSA	Variable Stiffness Actuator

W

WiFi	Wireless Fidelity
WPT	Wavelet Packet Transform
WTB	Wire Train Bus

Y

YOLO	You Only Look Once

CHAPTER 1

Introduction

1.1 Overview of rail transit robots

Railway transit is vital to the national economy. Research in Japan and China indicates that development of the railway can lead to economic growth in the railway field [1,2]. To stimulate economic development, governments are actively developing the railway transport industry. In such a large industrial system, the development of automation can improve efficiency and reduce costs, so the level of automation in railways should be increased. The increase in automation requirements in the rail transit system generates a pursuit for rail transit robot systems.

In worldwide, many countries have proposed similar strategic plans to encourage the development of robots in the railway transit system. Taking China as an example, "A Country With a Strong Transportation Network" and "Smart Railway" are two important plans that encourage the development of automation of the rail transit system. Driven by these plans, many Chinese rail equipment manufacturers and operators have carried out extensive research in robotics. Zhuzhou CRRC Times Electric Co., Ltd. applied an automatic generating line to produce high-speed train converters [3]. CRRC Qishuyan Institute Co., Ltd. developed an intelligent manufacturing workshop for gear transmission systems for high-speed trains. CRRC Zhuzhou Institute Co., Ltd. combined automatic guided vehicle (AGV) technology with urban transportation equipment and designed autonomous rail rapid transit (ART) [4].

A variety of robot systems are employed. The use of robot systems in the rail transit system can be divided into three aspects: manufacturing, dispatch, and maintenance. In these three parts of rail transit, different kinds of robots have completely different roles, as shown in Fig. 1.1.

1.1.1 Rail transit robots in manufacturing

"A Country With a Strong Transportation Network" points out that intelligent manufacturing is required for the rail transit system. In the modern production line, robots can greatly improve processing efficiency. Taking the production of a high-speed train gearbox as an

Robot Systems for Rail Transit Applications. https://doi.org/10.1016/B978-0-12-822968-2.00001-2

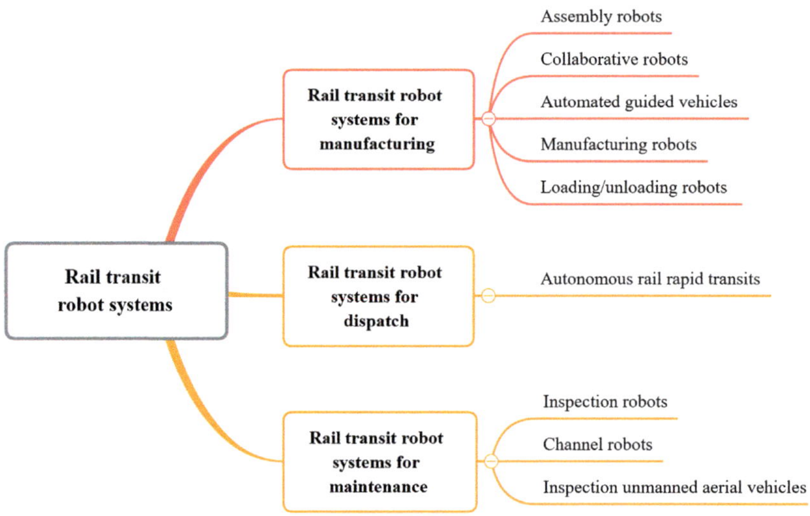

Figure 1.1
Robots in the manufacturing, dispatch, and maintenance of rail transit.

example, in the process flow of a transmission gear, robot arms can result in the efficient transmission of gears between different machine tools; when welding the gearbox, the welding robots can improve machining efficiency; when assembling the gearbox, assembly robots can work with high-precision; when the train is assembled, AGV enables the gearbox to be transported quickly between workshops. Applying robots in manufacturing not only saves processing time and labor costs, it improves manufacturing quality.

1.1.1.1 Assembly robots

As for assembly robots, the primary mission is to achieve high-precision positioning of the workpiece. According to previous research, assembly costs account for 50% of total manufacturing costs [5]. Assembly robot systems can also be divided into rigid assembly and flexible assembly robots. Rigid assembly robots are customized processing systems for specific workpieces in the traditional industrial environment.

Rigid assembly robots have poor generalization. If the production line is replaced with processed parts, the equipment needs to be customized. Replacement of equipment will cause a great economic burden. Compared with rigid assembly robots, flexible assembly robots can design customized processing programs according to the workpiece. Flexible assembly robots are programmable, which can result in different assembly schemes for different workpieces. Flexible assembly robots are significant for a flexible assembly system. In current industrial development, flexible assembly robots are the focus of development [6]. In the following discussion, assembly robots refer to flexible assembly robots.

An assembly robot consists of four components: machinery components, sensors, controllers, and actuators. To bring about a complex workpiece track in the real assembly environment, assembly robots usually have more than four degrees of freedom (DOFs). Mainstream assembly robots can be divided into two types: selective compliant assembly robot arms (SCARAs) and six-DOF robots.

SCARAs have four DOFs, which are commonly used in electronic assembly, screw assembly, and so on [7]. SCARAs are specially designed for assembly applications by Yamanashi University. SCARAs contain two parallel joints, which can assemble a workpiece in a specified plane. Compared with six-DOF robots, advantages of SCARAs are a higher assembly speed and precision; disadvantages are limited workspace. Commonly used control strategies for SCARAs contain adaptive control, force control, robust control, and so forth [8]. In state-of-the-art research on robot control, intelligent algorithms are employed to improve control performance [9]. Dulger et al. applied a neural network to control the SCARA [10]. The neural network was optimized by particle swarm optimization to improve performance. Son et al. adopted an optimized inverse neural network for feedback control [11]. To deal with disturbances in running, the parameters of the inverse neural network are updated by a back propagation algorithm. Luan et al. used the radial basis function (RBF) neural network to achieve dynamic control of the SCARA [12].

Six-DOF robots can locate the workpiece at almost any point. Thus, six-DOF robots can handle the assembly task of complex three-dimensional (3D) workpieces. The dynamics of six-DOF robots are basic for operating the robots. Zhang et al. considered the friction of the robots and used a hybrid optimization method to model the dynamics of the six-DOF robot [13]. After optimization, dynamic accuracy increased significantly. Yang et al. proposed a simulator for the dynamics of the six-DOF robot [14]. Robots with a large degree of freedom have large feasibility. However, too much freedom is uneconomical. To handle the trade-off between economy and feasibility, the DOF can be optimized for specific tasks. Yang et al. proposed an optimization method to minimize the DOF [15]. This optimization method can reduce the DOF and improve the use of the DOF.

Assembly robots should cooperate with the ancillary equipment. The fixtures are vital equipment to ensure cooperation in performance. The fixtures can fix the relative position between the workpiece and the robot under load. If the precision of the location of the fixtures is low, no matter how accurate the positioning precision of the robot is, it cannot achieve high-precision assembly. Currently, the flexible fixture is a future development [16]. Lowth et al. proposed a unique fixture that can adjust the radial and angular adaptively [17]. Although auxiliary devices are applied for assembly robots, the results of assembly robots may still be unsuccessful. Avoiding unsuccessful assembly is particularly important in electric connector assembly, because the electric connector is not a rigid component. To detect the unsuccessful assembly of the electric connector assembly, Di et al. proposed a hybrid detection system with a force sensor and camera [18].

The fault diagnosis and prognosis system of assembly robots guarantees assembly accuracy. There are many studies about fault diagnosis and prognosis systems. Huang et al. designed a classifier for the wiring harness robot [19]. That study modeled the manufacturing process was and calculated the fault with a fuzzy model. Baydar et al. introduced a diagnosis model with error prediction [20]. The proposed model integrated the Monte Carlo simulation, genetic algorithm, and so forth. The functions of the fault diagnosis and prognosis system for assembly robots should contain the main aspects as given in Choo et al. [21]:

(a) The health states of the assembly robots are monitored in real time. The monitored data are logged into the dataset. The health features are extracted from the health states of the assembly robots. The faults and remaining useful life can be calculated according to the features.
(b) According to the fault diagnosis and prognosis results, the assembly tasks are reassigned to make sure the failed assembly robots are replaced by the fully functioning robots. The maintenance plans can be made to repair the failed robots.

1.1.1.2 Collaborative robots

When applying the resulting classical industrial robot systems, interaction between robots and humans is limited. There are three reasons for this phenomenon:

(a) Traditional industrial robots do not consider moving humans. If a collision occurs along a certain trajectory, it may cause great damage to humans. Therefore, the working area of the industrial robot is mostly separated from the working area of the human.
(b) The weight and volume of traditional industrial robots are large, and it is difficult for humans to operate robots.
(c) Reprogramming of robots is difficult and requires special programming tools for tuning.

However, human–robot collaboration can combine human creativity with the efficiency of robots and can amplify the flexibility of robots and further improve work efficiency. Under this demand, collaborative robots are born. Compared with traditional robots, collaborative robots have three main advantages: safety, ease of operation, and ease of teaching. Some robot manufacturers have launched collaborative robotics products. The robot company Universal Robot launched the UR3 collaborative robot [22]. This robot is the first truly collaborative robot. The UR3 collaborative robot is based on a six-DOF robotic arm and is flexible enough to achieve complex motion trajectories. In terms of safety, the UR3 collaborative robot has a collision monitoring system that protects human safety by monitoring the joint position, speed, and power of the robot.

KUKA Robotics has launched a collaborative robot, the LBR iiwa [22]. The robot's seven-DOF design provides greater flexibility than traditional six-DOF robots, enabling more complex trajectories to cope with complex environments that work with humans. The shape of the robot is designed to be ergonomic and easy for humans to operate. The outer casing is made of aluminum alloy, which can reduce weight and improve operability. The robot is equipped with torque sensors at each joint to monitor collisions in real time. The teaching method of the robot is dragging, which reduces the technical threshold of the robot operator.

ABB launched the robot YuMi [22]. The robot is highly safe and can achieve human–robot interaction in a small space. To improve the performance of collaborative robots, AAB acquired Gomtec Robotics, which launched the collaborative robot Roberta [22]. The Roberta can handle higher load applications compared with YuMi.

Franka Emika launched the Franka Collaboration [23]. Like the LBR iiwa, the robot has seven DOFs. It is also equipped with a torque sensor on each joint to enable collision monitoring.

Rethink Robotics launched the two-arm collaborative robot Baxter and the one-arm collaborative robot Sawyer [24]. The two robots are exquisitely designed with high positioning accuracy and can be assembled with high precision.

Safety in human–robot interaction is essential for collaborative robots. Threats to the safety of collaborative robots can be divided into two aspects [25]:

(a) The first kind of threat is from robots. During operation, the robot may collide with workers and cause injuries. To ensure the safety of employees, the robot needs to detect the location of workers in real time and determine whether the location of workers is in a safe position. If workers intrude on the safe area, the robot should immediately stop to avoid a collision. In case a collision between a person and a robot occurs, the robot needs to detect the collision in time and change torque to minimize damage to workers. Mohammed et al. proposed a collision avoidance system for collaborative robots [26]. This system used depth vision sensors to detect the position of the worker. Considering the virtual model of the collaborative robot, a collision could be detected. The collaborative robot could take measures to avoid collisions. In addition to collision detection, it is necessary to ensure the integrity of the collaborative robot control system during operation. Failure of any part of the sensors, controllers, or actuators will lead to the failure of human–computer interaction, thus threatening the safety of workers. In addition, interaction with robots may cause mental stress to workers [27], which will increase the risk for a collision.

(b) The second kind of threat is from the industrial process. In the process of human–robot interaction, workers need close contact with the manufacturing process. The temperature of the manufactured workpieces can cause damage to workers. Fallen workpieces can also endanger workers. Therefore, it is necessary to fully consider the impact of machining parts on workers in the design process of collaborative robots. In addition, the unreasonable ergonomic design of the collaborative robot during maintenance will have an impact on safety.

1.1.1.3 Automatic guided vehicles

According to the level of automation, manufacturing systems can be divided into three levels [28]. Manufacturing systems in the first level are manual. Those in the second level have small-scale automated manufacturing in which transportation is carried out manually. Manufacturing systems in the third level have large-scale automated manufacturing and use automated transportation. Manufacturing systems in the third level are also called flexible manufacturing systems (FMSs). According to the Material Handling Industry of America, only 20% of the time is spent on processing and manufacturing; the remaining 80% is used for storage, handling, waiting for processing, and transportation [29]. As factory automation increases, transportation efficiency between workstations needs to improve. In the FMS, the AGV can improve the use of space in the factory and the efficiency of transportation in the material handling system. Therefore, transportation costs can be reduced.

The rail transit manufacturing system is a typical FMS. Transportation is an important part of the rail transit manufacturing system. In the rail transit processing environment, not only the transfer of workpieces between processes within the plant but also the free flow of workpieces between plants is required. In the traditional rail transit manufacturing environment, transportation inside the factory is carried out by gantry cranes and the transportation between factories is carried out by trucks. These modes of transportation have some disadvantages. There is a safety hazard when using gantry cranes to lift. If the lifting workpiece falls, it may cause serious safety accidents. The use of trucks carries a higher cost and is less efficient to transport. Improving the safety and economy of products in the transportation process is important for the development of the rail transit industry. In the current manufacturing environment, the AGV is an effective mode of transportation. The AGV is safer compared with gantry cranes and is more automated and more efficient for transportation than trucks.

The typical structure of an AGV consists of sensors, chassis, a control unit, and so on [30,31]. During work, the sensor can determine the position of the AGV and transmit the current position to the control decision system, and the control decision system plans an optimal path. The chassis is driven by the controller's control command to transport the workpiece to the designated position.

To improve transportation performance, the model predictive control of the AGVs should be achieved. The remaining power and the working state of the AGV should be predicted to calculate the remaining life of the AGV, and the dispatch plan of the AGV can be optimized with consideration of these factors to improve the operational safety of the AGV. Popular data-driven forecasting methods contain statistical methods, intelligent methods, and hybrid methods. The statistical methods can discover the statistical rule of the data and generate an explicit equation for prediction. Commonly used statistical methods contain autoregressive moving average, Winner process, Gaussian process, and so on. Intelligent methods can generate better forecasting performance than statistical methods with the help of the strongly fitting capacity of the neural network. The Elman neural network, multilayer perceptron, and extreme learning machine (ELM) are the three most popular intelligent methods. However, the training process of these neural networks depends on the initial values in some way. If the initial value is unsuitable, training of the neural network may stop at the locally optimal solution. To improve the performance of intelligent prediction methods, the initial values can be optimized by optimization algorithms [32]. Hybrid prediction methods combine data processing algorithms with statistical or intelligent methods. The decomposition algorithms are proved to be effective [33]. The decomposition algorithms can divide the raw series into several more stationary subseries. Each group of subseries has a simpler fluctuation mode than the raw series, so it is more predictable.

After optimization, the control command can be assigned to the AGVs in two different ways: static control and dynamic control [34]:

(a) Static control. The control commands are assigned before the task. Once the AGVs receive the control command, the transportation path will not be changed until the AGVs receive another control command. This control scheme is simple and easy to operate. However, flexibility is weak.
(b) Dynamic control. This control method can adjust the control commands according to the real-time state of the AGVs, so the task scheduling strategy is complex.

The multi-AGV system is being studied worldwide. Compared with the single AGV, the advantages of the multi-AGV system are:

(a) The multi-AGV system can cover a large area. The single AGV can achieve transportation only between points. In the modern manufacturing environment, the transportation task is far more complex than the point-to-point transportation. The multi-AGV system can build a transportation network and improve transportation efficiency.

(b) The multi-AGV system can execute the transportation task in parallel. A multi-AGV system can perform tasks simultaneously for a complex task. Thus, the multi-AGV system can greatly improve the efficiency of transportation.

The dispatch and routing of the multi-AGV are important. The conflict-free function of the multiple AGVs is the bottleneck for the multi-AGV system. Draganjac et al. proposed a control algorithm for the multi-AGV system [35]. The proposed algorithm can detect the conflict between the AGVs and guarantee the safe operation of the multi-AGV by the priority mechanism. Miyamoto et al. proposed a conflict-free routing algorithm for the multi-AGV system [36]. Because of the limited memory space of each AGV, a heuristic algorithm was adopted for routing. Małopolski et al. considered the transportation system in the factory as a combination of squares [37]. Based on the square topology, a novel conflict-free routing algorithm for the multi-AGV system was proposed.

The dispatch plan of the multi-AGV should be calculated to optimize the task waiting time, collision, load use, and so forth [38]. The optimization methods can be divided into single- and multiple-objective optimization. Single-objective optimization can optimize only one objective function. If the objective function consists of several objective subfunctions, these subfunctions should be combined as the weighted sum [39]. However, it is difficult to design these weights. Therefore, the generated optimization results might not be the global optimal solution. The multiple-objective optimization can balance the trade-off between different subfunctions and generate a Pareto front [40]. The Pareto front contains many solutions, each of which has both advantages and disadvantages. The final optimization results should be selected according to the expert. In this manner, the intelligent decision-making ability of humans can be used to improve the dispatch performance of the AGV.

The environment of the factory is dynamic. The AGVs should be flexible enough to cope with the dynamic manufacturing system. There are many studies on the dynamic transportation system. Brito et al. proposed a dynamic obstacle avoidance algorithm for the dynamic unstructured environment [41]. In this algorithm, model predictive control is applied to improve control performance. This algorithm was verified in the environment with walking humans. Li et al. proposed an integrated algorithm for obstacle avoidance [42]. This algorithm can generate a path for the AGV by a model-predictive algorithm. The AGV can track the generated path reliably.

1.1.1.4 Manufacturing robots

Manufacturing robots contain many types including welding robots, drilling robots, grinding robots, milling robots, and so on. Manufacturing robots can produce better machining performance than a classical computerized numerical control (CNC) machine.

For example, it is proven that a workpiece polished by a manufacturing robot has a better surface quality than any CNC machine [43]. The better performance of the manufacturing robots is because the robots have more flexibility to make sure the tool is in the right position.

Welding robots are one of the most widely used manufacturing robots. Difficulties of welding robots are [44]: (1) it is hard to observe the welding seam in a complex manufacturing task, (b) it is hard to obtain the absolute and relative locations of the workpieces, and (c) the trajectory is hard to track. Current commonly used location methods for the welding seam are based on optical sensors such as a depth camera and laser sensor [45,46]. Jia et al. proposed a welding seam location method and trajectory tracking algorithm [44]. The proposed location method was achieved using a laser scanner, which could obtain the location and direction of the pipe. The cubic spline was applied to fit the obtained welding seam. The velocity control was used to track the welding trajectory. Liu et al. proposed a trajectory planning algorithm for welding robots to cope with a single Y-groove welding task [47]. This study provided two different velocity planning algorithms.

A key concern with drilling robots is positioning accuracy. An inaccurate hole can reduce the mechanical performance of the equipment. The positioning compensation method can be divided into two types: model-based and model-free [48]. Model-based methods can guide the robot to move according to the measured positioning error. The essentials of model-based methods are the dynamics and kinematics of the drilling robots. Model-based methods are time-consuming. Model-free methods can solve this drawback. Model-free methods build a model to describe the relationship between the positioning error and the robot's joints' parameters, which do not consider the dynamics and kinematics of the robots. Commonly used model-free methods contain the interpolation method [49], cokriging method [50], and so on. Neural networks are applied to compensate intelligently for positioning . With the help of the strongly fitting capacity of neural networks, these intelligent positioning compensation methods obtain good performance. Yuan et al. used ELM to predict positioning error and guide the robot to compensate for it [51]. Chen et al. used the RBF neural network to estimate positioning error. The bandwidth of the adopted RBF neural was fine-tuned. Positioning accuracy can be improved by more than 80% with these intelligent positioning compensation methods [52].

The grinding manufacturing is highly precise, so the positioning accuracy of the grinding robots requires attention. In real applications, the grinding robots may deviate from the preset track because of the disturbance and wear to the tools. Thus, grinding robots should have the ability to self-adjust, to ensure manufacturing quality. Control of grinding robots is more difficult than for CNCs because grinding robots have more DOFs. Huang et al. proposed an intelligent gear grinding system [53] in which the robot arm can detect the actual trajectory by a vision sensor and adjust the trajectory adaptively. Two cameras are adopted for tool center point calibration. When coping with the complex grinding task, multiple grinding robots are necessary, because multiple grinding robots can cope with the complex manufacturing task more easily than a single grinding robot. Han et al. [54] proposed a multiple grinding robot system and introduced the trajectory planning method for the multiple-robot system. Experimental studies indicated that the multiple grinding robot system can generate a steadier manufacturing trajectory.

The robot only needs five DOFs for the milling task; the reserved one is the DOF of the spindle of the milling cutter [55]. In the real application, six-DOF robots are applied for milling to improve flexibility. The stiffness of milling robots is a challenge in manufacturing. Milling robots usually have low stiffness. Robots may deform or vibrate when milling workpieces. There have been many studies on improving milling performance. Peng et al. optimized the stiffness of the robots in the feed direction with a seven-DOF robot [56].

Commonly used manufacturing robots are developed from six-DOF robots. Trajectory planning is a common technology for manufacturing robots. The trajectory planning algorithm can generate the optimal trajectory and the robot control algorithm can drive the robot to follow a predetermined trajectory [57]. The optimization targets of the trajectory are the position, velocity, accelerated velocity of each joint. The trajectory planning algorithm is the basis of the industrial robot control system. The optimization functions of the trajectory contain many aspects, including reduce execution time, consumed energy, and impact.

1.1.1.5 Loading–unloading robots

In the existing transportation manufacturing environment, processing each workpiece requires different procedures. The connection between processes requires the workpiece to be moved from one machine to another [58]. This repetitive work can be replaced by loading–unloading robots. The application of loading–unloading robots has many benefits. On the one hand, loading and unloading of robots is repetitive and consistent, which can avoid a decrease in work efficiency caused by worker fatigue. On the other hand, using the robots can parameterize the loading and unloading operation and the fault source can be quickly located when troubleshooting problems, which is convenient for improving the processing quality.

Loading–unloading robots have a wide range of applications. Liu et al. adopted a loading–unloading robot for the CNC and designed a control system based on a programmable logic controller [59]. Zhang et al. developed a robot system for loading–unloading [60]. The designed robot was able to separate good and bad workpieces. Fan et al. used the loading–unloading robot for electric equipment detection

[61]. With the help of the loading–unloading robot, automatic and high-precision detection is achieved.

The development of loading–unloading robots shows a trend in intelligence. Computer vision and neural network technology have been widely applied [62]. The adoption of computer vision can increase the flexibility of loading–unloading robots. Loading–unloading robots can observe the posture of the workpiece and adjust the robots to grasp the workpiece from the appropriate position. The neural network is widely used in controller design with its strong generalization ability. Gu et al. proposed a visual servo loading–unloading robot [63]. The camera on the manipulator can detect the location of the workpiece and help the robot capture the workpiece precisely. The design contains two steps: (1) extract the features of the observed image and target image and calculate the error between these features; and (2) input the error into the neural network and output the control command for the robot. In the proposed robot control structure, the fuzzy neural network was adopted as the controller.

1.1.2 Rail transit robots in dispatch

The running stage of the rail transit system takes up the longest length and generates the highest cost in the whole life cycle of the rail transit system. With the help of robot technology, the rail transit system can achieve automatic operation, thus improving operational safety and reducing transportation costs. Autonomous driving is the most important form of automation in rail transit. Autonomous driving is the application of robot control technology to the railway train. The automatic train control of the railway train contains automatic train supervision (ATS), automatic train protection (ATP), and automatic train operation (ATO) [64]. The ATS can supervise the running states of the railway train. The ATP system can monitor the train's running position and obtain its speed limit to ensure its running interval and safety. The ATO can accept output information of the ATS and ATP and generate control instructions [65].

The international standard International Electrotechnical Commission 62,290 defines four grades of automation (GOA) [66]:

(a) GOA1. The train is able to monitor the train's operating status continuously. Operation needs to be carried out by drivers.
(b) GOA2. This level is self-driving with driver duty. The train is able to drive automatically through the signal system, but the driver is required to close the door and issue the command.
(c) GOA3. There is no need to equip drivers on this level of the train; the train can automatically complete the whole process of operation, including outbound, pit stop,

switch door, and so on. However, there is still a need for onboard personnel to deal with emergencies on this level of train.

(d) GOA4. No operator is required on this class of trains, and the train's control system can automatically respond to unexpected situations.

ART is an important example of the automation of rail transit operation; it is developed by CRRC Zhuzhou Institute Co., Ltd. The core technology of this train is virtual orbit tracking control technology. That is to say, the ART train does not need a true track, but rather, the marks on the road serve as tracks. Because the intelligent rail train does not have a physical track, existing positioning devices used in the subway train and high-speed train cannot be used. The ART adopts the global satellite positioning system as the main positioning method [4]. During the operation, inertial sensors and angle sensors installed in intelligent trains are used to monitor the movement posture and position of the train, thereby enabling a comprehensive perception of the state of the ART train. The monitored train status information is fed back to the controller of the intelligent rail train to control the intelligent rail train to run along the virtual track. Using this control technology, all wheels of the ART train can be driven along a virtual track [4]. This position accuracy can ensure the passing performance of the ART rail train in urban traffic. To improve operational efficiency, the ART train is also equipped with automatic driving technology [4]. The ART train can intelligently sense the environment and control the automatic operation of the train.

1.1.3 Rail transit robots in maintenance

Robots have become the focus of modern rail transit equipment manufacturing and operation. With the increase in railway operating mileage and operational density, the workload of railway maintenance support is also increasing, and higher requirements are put forward for the maintenance and repair of railway infrastructure equipment. Currently, the maintenance of railway infrastructure equipment is mainly carried out by workers [67]. The main problems

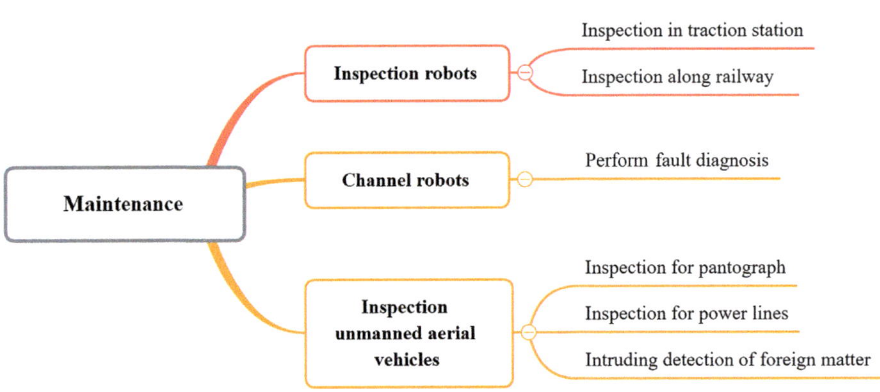

Figure 1.2
Role of robots in rail transit maintenance.

are that (1) diversified test operations have higher requirements for operators, (2) it is inconvenient to have operators carry measurement equipment, and (3) manual inspections maintain labor intensity and quality is uncontrollable. As shown in Fig. 1.2, robots can have an important role in the maintenance stage of rail transit in this case.

1.1.3.1 Inspection robots

The objectives of railway maintenance include tracks, instruments, and equipment next to railways, traction substations, pantographs, temperature of instruments, bogies, and foreign matter intrusion. According to the work scenario, the inspection area of rail transit inspection robots is mainly divided into traction substations and railway.

1.1.3.1.1 Inspection in traction substation

The traction substation converts electric energy from the regional power system into electric energy suitable for electric traction according to the different requirements of electric traction on current and voltage. The resulting electricity is then sent to catenary lines set up along the railway line to power electric locomotives, or to the urban rail transit system to power electric subway vehicles. The inspection work is the top priority for the daily maintenance of the traction substation. However, because of insufficient staffing and other reasons, the inspection cannot meet the needs. In addition, with the increase in substations, operation, and maintenance, personnel are struggling to comply. It is also difficult to keep satisfactory working conditions and a strong sense of responsibility.

Currently, the technology of intelligent robots is increasingly developing and the use of robots to replace manual inspections of equipment partially or completely has become the trend of future substations [68]. Substation intelligent inspection robots are equipped with advanced equipment such as infrared thermometers, high-definition cameras, and audio signal receivers. According to the preset inspection tasks, the entire station equipment and partitions are inspected in time. All task modules of meter reading, temperature measurement, sound collection, and functional inspection are processed together to identify and alert regarding discovered equipment abnormalities, defects, and hidden dangers in a timely manner. In addition, robots intuitively form charts and report weekly on operational data indicators of devices stored in the inspection so that maintenance personnel can perform equipment maintenance and fault analysis better.

By using the inspection robots in the substation, the equipment in the substation can be inspected more reliably and the data analysis returned by the system can give the maintenance personnel a better understanding of the working state of the equipment. Recording and analyzing historical data can also significantly improve the predictability of equipment in the substation, providing more effective data support for the condition of maintenance and a status assessment of the equipment, which will directly reduce the possibility of equipment accidents [69]. Using inspection robots to inspect substation equipment can not only keep field maintenance personnel away from the field equipment, which improves the safety and reliability of the inspection work, it provides a new method for spot inspection for the popularization of an unattended substation and truly improves the safety and reliability of substation equipment.

1.1.3.1.2 Inspection along the railway

With the rapid development of rail transit and the continuous improvement in passenger high-speed processes, railway transport has been gradually heading toward function integration, information sharing, command concentration, and highly automated transition and transformation. The urgent demand for railway transportation safety poses new challenges to the railway traffic safety assurance system. To adapt to the new needs of railway development, it has become the focus of the construction of the railway traffic safety assurance system to build a fully covered and highly reliable railway intelligent monitoring system based on high information sharing [70]. The higher speed brings convenience to people, but it also increases the difficulty of preventing railway disasters.

Although a high-speed train is equipped with a more advanced safety system, the large braking distance caused by high-speed driving reduces the flexibility of the train to respond to emergencies. In the event of an accident, a high-speed train often has more serious consequences than an ordinary train. Therefore, various factors such as the intrusion of foreign matter, natural disasters, and line failures may cause major railway accidents. Casualties and property loss caused by the intrusion are enormous and frequently occur. Therefore, research and development into railway foreign matter intrusion detection are of great significance to ensure the safe operation of trains [71].

The intrusion of foreign matter has the characteristics of irregularity, suddenness, and unpredictability. The fixed transportation mode of the operation route makes emergency measures that can be adopted limited when the train detects an intrusion on the road. If real-time detection of invasive foreign matter can be achieved and necessary measures are taken in time, property loss can be reduced as much as possible. The traditional foreign matter intrusion approach is to place sensors at key positions along the railway to monitor intrusion conditions in real-time. This method has many drawbacks, such as an unreasonable arrangement of the sensor and fixed monitoring position. With the rapid development of robotics, robots have entered the field of railway foreign matter invasion detection. The railway is inspected by an infrared sensor, a laser sensor, a smart camera, and an ultrasonic device.

1.1.3.2 Channel robots

High-speed trains may experience poor performance or even failure of key components during long-term high-speed service. If a safety accident occurs, it will cause severe economic loss and even casualties. Therefore, how to monitor and diagnose its operating status has become an important research topic [72]. Among them, the bogie is a main components of the train. The basic structure of the bogie mainly includes the wheel set, suspension system, frame, auxiliary suspension system, and so on. Therefore, it is very important to research fault diagnosis for the high-speed train, especially the diagnosis of bogie faults [73].

With the complex diversification of high-speed rail operating conditions and the increase in operating mileage, mechanical wear and aging of vehicles are gradually accelerating, which causes safety hazards for train operation. Among them, the train axle is an important component to support the running of the train. During the driving process, almost all of the weight of the train and the impact caused by vibration are assumed. Therefore, the axle is one of the most vulnerable parts of the train. As a result of improper assembly of the axles and excessive operation of the axles during maintenance, the weight of axles will be aggravated. When the axle temperature rises abnormally, movement between the axle and bearings will be worse, friction and wear will be aggravated, and smoothness will be reduced. In this case, if the vehicle continues to run without emergency treatment, hidden safety hazards will occur and the train will be delayed or even derailed. Therefore, the health status of the bogie is directly related to the safety of the vehicle operation. The fault warning and diagnosis of the axle of the high-speed train are an urgent problem to be solved.

With the rapid development of sensor technology, a large amount of real-time monitoring data generated during the operation of high-speed trains has been collected. Through the analysis and mining of historical monitoring data, the potential correlation and periodicity can be explored to provide a basis for train fault diagnosis. By analyzing the monitoring data collected during the train's operation, it is important to identify the health status and the potential safety hazards that may exist at the time of the early warning, which is greatly significant for ensuring the safe operation of the train and improving the reliability of the train's operation.

Bogie fault diagnosis includes fault vibration signal processing and fault identification. Among them, fault vibration signal processing includes time-domain analysis, frequency-domain analysis, and time-frequency analysis. Fault identification uses the expert system, pattern recognition, neural networks, deep learning, and so forth [74,75]. Because the high-speed train bogie has nonlinear characteristics, its vibration signal is a nonlinear, nonstationary signal. Therefore, applying effective fault vibration signal processing methods to extract state information and constructing a fast and effective bogie fault diagnosis method is critical for accurately evaluating the safety state of the train.

With the rapid development of robotics, more and more trains are on the line and rail transit channel robots have emerged. Compared with industrial robots and inspection robots, rail transit channel robots have been in development for a short time, but they form a certain scale. Channel robots are mainly based on intelligent technology and integrate various sensing equipment. It is dedicated to the collection and diagnosis of fault information. The biggest contribution of rail transit channel robots is to replace on-site workers to check the train's status. The staff of the inspection station can carry out remote control operations through network technology and can reliably monitor the site in real time, which can objectively monitor the equipment and meet the requirements of flexibility, high precision, and antiinterference. In general, rail transit channel robots need to meet the following requirements:

(a) The channel robot belongs to the mobile robot. Unlike the fixed industrial robot, it needs to move according to the actual task. Therefore, the navigation control system needs to have a control module to perform robot motion control. Of course, the movement of the channel robot can be along the guide rail or autonomous navigation.
(b) The robot needs to carry a variety of sensors to obtain information about the detected target, laying the foundation for intelligent fault diagnosis.
(c) The navigation system needs to be modularized to make it easy to replace modules and easy to maintain. Communication between modules must be completed.
(d) The channel robot needs to upload collected information and the corresponding fault type, fault location, and so on to the prebuilt historical fault database. With the continuous increase in fault information, the database has gradually grown. Data mining can be used to predict the fault and life of train equipment and to identify potential dangers in advance.

Rail transit channel robots mainly perform a fault diagnosis of trains. Generally speaking, the workflow is as follows: (1) the maintenance task is carried out along the ground track to the checkpoint; (2) the vehicle maintenance information is collected by the onboard sensor including the visible light camera, the laser sensor, and the infrared camera, and some sensors are installed on the robot arm; (3) the collected image for fault diagnosis analysis is uploaded; and (4) the fault diagnosis results are reported and stored, including the fault types, fault problems, fault degrees, and so on. In addition to the growing database, the big data analysis module will perform data mining on train bogie equipment failure information to predict potential equipment failures. In addition, the robot continuously detects its own power and performs a recharging task once the minimum battery threshold is reached.

Channel robots with a trouble of moving electric multiple units (EMU) detection system (TEDS) intelligent sensing system have been used for EMU fault diagnoses, aiming to use the rail transit channel robots to complete train status detection and fault diagnosis [76].

The hardware structure of a channel robot mainly includes a ground rail, dual-arm robot, laser sensor, visible light camera, infrared camera, server, and recharging device. Among them, the function of the robot guide is to drive the robot and make it move along the preplanned route to expand the working radius of the industrial robot. The arm of a two-armed robot is similar to that of an industrial robot. The difference is that a two-armed robot can perform more complex tasks and control is more difficult. Laser sensors, visible light cameras, infrared cameras, and other in-vehicle sensors belonging to a channel robot are used to collect different types of data. The background server is mainly used for the channel robot control, data storage, fault diagnosis, and big data processing. The self-recharging technology of the recharging device is an important means to ensure the normal operation of the channel robot. Currently, the more advanced one is noncontact recharging. Compared with contact recharging, noncontact recharging has the advantages of a small footprint and low cost.

The software of a rail transit channel robot mainly includes intelligent visible light image analysis, infrared thermal imaging temperature analysis, location detection, and big data analysis. Among them, visible image intelligent analysis and infrared thermal image analysis are used for fault diagnosis of the train. Visible light image analysis mainly includes image registration and change detection algorithms, and the train fault is found by comparing images. Infrared thermal imaging temperature analysis mainly includes fault location and image segmentation, fault location image recognition, and classification. The former uses image processing technology and the latter relies on machine learning methods. Location detection mainly uses laser ranging technology to determine the location of train faults. Big data analysis is mainly based on the train bogie equipment fault information database; it uses data mining and machine learning and predicts the potential fault of the equipment to avoid accidents.

1.1.3.3 Unmanned aerial vehicles for inspection

Railway accidents are sudden and random, so regular inspections of railways must be rigorous and critical. With the continuous development of unmanned aerial vehicle (UAV) technology, the demand for civilian UAVs in many countries is growing and UAVs have an increasingly important role in many fields. Research scholars have a major interest in developing new types of UAVs that can fly autonomously in different environments and locations and perform a variety of tasks [77]. Therefore, various types of the UAVs with different sizes and weights have been invented and designed for different situations and scenes. For example, UAVs slowly began to replace some manual work of railway operations and maintenance and have been successfully applied in the field of railway safety. Combining existing technical skills of UAVs with the daily safety requirements of railway tracks, UAVs can inspect for daily monitoring of railway tracks. Rail transit inspection UAV inspection refers to the use of the UAVs equipped with visible light cameras and infrared sensors to detect the operation status of railway equipment and real-time weather and discover the hidden dangers of railway train operation. Equipment along the railway includes pantographs, power lines, and so on.

1.1.3.3.1 Pantograph

The pantograph and catenary technology can connect electrical energy from a contact network to an electric locomotive [78]. As the speed of the train increases, the vibration characteristics of the catenary and the pantograph are particularly important for the stability and safety of the train. To obtain a stable current, the contact pressure should be maintained between the contact slide of the pantograph and the contact line. Receiving current from the traction net by the pantograph is a process in which several mechanical movements occur simultaneously during the high-speed running of a high-speed train. The pantograph and the locomotive are closely connected. When the locomotive is driving fast, the pantograph is also rubbing quickly and moving forward relative to the contact line.

Because of the structural characteristics of the pantograph, it maintains tension, so it generates vibration in the vertical direction. To ensure the pressure value, the pantograph is also in contact with the catenary, and high-speed lateral vibration occurs. The catenary has an inherent vibration frequency that creates a traveling wave propagating along the catenary. These kinds of movements occur at the same time and are likely to cause separation between the nets [78]. In the case of high-speed driving, these types of motion are superimposed, and the accuracy of construction of each line is different. This makes it difficult to maintain stable contact pressure between the pantograph and the contact line. When the contact pressure is zero, it is likely that mechanical disengagement will occur, which is the pantograph-catenary offline. When the electric locomotive runs at high speed, contact between the pantograph and the high-voltage catenary is good, which is equivalent to a closed circuit with an inductive load, and the load operates normally. When the net is separated, it is equivalent to disconnecting the inductive load circuit.

Because of its circuit characteristics or overvoltage, overvoltage between networks is high enough to rupture the air around the gap, which will cause a spark discharge and then an arc discharge. Whether it is spark discharge or arc discharge, the locomotive current at this stage will be greatly affected. At this stage, not only will overvoltage be generated, electromagnetic noise will be disturbed. In addition, high heat released by the arc will melt the metal and cause wear on the pantograph and catenary. When there is an offline situation between the pantograph and catenary, safe and stable operation of the high-speed railway will face great hazards: unstable operation of the electric locomotive, severe wear of the pantograph slide, deterioration of the traction motor rectification conditions, and generated overvoltage in the main circuit.

According to statistics, railway accidents caused by failure of the traction power supply system accounted for 40% and 30% of accidents in 2009 and 2010, respectively [79]. The quality of the pantograph-catenary system has become an important prerequisite for speeding and improving operational reliability [79]. Currently, the real-time dynamic monitoring system of the pantograph-catenary system has been installed on newly built trains, and the test system has been installed in the operation EMUs. The system is installed on top of the train and can detect the working state of the pantograph in real time. The main method is to collect image information through the camera for analysis and processing, and the pan/tilt can adjust the shooting angle of the camera. Nevertheless, the monitoring effect of the bow monitoring system will be limited by the fixed installation position of the camera and the range of camera angles, so that the pantograph can be monitored from only one direction and the pantograph cannot be fully fault-monitored. With the gradual rise of UAV technology, a train pantograph monitoring system based on the UAV inspection has emerged. Depending on the flexibility of the UAVs, it is theoretically possible to achieve all-around detection of the state of the pantograph and avoid blind spots in the pantograph failure. Moreover, considering simultaneous work in different areas, multiple sets of parallel UAVs can be used for detection to improve efficiency.

1.1.3.3.2 Power lines

The overhead transmission line is a kind of steel frame structure that erects the wire and keeps it away from the ground. The safety of overhead transmission lines affects people's normal life, is related to economic production in the country, and even poses a threat to national security. Long-distance transmission lines are characterized by long line distances and high voltage levels. Transmission lines are also called overhead transmission lines. The transmission capacity of the lines is large, the transmission distance is long, and they are in the open air for a long time. They are often affected by the surrounding environment and natural changes such as snow, lightning, ice, external forces, birds, and so forth. An ultrahigh voltage (UHV) transmission line has high voltage, a long line, a high pole tower, a large conductor, and a complex geographical environment. It puts high requirements on operation and maintenance [80]. Traditional line maintenance and overhaul operations have complicated processes, high risks, a poor monitoring work environment, and low labor efficiency. If there is an emergency trip, power failure, and nature force majeure, only low-end inspection equipment such as a telescope and night vision device can be used. Therefore, the manual operation and maintenance mode are unable to cope with the safe operation and maintenance of the UHV over a long distance.

1.1.3.3.3 Foreign matter invasion detection

Compared with the land inspection robots introduced earlier, the UAVs can detect foreign matter intrusion with the obvious advantages of wide detection range and emergency response [81]. When the railway is hit by a major natural disaster, the advantages of inspection UAVs are highlighted. For example, serious mudslide disasters may occur along the railway line, posing a serious hazard to the railway subgrade. If a mudslide is not found and taken emergency measures in time, it will lead to serious traffic accidents. In general, the occurrence of mudslide disasters is accompanied by strong rainfall. In this case, the train driver's line of sight is seriously degraded and it is difficult to detect the danger in time during heavy rain. Historically, there have been train crashes in which the accident occurred because heavy rains caused mudslides to ruin railway bridges [82]. Railway inspection UAVs can take pictures of the railway along with the high altitude and determine whether the railway is normal along the way through wireless transmission technology and image processing technology.

1.2 Fundamental key problems of rail transit robot systems

Key problems of rail transit robots are positioning navigation and path planning, human–robot interaction, and power management, (Fig. 1.3).

1.2.1 Navigation

1.2.1.1 Development progress

The development of robotics has gradually expanded from the industrial field to rail transit, services, medical, entertainment, fire protection, and other fields. The application of robot technology has brought tremendous productivity and economic benefits to society. How to make robots more efficient and serve humans more intelligently is the focus of scientific research. The positioning and navigation of robots have been the core emphases of content since the development of robotics.

The positioning of mobile robots is a prerequisite for all of their actions. The accuracy of positioning determines the accuracy of the robots' feasible operation. It is the basis for ensuring the robot navigation system. The positioning of rail transit robots is to use sensors to sense the surrounding environment and then determine its position and posture in the workspace. In the case of positioning, robots start the navigation operation and move to the target autonomously through a certain guiding method. Path planning means that the autonomous mobile robots can intelligently plan the relatively shortest and least

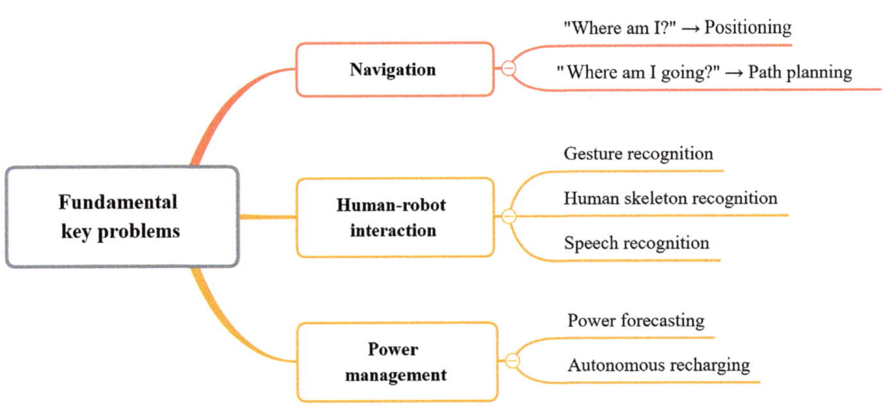

Figure 1.3
Key problems of rail transit robot systems.

time-consuming path for the target task. Path planning can independently interact with the intelligent environment through onboard equipment such as an automatic elevator ride, intelligent access control, and automatic obstacle avoidance function. Under the premise of ensuring the safety of robots, people, and the work environment, the target task is completed.

1.2.1.2 Methodologies

1.2.1.2.1 Positioning

The key to the positioning and path planning of mobile robots lies in two points: "Where am I?" and "Where am I going?'". The former is positioning and the latter is path planning. Mobile robot positioning technology solves the first problem: that is, it achieves the precise positioning of robots. If a robot can self-navigate, the first condition is that the robot knows where it is. There are several basic tasks for navigation: (1) global positioning, sense the position, posture, and environment information of the robot; (2) path planning, plan an accessible path to the target point from the environmental model; and (3) motion control, select the motion control method, and move to the destination along the planned safe path.

Many scholars have made great progress in the field of positioning [83]. The main research directions of mobile robot positioning technology include simultaneous positioning and map construction, path planning, and motion control. In the process of robot navigation, positioning is the first problem to be solved; the posture information and current state of robots need to be known. Existing positioning methods of intelligent mobile robots mainly include relative and absolute positioning [84]. Absolute positioning can obtain the posture and position information of a robot directly through the sensor carried by the mobile robot itself. Relative positioning requires the determined original position. Sensors for a robot can obtain the concrete position of the distance and direction to the target point. According to the current point moving distance and the direction of rotation angle, the robot can be aware of its posture and direction. Absolute positioning methods include a magnetic compass and global positioning system (GPS). Relative positioning can be divided into the dead reckoning method based on the inertial sensor and the dead reckoning method based on the odometer. According to the different types of sensors used to perceive the environment, navigation of mobile robots mainly includes GPS, gyroscope, inertial, photoelectric coding, magnetic compass, laser, and ultrasonic.

1.2.1.2.2 Path planning

The path planning problem of mobile robots is a hot spot in the field of mobile robot navigation research [85]: mobile robots can find an optimal or near-optimal path from the starting state to the target state that avoids obstacles based on one or some performance indicators (such as the lowest working cost, the shortest walking route, the shortest walking time, etc.) in the motion space. Generally speaking, the path planning selects the shortest distance of the path: that is, the shortest length of the path from the starting point to the target point as a performance indicator. The mobile robot path planning method can be divided into two types according to the known degree of environmental information: path planning based on global map information or local map information [86]. These two path planning methods are referred to as global path planning and local path planning. The global path planning method can generate the path under the completely known environment (the position and shape of the obstacle are predetermined). With the global map model of the environment where the mobile robots are located, the search is performed on the established global map model. The optimal algorithm can obtain the optimal path. Therefore, global path planning involves two parts: establishment of the environmental model and the path planning strategy.

Path planning requires building an environmental map. Environmental map construction refers to the establishment of an accurate spatial location description of various objects in the environment in which the robot is located, including obstacles, road signs, and so on: that is, the establishment of a spatial model or map. The purpose of constructing the environmental map is to help the mobile robot plan an optimal path from the starting point to the target point in the established environment model with obstacles.

There are many mature methods for establishing an environment model for mobile robot path planning. Existing basic environment models mainly include a grid decomposition map, quad split graph, visibility graph, and Voronoi diagram. After the environmental map is built, global path planning is carried out. Algorithms of global path planning are mainly divided into two types: heuristic search methods and intelligent algorithms. The initial representation of the heuristic search is the A* algorithm developed by the Dijkstra algorithm. The A* algorithm is the most commonly used heuristic graph search algorithm for state space. In addition to solving problems based on state space, it is often used for the path planning of robots. Many scholars have improved the A* algorithm and obtained other heuristic search methods [87,88]. Intelligent algorithms have lots of studies, including ant colony [89], particle swarm [90], genetic [91], bat [92], simulated annealing [93], and so forth.

Different from the global path planning method, the local path planning method assumes that the position of the obstacle in the environment is unknown, and the mobile robot perceives its surrounding environment and its state only through the sensor. Because the global information of the environment cannot be obtained, the local path planning focuses on the current local environment information of the mobile robot and uses the local environment information obtained by the sensor to find an optimal path from the starting point to the target point that does not touch the obstacle in the environment. The path planning strategy needs to be adjusted in real time. Commonly used methods for local path planning include the rolling window [94], artificial potential field [95], and various intelligent algorithms [96].

1.2.1.3 Applications

Positioning and path planning are applied to almost all mobile robots, the most widely used of which are the AGV robots and rail transit inspection robots.

1.2.1.3.1 Carrying automatic guided vehicle robots

To reduce labor costs and storage costs, logistics sorting has gradually shifted from manual to automated. Intelligent unmanned storage has gradually become a hot spot for many scholars [97]. Unmanned factories and unmanned warehousing have become a trend in development. The intelligent logistics system began to appear, which aims to reduce handling, management, communication, and labor costs. All links in picking and unloading are carried out in an orderly manner without manual intervention. AGV robots have become the main method of automated warehousing. Performing collision-free path planning for multiple AGVs and performing tasks is a difficult problem for the AGV group.

1.2.1.3.2 Rail transit inspection robots

The main method of inspection in rail transit is to use human power to conduct inspections. However, the work environment of rail transit equipment is complex and conditions are harsh. In case of bad weather, there is a personal safety risk for personnel conducting a manual inspection. At the same time, the efficiency of manual inspections is low, depending on business proficiency and the subjective initiative of the inspection personnel. The long-term labor of mechanical repetition easily causes negligence, so rail transit inspection robots have emerged as the times require, and will be used in the rail transit line and station equipment. New methods of inspection are presented and manual inspections will be phased out. Inspection robots can use various sensors equipped to detect the rail transit and rail transit vehicle equipment through various advanced technologies autonomously or with assisting inspectors. The use of inspection robots can detect the work conditions of rail transit equipment under severe conditions and ensure the normal and orderly operation of rail transit. This is important for the future realization of the full automation of rail transit. [98].

1.2.2 Human—robot interaction

1.2.2.1 Development progress

Human—robot interaction is a technology that studies people, computers, and their interactions [99]. It is a focus of competition in the information industry. Its extensive application will profoundly change the manufacturing, construction, and transportation industries. So far, human—robot interaction technology has evolved to the interface of gestures and language, combining the powerful processing power of computers. The design of the current human–robot interaction system has achieved a lot of results, such as the handwritten Chinese character recognition system, Chinese speech recognition system, and gesture recognition. In terms of handwritten Chinese character recognition, the most advanced system has been able to identify 28,000 Chinese characters after years of research and development [100]. If written at a speed of 12 Chinese characters per minute, the recognition rate of the system is nearly 100% [100]. In the aspect of sign language recognition and synthesis, the Institute of Computing Technology, Chinese Academy of Sciences has successfully developed a Chinese sign language recognition and synthesis system based on multifunctional perception [101].

1.2.2.2 Methodologies

1.2.2.2.1 Human–robot interaction based on gestures

Gestures are a natural and intuitive mode of communication. Vision-based gesture recognition is a key technology that is indispensable for the realization of a new generation of human–robot interaction. Gestures are a variety of actions produced by a human hand or arm. They include static and dynamic gestures. Because of the diversity and ambiguity of gestures and the complex deformation of human hands, vision-based gesture recognition is a multidisciplinary and challenging research topic. To find a breakthrough, the use of gestures in interpersonal communication is required. In the vision-based gesture recognition system: (1) the video data stream is acquired by one or more cameras; (2) the system detects whether there is a gesture in the data stream according to the interaction model; and (3) if so, the gesture is segmented from the video signal, and the gesture model is selected for gesture analysis. The analysis process includes feature detection and model parameter estimation. In the recognition process, gestures are classified according to model parameters and gesture descriptions are generated as needed. Finally, the system drives specific applications according to the generated descriptions.

The gesture model is important for the gesture recognition system, especially for determining the recognition range. The model selection depends on the specific application. A simple and rough model may be sufficient for a given application. However, to achieve natural human–computer interaction, a complex gesture model must be established. In this way, the recognition system can respond correctly to most gestures made by users. From the current literature, most gesture modeling methods can be summarized as gesture modeling based on the apparent model and gesture modeling based on the 3D model.

The task of the gesture analysis is to estimate the parameters of the selected gesture model. The analysis generally consists of feature detection and parameter estimation. In the feature detection process, the subject that makes the gesture must be located.

Depending on the conditions used, positioning techniques can be classified as color-based, motion-based, and multimode. Most color-based positioning techniques rely on histogram matching or the use of skin training data to create a lookup table. Color-based positioning techniques have significant drawbacks (i.e., skin color changes under different lighting conditions). Gesture recognition is the process of classifying a trajectory (or point) in a model parameter space into a subset of that space.

The static gesture corresponds to a point in the model parameter space, and the dynamic gesture corresponds to a trajectory in the model parameter space. Therefore, their identification methods are different. Static gesture recognition algorithms include recognition based on classical parameter clustering techniques and recognition based on nonlinear clustering techniques. Unlike static gestures, dynamic gestures involve time and space. Most dynamic gestures are modeled as a single trajectory in the parameter space. Because of the different habits of different users, the difference in speed and proficiency in gestures will cause nonlinear fluctuations on the time axis of the trajectory. Eliminating these nonlinear fluctuations is an important problem that must be overcome by dynamic gesture recognition technology. Considering the different processing of the time axis, existing dynamic gesture recognition technology can be divided recognition based on hidden Markov model [102], recognition based on dynamic time warping [103], and so forth.

1.2.2.2.2 Human–robot interaction based on human skeleton

The development of human pose estimation has become increasingly practical. In the fields of gait analysis, human–robot interaction, and video surveillance, human pose estimation have broad application prospects. The mainstream human body pose estimation algorithm can be divided into traditional and deep learning-based methods.

Traditional methods are generally based on the graph structure; they design a 2D body part detector, use the graph model to establish the connectivity of each component, and continuously optimize the graph structure model to estimate the human body posture combined with the relevant constraints of human kinematics. Although traditional methods have high time efficiency, the extracted features are mainly histogram of oriented gradient [104] and scale-invariant feature transform features [105]. These features cannot fully use image information; thus, the algorithm is subject to different viewing angles, occlusions, and inherent geometric ambiguities of the image.

At the same time, because of the single structures of the traditional models, when the posture of the human body changes greatly, the traditional models cannot be accurately characterized and expressed. There are multiple feasible solutions to the same data; that is, the results of attitude estimation are not unique, which limits the scope of application of traditional methods. On the other hand, most traditional methods are based on digital images such as depth images for feature extraction. However, because the acquisition of depth images requires professional acquisition equipment, the cost is high, so it is difficult to apply to all application scenarios. In addition, the acquisition process needs to synchronize the depth cameras of multiple perspectives, which reduces the impact of the occlusion problem. These factors will make the acquisition process of human posture data complicated and difficult. In contrast, monocular cameras are more common, although the color images they collect are susceptible to environmental factors such as illumination.

Neural networks can be used to extract more accurate convolution features than artificial features and improve the performance of the forecast. Because of this, human body pose estimation methods based on deep learning have been deeply studied [106–108]. The human body pose estimation method based on deep learning mainly uses a convolutional neural network (CNN) to extract the human body pose features from the image. Compared with artificial design features of the traditional method, the CNN can obtain more rich features of semantic information as well as obtain different types of multiscale and multiple-type human joint point feature vectors.

1.2.2.2.3 Human–robot interaction based on speech recognition

Speech recognition is the process of converting human sound signals into words or instructions. Speech recognition is based on speech. It is an important research direction of speech signal processing and a branch of pattern recognition. The research of speech recognition involves many subject areas such as computer technology, artificial intelligence, digital signal processing, pattern recognition, acoustics, linguistics, and cognitive science. It is a multidisciplinary comprehensive research field. Different research areas have emerged based on research tasks under different constraints. According to the requirements of the speaker's way of speaking, these areas can be divided into isolated words, connected words, and continuous speech recognition systems. According to the degree of dependence on the speaker, these areas can be divided into speech recognition systems for the specific person and nonspecific person. According to the size of vocabulary, they can be divided into small vocabulary, medium vocabulary, large vocabulary, and infinite vocabulary speech recognition systems.

From the perspective of the speech recognition model, the theory of the speech recognition system is based on pattern recognition. The goal of speech recognition is to transform the input speech feature vector sequence into a sequence of words using phonetic and linguistic information. According to the structure of the speech recognition system, a complete speech recognition system includes a feature extraction algorithm, acoustic model, and language model and search algorithm. The speech recognition system is essentially a multidimensional pattern recognition system. For different speech recognition systems, the specific recognition methods and techniques used by people are different, but the basic principles are the same. The collected speech signals are sent to feature extraction. The module processes and the obtained speech features are sent to the model library module. The speech pattern matching module identifies the segment speech according to the model library, and finally obtains the recognition result.

1.2.2.3 Applications

In the field of rail transit, rail transit equipment manufacturing and rail transit operation and maintenance are the main components that determine whether the rail transit system can operate safely, efficiently, and economically for a long time. Introducing human–robot interaction technology into rail transit equipment manufacturing will change the traditional equipment manufacturing production mode and transform the original staff to work together with production tools, which will greatly improve the quality and efficiency of equipment manufacturing. In the field of operation and maintenance, the friendly interface achieved by high-level human–robot interaction will enhance the passenger's ride experience from the inbound to the outbound service cycle and will greatly improve the management level of the rail transit system [109].

1.2.2.3.1 Intelligent rail transit equipment manufacturing

Intelligent industrial equipment has become a trend in the development of the global manufacturing industry. The robotic group used for welding replaces a separate welder to perform collaborative and high-precision welding and assembly operations. The automated guided carrier platform connects manufacturing modules in an intelligent manufacturing environment, based on unmanned and intelligent obstacle-avoidance delivery platforms. It can be shuttled within a complex environment, and the safety of production is greatly improved, which also increases the automation of the assembly line. In a complex intelligent manufacturing environment, human–robot interaction control technology will have an important role in the coordinated operation of personnel and carrier platforms and the operator control of the delivery platform. On the one hand, the automatic guided carrier platform requires the operator to provide instructions to complete the loading, transporting, and unloading operations of the carrier platform. On the other hand, the automatic guided carrier platform needs to obtain operator instructions in a small space to complete high-performance obstacle avoidance [110].

1.2.2.3.2 Intelligent operation and maintenance

To ensure the stability, safety, and reliability of the urban rail transit system, it is necessary to achieve the intelligent manufacturing of rail vehicles, but also to ensure its intelligent operation and maintenance and improve system efficiency. The intelligent operation and maintenance framework includes parts such as the automatic identification of vehicles, automatic equipment control, processing data collection and analysis, and vehicle status detection, and is interconnected with the network.

In 2017, Tao et al. proposed a method for intelligent state-of-sales awareness and process control through intelligent instrument and detection technology [111]. They discussed the application mode of intelligent instrument and detection technology in intelligent manufacturing systems, digitalized in combination with actual projects. The innovative practice of production line transformation and construction puts forward the development direction of intelligent instruments and detection technologies that meet the requirements of intelligent manufacturing. Through the practical application of the project, this model can effectively promote the realization of data acquisition and intelligent perception in intelligent manufacturing systems and can provide effective data resources. In 2019, Xu et al. proposed an AGV intelligent logistics scheduling model and response method based on the request scheduling response model for the dynamic scheduling of intelligent manufacturing workshop logistics [112].

1.2.3 Power management

Robots can do a lot of work instead of humans; they have an important role in the fields of manufacturing, service, and rail transportation. Power management is mainly divided into the robot's power forecasting and autonomous recharging.

1.2.3.1 Power forecasting

The robot is powered by a lithium battery during the work process. How to ensure that the robot does not fail owing to a low battery during the mission is a serious problem. Traditional robots continuously detect their own power and perform recharging tasks when they find that the battery is below the minimum threshold. This method is inapplicable to the robot that is performing the task. For example, whether the robot can have enough power to reach the next tower or the next-level recharging base station during the inspection is an urgent problem to be studied. In addition, when the inspection personnel uses the robot to inspect the transmission line, it is necessary to grasp the lifetime and cruising range of the robot in real time, and then the next inspection plan can be formulated according to the robot's endurance capacity. Most research on the cruising range of the inspection robot uses a rough estimation of the amount of power consumed by the robot at different stages; thus, the accumulation method is not scientific. Many scholars have begun to use the power time series data of the robot to establish a power forecasting model, to estimate the power in the subsequent time period [113,114]. The power forecasting model is built by data processing methods and machine learning.

1.2.3.2 Autonomous recharging

Most intelligent robots use battery packs to power them and their working hours are generally shorter. The mobile robot that can self-charge to achieve long-term continuous work when the battery power is reduced to a certain level has attracted the interest of researchers. Research in the field of robotic autonomous recharging has begun. In the mid-20th century, Walter developed the world's first self-aware electric robot [115], which was able to achieve autonomous recharging. Walter designed a recharging station with a charger and illuminator installed. It not only avoids obstacles, it uses a light beam to reach the recharging station for recharging. Autonomous recharging for mobile robots can be mainly divided into contact and noncontact, which is major topic in the field of robotics [116].

1.3 Scope of this book

Today, robotics and advanced rail transit are two major areas of development in the world. This book starts with the field of rail transit manufacturing, operation, and maintenance. It also introduces different types of robots and related technologies in detail. The book includes eight chapters:

Chapter 1: Introduction.

This chapter first outlines rail transit robots. Then, it describes three basic issues of robotics, including navigation, human–robot interaction, and power control.

Chapter 2: Rail transit assembly robot systems.

This chapter first introduces progress in development and the key technologies of assembly robots. Then, it introduces the main components of assembly robots: machinery components, sensors, controllers, and actuators. Arm dynamics is a major research topic of assembly robots. This chapter introduces mathematical models of forward dynamics and inverse dynamics, discusses various trajectory planning algorithms in joint and Cartesian space, and proposes an artificial neural network algorithm for the inverse dynamic optimization calculation of the robot arm. Finally, the chapter presents the future direction of intelligent assembly robots.

Chapter 3: Rail transit collaborative robot systems.

This chapter first gives a basic definition of collaborative robots. The chapter then summarizes the development history, application field, and mode of collaborative robots. The hardware composition and characteristics of the collaborative robot are summarized. Finally, the basic concepts of feature extraction, object detection, and object tracking in visual perception are introduced.

Chapter 4: Automatic guided vehicles in rail transit intelligent manufacturing environment.

This chapter introduces progress in the development and types of AGVs in the rail transit intelligent manufacturing environment. Then, the main components of AGVs are introduced. Next, the key technologies in AGVs are introduced. This part includes the navigation methods of AGVs, the path planning of AGVs, human–robot interaction methods, and the AGV path planning application in rail transit intelligent manufacturing environment.

Chapter 5: Autonomous rail rapid transit systems.

This chapter introduces the hardware of the ART, which can be divided into railway-based vehicle basic components and various sensors based on the ART trains. Then, the technologies of the ART are introduced, including road traffic interaction, navigation, communication, and scheduling management. Finally, pedestrian detection algorithms for the ART are introduced in detail, including traditional pedestrian detection algorithms and those based on deep learning.

Chapter 6: Rail transit inspection robots systems.

This chapter introduces the development history, function, and hardware structure of rail transit inspection robots. The hardware structure is described in detail and explained in principle, which mainly includes the drive wheel, laser sensor, pan/tilt, infrared thermal imager, binocular vision camera, ultrasonic sensor, radio-frequency identification sensor, sound collection device, and wireless recharging device. Then, two key technologies to ensure inspection robots normally complete the inspection work, navigation methods, and the hand–eye system are introduced in detail.

Chapter 7: Rail transit channel robot systems.

This chapter introduces the development history and hardware composition of rail transit channel robots, including the ground rail, dual-arm robot, infrared thermometer, and laser sensor. Then, the chapter describes the TEDS intelligent sensing system in detail, including visible light image intelligent analysis, infrared thermal image temperature warning, location detection, big data analysis, and autonomous recharging. Finally, fault diagnosis algorithms based on deep learning model and related principles are introduced.

Chapter 8: Rail transit inspection unmanned aerial vehicle systems.

This chapter introduces the development history of UAVs and their applications in various fields. Next, it introduces the basic structure of fixed-wing UAVs, unmanned helicopters, and rotary-wing UAVs. Various sensors are applied to rail transit inspection. Then, UAV technologies are described in detail, including communication methods, data collection methods, and scheduling methods. The communication mode, UAV condition monitoring, positioning and navigation algorithm, path planning algorithm, and mission assignment method are introduced. Finally, applications of UAVs in detecting of rail transit intrusion are introduced.

References

[1] K. Kobayashi, M. Okumura, The growth of city systems with high-speed railway systems, Ann. Reg. Sci. 31 (1997) 39–56.

[2] Y. Huang, The correlation between HSR construction and economic development—Empirical study of Chinese cities, Transport. Res. Pol. Pract. 126 (2019) 24–36.

[3] D. Xv, H. He, SPC method research based on rail transit product manufacturing process, Elec. Locom. Mass Transit Veh. 41 (2018) 72–77.

[4] S. Zhao, Research on key technical scheme of ART signal system, Railway Commun. Signal Eng. Technol. 16 (2019) 56–58+61.

[5] H. Mosemann, F.M. Wahl, Automatic decomposition of planned assembly sequences into skill primitives, IEEE Trans. Robot. Autom. 17 (2001) 709–718.

[6] N. Papakostas, M. George, S. Makris, et al., Industrial applications with cooperating robots for the flexible assembly, Int. J. Comput. Integrated Manuf. 24 (2011) 650–660.

[7] Sonick Suri, A. Jain, N. Verma, et al., SCARA industrial automation robot, in: International Conference on Power Energy, Environment and Intelligent Control, 2018, pp. 173–177, 2018.

[8] C. Urrea, Juan Cortés, J. Pascal, Design, construction and control of a SCARA manipulator with 6 degrees of freedom, J. Appl. Res. Technol. 14 (2016) 396–404.

[9] A. Saxena, A. Saxena, Review of soft computing techniques used in robotics application, Int. J. Inf. Comput. Technol. 3 (2013) 101–106.

[10] L.C. Dulger, M.T. Das, R. Halicioglu, et al., Robotics and servo press control applications: experimental implementations, in: International Conference on Control, Decision and Information Technologies (CoDIT), 2016, pp. 102–107, 2016.

[11] N. Ngoc Son, C. Van Kien, H. Pham Huy Anh, A novel adaptive feed-forward-PID controller of a SCARA parallel robot using pneumatic artificial muscle actuator based on neural network and modified differential evolution algorithm, Robot. Autonom. Syst. 96 (2017) 65–80.

[12] F. Luan, N. Jing, Y. Huang, et al., Adaptive neural network control for robotic manipulators with guaranteed finite-time convergence, Neurocomputing 337 (2019) 153–164.

[13] L. Zhang, J. Wang, J. Chen, et al., Dynamic modeling for a 6-DOF robot manipulator based on a centrosymmetric static friction model and whale genetic optimization algorithm, Adv. Eng. Software 135 (2019) 102684.

[14] J. Yang, Z. Xu, Q. Wu, et al., Dynamic modeling and control of a 6-DOF micro-vibration simulator, Mech. Mach. Theor. 104 (2016) 350–369.

[15] G. Yang, I.-M. Chen, Task-based optimization of modular robot configurations: minimized degree-of-freedom approach, Mech. Mach. Theor. 35 (2000) 517–540.

[16] A. Gameros, S. Lowth, D. Axinte, et al., State-of-the-art in fixture systems for the manufacture and assembly of rigid components: a review, Int. J. Mach. Tool Manufact. 123 (2017) 1–21.

[17] L. Stewart, A. Dragos, Axinte. An assessment of "variation conscious" precision fixturing methodologies for the control of circularity within large multi-segment annular assemblies, Precis. Eng. 38 (2014) 379–390.

[18] Di Pei, J. Huang, F. Chen, et al., Hybrid vision-force guided fault tolerant robotic assembly for electric connectors, in: International Symposium on Micro-NanoMechatronics and Human Science, 2009, pp. 86–91, 2009.

[19] J. Huang, T. Fukuda, T. Matsuno, Model-based intelligent fault detection and diagnosis for mating electric connectors in robotic wiring harness assembly systems, IEEE/ASME Trans. Mechatron. 13 (2008) 86–94.

[20] C. Baydar, K. Saitou, Off-line error prediction, diagnosis and recovery using virtual assembly systems, J. Intell. Manuf. 15 (2004) 679–692.

[21] B.Y. Choo, P.A. Beling, A.E. Laviers, et al., Adaptive multi-scale phm for robotic assembly processes, in: Proceedings of the Annual Conference of the Prognostics and Health Management Society Prognostics and Health Management Society Conference, vol. 6, 2015, pp. 1−23.
[22] R. Bogue, Europe continues to lead the way in the collaborative robot business, Ind. Robot: Int. J. 43 (2016) 6−11.
[23] J. Sandoval, M.A. Laribi, S. Zeghloul, A low-cost 6-DoF master device for robotic teleoperation, in: IFToMM International Symposium on Robotics and Mechatronics, 2019, pp. 473−480.
[24] C. Thomas, B. Matthias, B. Kuhlenkötter, Human-robot-collaboration-new applications in industrial robotics, in: International Conference on Competitive Manufacturing, 2016, pp. 293−299.
[25] S. Bragança, E. Costa, I. Castellucci, et al., A brief overview of the use of collaborative robots in Industry 4.0: human role and safety, Occup. Environ. Safety Health 202 (2019) 641−650.
[26] A. Mohammed, B. Schmidt, L. Wang, Active collision avoidance for human−robot collaboration driven by vision sensors, Int. J. Comput. Integrated Manuf. 30 (2016) 970−980.
[27] A. Khalid, P. Kirisci, Z. Ghrairi, et al., Towards implementing safety and security concepts for human-robot collaboration in the context of Industry 4.0, in: Proceedings of 39th International MATADOR Conference on Advanced Manufacturing, 2017, pp. 1−7.
[28] S.R. González, I. Mondragón, Z. Gabriel, et al., Manufacturing control architecture for FMS with AGV: a state-of-the-art, in: Advances in Automation and Robotics Research in Latin America, 2017, pp. 157−172.
[29] A. Gaur, M. Pawar, AGV based material handling system: a literature review, Intl. J. Res. Scient. Innovat. 3 (2016) 33−36.
[30] X. Wang, C. Wang, X. Xu, et al., Research on an omni-directional AGV with differential wheels, in: IEEE International Conference on Mechatronics and Automation, 2016, pp. 1566−1571, 2016.
[31] S.W. Yoon, S.-B. Park, J.S. Kim, Kalman filter sensor fusion for Mecanum wheeled automated guided vehicle localization, J. Sensors 2015 (2015).
[32] S. Córdova, H. Rudnick, Á. Lorca, et al., An efficient forecasting-optimization scheme for the intraday unit commitment process under significant wind and solar power, IEEE Trans. Sustain. Energy 9 (2018) 1899−1909.
[33] C. Xu, H. Chen, W. Xun, et al., Modal decomposition based ensemble learning for ground source heat pump systems load forecasting, Energy Build. 194 (2019) 62−74.
[34] Y. Jiang, Q. Wang, J.-W. Gong, et al., Research on temporal consistency and robustness in local planning of intelligent vehicles, Acta Autom. Sin. 41 (2015) 518−527.
[35] I. Draganjac, D. Miklic, Z. Kovacic, et al., Decentralized control of multi-AGV systems in autonomous warehousing applications, IEEE Trans. Autom. Sci. Eng. 13 (2016) 1433−1447.
[36] T. Miyamoto, K. Inoue, Local and random searches for dispatch and conflict-free routing problem of capacitated AGV systems, Comput. Ind. Eng. 91 (2016) 1−9.
[37] W. Małopolski, A sustainable and conflict-free operation of AGVs in a square topology, Comput. Ind. Eng. 126 (2018) 472−481.
[38] T. Itou, K. Konishi, H. Wakayama, et al., Design and analysis of AGVs dispatch timing controller for self-synchronized production, in: IEEE Conference on Decision and Control (CDC), 2018, pp. 2797−2802, 2018.
[39] I.M. Galván, J.M. Valls, A. Cervantes, et al., Multi-objective evolutionary optimization of prediction intervals for solar energy forecasting with neural networks, Inf. Sci. 418 (2017) 363−382.
[40] R.T. Marler, J.S. Arora, Survey of multi-objective optimization methods for engineering, Struct. Multidiscip. Optim. 26 (2004) 369−395.
[41] B. Brito, B. Floor, L. Ferranti, et al., Model predictive contouring control for collision avoidance in unstructured dynamic environments, IEEE Robot. Autom. Lett. 4 (2019) 4459−4466.
[42] X. Li, Z. Sun, D. Cao, et al., Development of a new integrated local trajectory planning and tracking control framework for autonomous ground vehicles, Mech. Syst. Signal Process. 87 (2017) 118−137.

[43] Y. Chen, F. Dong, Robot machining: recent development and future research issues, Int. J. Adv. Manuf. Technol. 66 (2012) 1489–1497.
[44] Z. Jia, T. Wang, L. Li, et al., Laser vision-based automatic trajectory planning technology for spatial intersecting joint weld, Int. J. Precis. Eng. Manuf. 1–11 (2019).
[45] S.M. Ahmed, Y.Z. Tan, G.H. Lee, et al., Object detection and motion planning for automated welding of tubular joints, in: IEEE/RSJ International Conference on Intelligent Robots and Systems (IROS), 2016, pp. 2610–2615, 2016.
[46] D. Mitchell, F. Gu, Autonomous weld seam identification and localisation using eye-in-hand stereo vision for robotic arc welding, Robot. Comput. Integrated Manuf. 29 (2013) 288–301.
[47] Y. Liu, L. Shi, X. Tian, Weld seam fitting and welding torch trajectory planning based on NURBS in intersecting curve welding, Int. J. Adv. Manuf. Technol. 95 (2017) 2457–2471.
[48] D. Chen, P. Yuan, T. Wang, et al., A compensation method based on error similarity and error correlation to enhance the position accuracy of an aviation drilling robot, Meas. Sci. Technol. 29 (2018) 085011.
[49] Y. Bai, D. Wang, On the comparison of an interval Type-2 Fuzzy interpolation system and other interpolation methods used in industrial modeless robotic calibrations, in: IEEE International Conference on Computational Intelligence and Virtual Environments for Measurement Systems and Applications (CIVEMSA), 2016, pp. 1–6, 2016.
[50] D. Chen, P. Yuan, T. Wang, et al., A compensation method for enhancing aviation drilling robot accuracy based on Co-kriging, Int. J. Precis. Eng. Manuf. 19 (2018) 1133–1142.
[51] P. Yuan, D. Chen, T. Wang, et al., A compensation method based on extreme learning machine to enhance absolute position accuracy for aviation drilling robot, Adv. Mech. Eng. 10 (2018) 1–11.
[52] D. Chen, T. Wang, P. Yuan, et al., A positional error compensation method for industrial robots combining error similarity and radial basis function neural network, Meas. Sci. Technol. 30 (2019) 125010.
[53] C.K. Huang, Y.B. Guu, C.Y. Yang, et al., Intelligent automation module-based gear edge grinding system, in: IFToMM International Symposium on Robotics and Mechatronics, 2019, pp. 311–322.
[54] S. Han, X. Zhao, Q. Fan, et al., Automatic programming for dual robots to grinding intersecting curve, in: International Conference on Intelligent Robotics and Applications, 2019, pp. 508–516.
[55] G. Xiong, Y.E. Ding, L. Zhu, A feed-direction stiffness based trajectory optimization method for a milling robot, in: International Conference on Intelligent Robotics and Applications, 2017, pp. 184–195.
[56] F. Peng, R. Yan, W. Chen, et al., Anisotropic force ellipsoid based multi-axis motion optimization of machine tools, Chin. J. Mech. Eng. 25 (2012) 960–967.
[57] Z. Mu, B. Zhang, W. Xu, et al., Fault tolerance kinematics and trajectory planning of a 6-DOF space manipulator under a single joint failure, in: IEEE International Conference on Real-Time Computing and Robotics (RCAR), 2016, pp. 483–488, 2016.
[58] M. Ramadiansyah, W. Wahab, Modelling, simulation and control of a high precision loading-unloading robot for CNC milling machine, in: 15th International Conference on Quality in Research (QiR): International Symposium on Electrical and Computer Engineering, 2017, pp. 204–208, 2017.
[59] L. Liu, J. Li, H. Ming, Research on control system of up-down materials truss robot for a CNC Machine, Manuf. Autom. 41 (2019) 108–110+138.
[60] Z. Zhang, New machine tool loading and unloading equipment based on robot, Digital Technol. Appl. 37 (2019) 10–13.
[61] D. Fan, Z. Shao, L. Li, et al., Research on the intelligent industrial detection system for distribution network equipment, in: 4th International Conference on Control and Robotics Engineering (ICCRE), 2019, pp. 49–53, 2019.
[62] K. Hashimoto, A review on vision-based control of robot manipulators, Adv. Robot. 17 (2003) 969–991.
[63] J. Gu, H. Wang, Y. Pan, et al., Neural network based visual servo control for CNC load/unload manipulator, Optik 126 (2015) 4489–4492.

[64] H. Qi, W. Xu, Design of maglev Automatic Train Operation system and research on predictive control algorithm, in: IEEE International Conference on Computer Science and Automation Engineering, vol. 3, 2011, pp. 463–470, 2011.
[65] H. Dong, L. Li, B. Ning, et al., Fuzzy tuning of ATO system in train speed control with multiple working conditions, in: Proceedings of the 29th Chinese Control Conference, 2010, pp. 1697–1700.
[66] D. Tokody, I.J. Mezei, G. Schuster, An Overview of Autonomous Intelligent Vehicle Systems. Vehicle and Automotive Engineering, Springer, 2017, pp. 287–307.
[67] F. Bonnin-Pascual, A. Ortiz, On the use of robots and vision technologies for the inspection of vessels: a survey on recent advances, Ocean. Eng. 190 (2019) 106420.
[68] J.K. Da Cunha Pinto, M. Masuda, L.C. Magrini, et al., Mobile robot for hot spot monitoring in electric power substation, in: IEEE/PES Transmission and Distribution Conference and Exposition, 2008, pp. 1–5, 2008.
[69] G. Seet, S. Huat Yeo, W.C. Law, et al., Design of tunnel inspection robot for large diameter sewers, Proc. Comp. Sci. 133 (2018) 984–990.
[70] G. Tian, B. Gao, Y. Gao, et al., Review of railway rail defect non-destructive testing and monitoring, Chin. J. Sci. Instrum. 37 (2016) 1763–1780.
[71] J. Qu, Y. Liu, J. Zhang, et al., A method to monitor railway tracks' foreign body invasion based on phase sensitive optical fiber sensing technology, in: International Conference on Smart Grid and Electrical Automation (ICSGEA), 2017, pp. 315–319, 2017.
[72] H. Hu, B. Tang, X. Gong, et al., Intelligent fault diagnosis of the high-speed train with big data based on deep neural networks, IEEE Trans. Indust. Inform. 13 (2017) 2106–2116.
[73] M. Hong, Q. Wang, Z. Su, et al., In situ health monitoring for bogie systems of CRH380 train on Beijing–Shanghai high-speed railway, Mech. Syst. Signal Process. 45 (2014) 378–395.
[74] V. Venkatasubramanian, R. Rengaswamy, S.N. Kavuri, et al., A review of process fault detection and diagnosis: Part III: process history based methods, Comput. Chem. Eng. 27 (2003) 327–346.
[75] R. Liu, B. Yang, E. Zio, et al., Artificial intelligence for fault diagnosis of rotating machinery: a review, Mech. Syst. Signal Process. 108 (2018) 33–47.
[76] S. Liu, L. Kong, K. Fang, et al., A novel trouble of moving EMU detection algorithm based on contextual semantic information, in: 12th International Conference on Signal Processing (ICSP), 2014, pp. 827–830, 2014.
[77] R. Rodrigues Santos De Melo, D.B. Costa, Juliana Sampaio Álvares, et al., Applicability of unmanned aerial system (UAS) for safety inspection on construction sites, Saf. Sci. 98 (2017) 174–185.
[78] E. Karakose, M.T. Gencoglu, M. Karakose, et al., A new experimental approach using image processing-based tracking for an efficient fault diagnosis in pantograph–catenary systems, IEEE Trans. Indust. Informat. 13 (2016) 635–643.
[79] M. Tan, N. Zhou, J. Wang, et al., A real-time impact detection and diagnosis system of catenary using measured strains by fibre Bragg grating sensors, Veh. Syst. Dyn. 57 (2019) 1924–1946.
[80] N. Pouliot, P.-L. Richard, S. Montambault, LineScout technology opens the way to robotic inspection and maintenance of high-voltage power lines, IEEE Power Energy Technol. Syst. J. 2 (2015) 1–11.
[81] Stephen Klosterman, E. Melaas, J.A. Wang, et al., Fine-scale perspectives on landscape phenology from unmanned aerial vehicle (UAV) photography, Agric. For. Meteorol. 248 (2018) 397–407.
[82] K. Liu, M. Wang, Y. Cao, et al., Susceptibility of existing and planned Chinese railway system subjected to rainfall-induced multi-hazards, Transport. Res. Pol. Pract. 117 (2018) 214–226.
[83] Q. Fan, B. Sun, Y. Sun, et al., Data fusion for indoor mobile robot positioning based on tightly coupled INS/UWB, J. Navig. 70 (2017) 1079–1097.
[84] P. Goel, S.I. Roumeliotis, G.S. Sukhatme, Robust localization using relative and absolute position estimates, in: Proceedings 1999 IEEE/RSJ International Conference on Intelligent Robots and Systems Human and Environment Friendly Robots with High Intelligence and Emotional Quotients (Cat No 99CH36289), vol. 2, 1999, pp. 1134–1140.

[85] B. Patle, A. Pandey, D. Parhi, et al., A review: on path planning strategies for navigation of mobile robot, Defence Technol. 15 (2019) 582–606.

[86] K.H. Sedighi, K. Ashenayi, T.W. Manikas, et al., Autonomous local path planning for a mobile robot using a genetic algorithm, in: Proceedings of the 2004 Congress on Evolutionary Computation (IEEE Cat No 04TH8753), vol. 2, 2004, pp. 1338–1345.

[87] C. Wang, L. Wang, J. Qin, et al., Path planning of automated guided vehicles based on improved a-star algorithm, in: IEEE International Conference on Information and Automation, 2015, pp. 2071–2076, 2015.

[88] A.M. Chaudhari, M.R. Apsangi, A.B. Kudale, Improved a-star algorithm with least turn for robotic rescue operations, in: International Conference on Computational Intelligence, Communications, and Business Analytics, 2017, pp. 614–627.

[89] C.A. Liu, X.H. Yan, C.Y. Liu, et al., Dynamic path planning for mobile robot based on improved ant colony optimization algorithm, Acta Electron. Sin. 39 (2011) 1220–1224.

[90] V. Roberge, M. Tarbouchi, G. Labonté, Comparison of parallel genetic algorithm and particle swarm optimization for real-time UAV path planning, IEEE Trans. Indust. Informat. 9 (2012) 132–141.

[91] A. Tuncer, M. Yildirim, Dynamic path planning of mobile robots with improved genetic algorithm, Comput. Electr. Eng. 38 (2012) 1564–1572.

[92] J. Guo, Y.U. Gao, G. Cui, The path planning for mobile robot based on bat algorithm, Int. J. Autom. Contr. 9 (2015) 50–60.

[93] T. Turker, O.K. Sahingoz, G. Yilmaz, 2D path planning for UAVs in radar threatening environment using simulated annealing algorithm, in: International Conference on Unmanned Aircraft Systems (ICUAS), 2015, pp. 56–61, 2015.

[94] Q. Huang, G. Zheng, Route optimization for autonomous container truck based on rolling window, Int. J. Adv. Rob. Syst. 13 (112) (2016).

[95] O. Montiel, R. Sepúlveda, U. Orozco-Rosas, Optimal path planning generation for mobile robots using parallel evolutionary artificial potential field, J. Intell. Rob. Syst. 79 (2015) 237–257.

[96] M. Elhoseny, A. Tharwat, A.E. Hassanien, Bezier curve based path planning in a dynamic field using modified genetic algorithm, J. Compu. Sci. 25 (2018) 339–350.

[97] J. Sankari, R. Imtiaz, Automated guided vehicle (AGV) for industrial sector, in: 10th International Conference on Intelligent Systems and Control (ISCO), 2016, pp. 1–5, 2016.

[98] X. Gibert, V.M. Patel, R. Chellappa, Deep multitask learning for railway track inspection, IEEE Trans. Intell. Transport. Syst. 18 (2016) 153–164.

[99] T. Bellet, C. Martin, M. Mullins, et al., From semi to fully autonomous vehicles: new emerging risks and ethico-legal challenges for human-machine interactions, Transport. Res. F Traffic Psychol. Behav. 63 (2019) 153–164.

[100] Y.-C. Wu, F. Yin, C.-L. Liu, Improving handwritten Chinese text recognition using neural network language models and convolutional neural network shape models, Pattern Recogn. 65 (2017) 251–264.

[101] B. Liu, S. Nie, Y. Zhang, et al., Boosting noise robustness of acoustic model via deep adversarial training, in: IEEE International Conference on Acoustics, Speech and Signal Processing (ICASSP), 2018, pp. 5034–5038, 2018.

[102] K. Sinha, R. Kumari, A. Priya, et al., A Computer Vision-Based Gesture Recognition Using Hidden Markov Model. Innovations in Soft Computing and Information Technology, Springer, 2019, pp. 55–67.

[103] G. Plouffe, A.-M. Cretu, Static and dynamic hand gesture recognition in depth data using dynamic time warping, IEEE Trans. Instrum. Meas. 65 (2015) 305–316.

[104] X. Ji, J. Cheng, W. Feng, et al., Skeleton embedded motion body partition for human action recognition using depth sequences, Signal Proces. 143 (2018) 56–68.

[105] J. Wan, Q. Ruan, W. Li, et al., 3D SMoSIFT: three-dimensional sparse motion scale invariant feature transform for activity recognition from RGB-D videos, J. Electron. Imag. 23 (2014) 023017.

[106] M. Schwarz, H. Schulz, S. Behnke, RGB-D object recognition and pose estimation based on pre-trained convolutional neural network features, in: IEEE International Conference on Robotics and Automation (ICRA), 2015, pp. 1329–1335, 2015.

[107] I. Melekhov, J. Ylioinas, J. Kannala, et al., Relative camera pose estimation using convolutional neural networks, in: International Conference on Advanced Concepts for Intelligent Vision Systems, 2017, pp. 675–687.

[108] X. Liu, W. Liang, Y. Wang, et al., 3D head pose estimation with convolutional neural network trained on synthetic images, in: IEEE International Conference on Image Processing (ICIP), 2016, pp. 1289–1293, 2016.

[109] W. Chen, L. Zhao, D. Tan, et al., Human–machine shared control for lane departure assistance based on hybrid system theory, Contr. Eng. Pract. 84 (2019) 399–407.

[110] M.-P. Pacaux-Lemoine, D. Trentesaux, Gabriel Zambrano Rey, et al., Designing intelligent manufacturing systems through Human-Machine Cooperation principles: a human-centered approach, Comput. Ind. Eng. 111 (2017) 581–595.

[111] F. Tao, Y. Cheng, L. Zhang, et al., Advanced manufacturing systems: socialization characteristics and trends, J. Intell. Manuf. 28 (2017) 1079–1094.

[112] W. Xu, S. Guo, X. Li, et al., A dynamic scheduling method for logistics tasks oriented to intelligent manufacturing workshop, Math. Probl Eng. 7237459 (2019) 2019.

[113] H. Liu, N. Stoll, S. Junginger, et al., A new approach to battery power tracking and predicting for mobile robot transportation using wavelet decomposition and ANFIS networks, in: IEEE International Conference on Robotics and Biomimetics (ROBIO), 2014, pp. 253–258, 2014.

[114] A. Hamza, N. Ayanian, Forecasting battery state of charge for robot missions, in: Proceedings of the Symposium on Applied Computing, 2017, pp. 249–255.

[115] W.G. Walter, A machine that learns, Sci. Am. 185 (1951) 60–64.

[116] S. Chernova, M. Veloso, Multi-thresholded approach to demonstration selection for interactive robot learning, in: Proceedings of the 3rd ACM/IEEE International Conference on Human Robot Interaction, 2008, pp. 225–232.

CHAPTER 2

Rail transit assembly robot systems

2.1 Overview of assembly robots

2.1.1 Definition of assembly robots

The industrial robot is defined by the Robot Institute of America as a programmable multifunction operator for handling mechanical parts or workpieces, or a special mechanical device that can perform various tasks by changing programs [1]. The assembly robot is a special kind of industrial robot used to assemble parts or components on the assembly line. It is the core equipment of the flexible automated assembly system. It is a high-tech product related to the fields of optics, machinery, microelectronics, automatic control, and communication [2]. The industrial robot can handle goods, process parts, weld mechanisms, cut workpieces, and so on, to achieve assembly line operations and automated production.

Assembly refers to the process of assembling parts according to specified technical requirements; the assembled equipment becomes a qualified product after debugging and inspection. The assembly process takes place in the subsequent stages of production and is an important part of machinery manufacture [2]. The proportion of assembly operations in the manufacture of modern industrial products is gradually increasing. For example, assembly engineering in countries such as the United States and Japan occupies about 40% of the labor force in the manufacturing industry. Almost all products have assembly links, and related products in the rail transit field are no exception [3].

Rail transit, including railways, subways, magnetic levitation, and light rails, has the advantages of large capacity, fast speed, energy savings, and environmental protection, which greatly facilitates people's travel [4]. With the continuous improvements in Chinese science and technology, rail transit has further developed, and improvements in the modernization of rail transit equipment have become a current demand.

The production of rail vehicles is generally carried out by a parts supplier; and then, a vehicle factory is responsible for producing large parts and assembling the whole vehicle. The quality of the assembly directly determines the quality of the vehicle, so the factory should pay attention to the quality control of the assembly process. It is not only time-consuming and laborious, it has certain unsafe factors if all rely on manual operation.

Robot Systems for Rail Transit Applications. https://doi.org/10.1016/B978-0-12-822968-2.00002-4

With the advent of Industry 4.0, the industrial robot has gradually created huge advantages for various industries and has gradually been applied to the field of rail transit [5]. The assembly robot is a special kind of industrial robot that has the advantages of high assembly quality, fast response, and automatic operation. Assembly robots can achieve high-precision assembly, high-speed assembly, and low total assembly costs, so the assembly robot is rapidly evolving.

In the growth of various industrial robots, the assembly robot is in a leading position and has become the most widely used one. Therefore, in the assembly process of vehicles, the application of assembly robots is increasingly common, and assembly robots have had a huge role in improving product quality, shortening the assembly time, and promoting the development of industrial intelligence.

There are three existing assembly methods: manual, special assembly machinery, and assembly robot [6]. In these methods, the assembly speed of the assembly robot is at an intermediate level. Although its speed is slightly slower than that of the special assembly machine, it is suitable for large-volume, multivariety production operations and it can cope with complicated assembly tasks. Moreover, the assembly robot has a large load capacity, good assembly quality, and fast response speed, and can handle repetitive and polluting operations, which is beneficial to improving conditions for workers.

Although assembly robots account for an increasing proportion of industrial robots, they have disadvantages. Because the assembly process has become increasingly complicated, the technical level of the assembly robot needs to be improved. Therefore, it needs to adopt a more scientific and effective way to handle various assembly operations and arrange the operation mode of each process reasonably to improve operation quality. Moreover, the assembly robot has higher requirements for the working environment, and the adaptive ability still needs to be improved. External environment or human factors often interfere with the assembly robot, which sometimes results in a reduction in the assembly precision and efficiency of assembly robots.

Advantages and disadvantages of three assembly methods are shown in Table 2.1.

Table 2.1: Advantages and disadvantages of three assembly methods.

Assembly methods	Manual assembly	Special assembly machinery assembly	Assembly robot assembly
Advantages	Low cost and good flexibility	Fast assembly speed and high assembly accuracy	Good assembly quality and fast response
Disadvantages	Slow assembly speed	Only suitable for a certain assembly process; flexibility is poor	High requirements for working environment; adaptive ability needs to improve

2.1.2 Development progress of assembly robots

In the 1960s, the world's first industrial robot was created in the United States [3]. Although Japan's robotics technology started 5 or 6 years later than in the United States, after more than 40 years of development, robotics technology in Japan has a leading position in the world. Robotics technology in some European countries started late but has had better development [7].

With the continuous development of the assembly manufacturing industry, since the 1970s, the assembly robot has gradually appeared in the assembly operation factory [7]. According to a survey, the total number of assembly robots in the United States doubled from 1985 to 2000 [7]. The number of assembly robots in Japan has risen sharply, and its growth rate ranks first in the world. It once reflected the largest number of industrial robots in Japan.

The development of robots has gone through three generations: the programmable robot, which can automatically perform some repetitive operations according to established procedures; the adaptive robot, with certain self-adaptability, which can be programmed offline, and for which the work content can be changed according to different working objects [8]; and the intelligent robot, which can use various sensors to sense changes in the environment and make corresponding operations according to information fusion, and finally achieve the set goals. The assembly robot is a kind of industrial robot, and the development process is basically the same.

China began to develop robotics around the 1970s [9]. With the continuous support of the government, after the 1990s, the speed of development of Chinese robotics technology has steadily increased, and the technical level has been continuously innovated [9]. The industrial robot industry base has been established in many places in China, and the robotic automated production line has also made significant progress, gradually occupying a place on the world robotics stage [9].

With the continuous development of the Chinese manufacturing industry, the development of assembly robots in China has also achieved certain results. China has mastered the main structural design, drive system design, and coordinated control technology of assembly robots, and has made breakthroughs in basic components such as controllers, sensors, and reducers [10].

2.1.3 Types of assembly robots

The assembly robot system consists of several assembly robots and their corresponding peripheral equipment.

Assembly work includes pressing, riveting, bonding, and insertion of the parts, The assembly robot performs these actions by moving the rotating joint, turning the wrist joint, and operating the end effector [2]. The precise positioning of the assembly robot for identifying and assembling components is achieved with vision, tactile, and force sensors.

The peripheral equipment of the assembly robot assists in assembly, including the parts supplying and conveying devices. The parts supplying device is further composed of components such as a feeder and a tray, which are responsible for placing parts in a predetermined position; the conveying device is generally a conveyor belt and is responsible for conveying the workpiece to a predetermined working position. The peripheral equipment of the assembly robot and the assembly robot itself are all necessary parts of the assembly robot system. Generally, the amount of investment in peripheral equipment is higher than that for the assembly robot, and even the area needed for peripheral equipment is larger.

General steps of the assembly robot at work are shown in Fig. 2.1: (1) clamp a part by the arm, (2) place the part in the fixture, (3) place the part in the mounting position, and (4) return the arm to the part placement area and repeat the steps [11]. During the entire assembly process, the positioning accuracy and trajectory of the assembly robot arm are particularly important, and are related to the accuracy and efficiency of the assembly.

As shown in Fig. 2.2, assembly robots can be divided into different types according to different criteria. Taking the classification according to the form of motion of arms as an example, assembly robots can be divided into Cartesian coordinate assembly, cylindrical coordinate assembly, spherical coordinate assembly, vertical articulated assembly, and planar articulated assembly robot [12]. Cartesian coordinate assembly robots are the simplest type of assembly robot; the arm can be moved along three orthogonal coordinate axes to move and insert parts, and operation is simple. The arm of the cylindrical coordinate assembly robot can be lifted, swung, and telescoped. The arm of the spherical coordinate assembly robot can perform operations such as turning, lying, and telescoping. The vertical articulated assembly robot generally has six degrees of freedom (DOFs) and can perform operations anywhere in three-dimensional space. The planar articulated

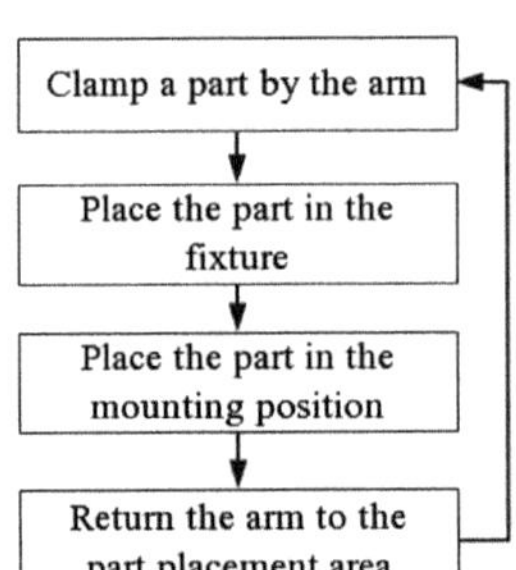

Figure 2.1
Working steps of the assembly robot.

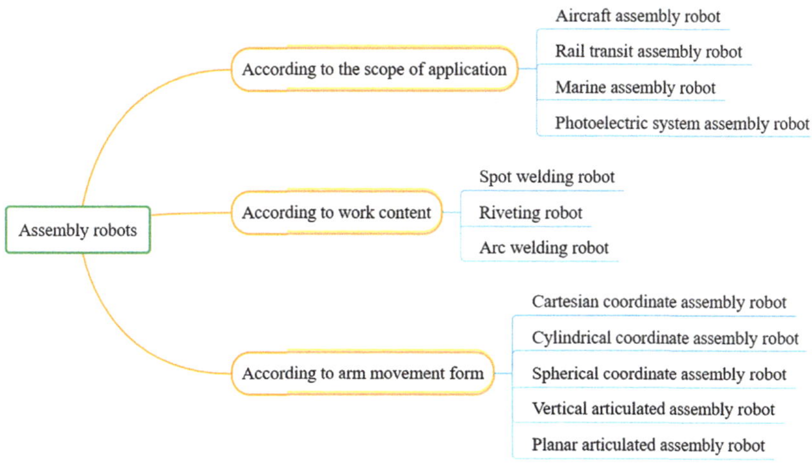

Figure 2.2
Different types of assembly robots.

assembly robot can complete automatic assembly, handling, and debugging of the product. It has the advantages of high precision, high speed, and good flexibility.

2.1.4 Key technologies of assembly robots

2.1.4.1 Precise positioning

To ensure the precise completion of assembly work, an assembly robot needs to be accurately positioned during operation. The static accuracy and dynamic response of the mechanical system will have a great impact on the positioning accuracy of the assembly robot's motion system [13]. The static accuracy of the mechanical system depends on the manufacturing geometric error, thermal deformation error, and load deformation error of the equipment, and the dynamic response.

The characteristics are related to those of the robot's control system, the damped natural frequency, and the external tracking signal. Therefore, to ensure the precise positioning of the assembly robot, it is necessary to minimize various deformation errors of the equipment, reduce the influence of environmental changes on the mechanical equipment, and continuously improve the dynamic response characteristics of the system.

2.1.4.2 Detection and sensing

The assembly robot can be disturbed by environmental changes at any time during operation; each assembly process needs real-time monitoring to ensure the quality of the entire assembly operation. This requires sensor-based detection and sensing technology to function. The sensor ensures that the assembly robot will have good automation function and achieve accurate assembly work.

A sensor is a detecting device that senses the transmission, processing, and display of measured information. It can sense the state of the internal, external, or operating object of the assembled robot, as well as process-related information such as direction, position, velocity, and acceleration. It uses electrical signals that are easy for humans to understand and record [14].

The function of the automatic detection and automatic control of assembly robots is inseparable from that of sensors. Commonly used sensors are photosensitive, acoustic, thermal, and pressure-sensitive sensors. The sensitivity and accuracy of the sensor largely determine the sensitivity of the assembly robot, and sensing technology has a huge role in various assembly operations. For example, when the robot arm of the assembly robot is operating, sensor detection technology can enable the specific positioning of the end position of the end effector. The tactile sensor can ensure that the end effector is opened or closed at an appropriate position, the assembly part does not fall off, and it can be accurately inserted into the appropriate location [14]. Moreover, the detection and sensing technology can also have a role in monitoring the temperature of the drive system and detecting the safety performance of the control system.

With continuous improvements in the level of detection and sensing technology, it is gradually developing toward a multidimensional detection sensor device and the integration of a detection sensor device, and will be better applied in assembly operations.

2.1.4.3 Assembly robot controller

Precise control of the assembly robot has a vital role in executing the assembly; research on the assembly robot operating system and kinematics control of the assembly robot has a decisive influence on the development of the assembly robot controller. The assembly robot controller includes a servo control module, a supervisory control module, and a local free control module [15].

The servo control is an effective method of controlling the amount of displacement, velocity, and acceleration of an object's motion. The servo control module of the assembly robot has good position, static force, and dynamic force control performance, and achieves hybrid servo control of the robot's operating position and space force, which is the basis of the whole system [15].

The supervisory control module of the assembly robot enables control of the safety of the system's operation, manual remote control, and other operations, and is the core of the whole system.

The local free control module is located above the servo module; it is relatively independent of the supervisory control module and can complete relatively complicated assembly operations.

The current assembly robot generally includes a special motion control card and a real-time operating system. They can accurately control some motions specified by the program and can also feed back some effective data in the field in real time.

2.1.4.4 Graphical simulation

Assembly includes the steps of conveying goods, welding, spinning, and bonding. The assembly path is complicated. To ensure the high-quality completion of assembly work, simulation of the assembly process is generally performed before formal operation.

For complex assembly operations, teaching programming is often inefficient. If a computer-aided design system is connected to the assembly robot system controller, information related to the assembly operation in the database can be used to enable off-line modeling of assembly equipment and the motion trajectory, and input of assembly process parameters, and then to conduct dynamic three-dimensional simulation [16]. This can generate assembly process control programs to verify the correctness, enforceability, and rationality of the assembly process.

Continuous testing can also improve the efficiency and reliability of assembly operations. Inspection and evaluation report generate intuitive visual effects for robot operation programming and debugging, and accumulate experience for the next operation, which makes the operation easier and more flexible.

2.1.4.5 Compliance wrist

The assembly robot is tasked with operating various products and different assemblies. To exceed the flexibility and sensitivity of the human hand and wrist during manual assembly, in addition to using the structural flexibility of the assembly robot, motion error in the assembly operation should be constantly corrected. It is necessary to design a compliance wrist specifically designed for assembly work. There are three common compliant assembly methods: passive, active , and active–passive.

As shown in Fig. 2.3, this chapter introduces the development progress and key technologies of rail transit assembly robots. Then, it discusses the main components of assembly robots: machinery components, sensors, controllers, and actuators. Arm dynamics is a research hot topic of assembly robots [17]. This chapter introduces mathematical models of forward and inverse dynamics and discusses various trajectory planning algorithms in joint and Cartesian space. Finally, the artificial neural network algorithm is introduced to perform the inverse dynamic optimization calculation of the robot arm.

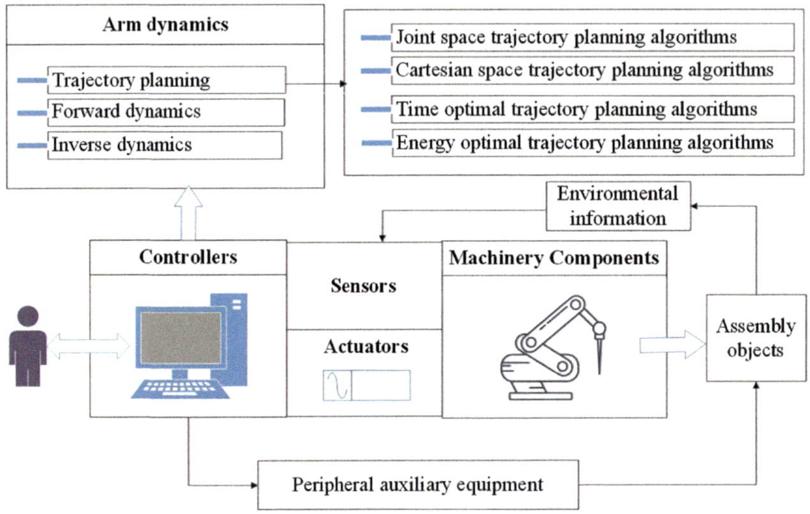

Figure 2.3
Overall frame diagram of rail transit assembly robot system.

2.2 Main components of rail transit assembly robots

The industrial assembly robot is equipped with a multifunctional, multi-DOF, and reprogrammable manipulator that can enable automatic positioning control. As shown in Fig. 2.4, the main components of an assembly robot are generally machinery components, sensors, controllers, and actuators [2].

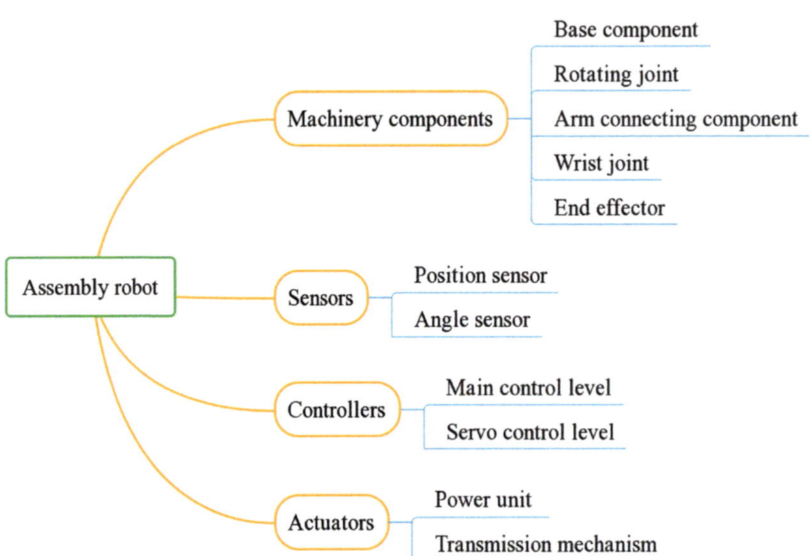

Figure 2.4
Main components of assembly robot.

Machinery components are the main part of assembly operation; they consist of the base component, rotating joint, arm connecting component, wrist joint, and end effector. Most assembly robots have three to six DOFs of movement, and the wrist typically has one to three DOFs of movement. There are many types of sensors, including position and angle sensors. The controllers are equivalent to the human brain and include the main and servo control levels. The actuators are the part that transmits power and drives the operating machine. They include the power unit and transmission mechanism.

From the structural form, assembly robots include the Selective Compliance Assembly Robot Arm (SCARA) robot and the Cartesian robot [3].

The SCARA robot is the most widely used assembly robot. It was first developed by Japan and is designed for vertical installation movements. The robot has four axes and four DOFs of motion, including translational freedom in the X, Y, and Z directions and rotational freedom around the Z axis [18]. The SCARA robot has greater flexibility in the horizontal direction and its plane-rotating joint can be loosened to make small adjustments to the position of the hole when inserting the component. However, it has large rigidity in the vertical insertion direction, which can be fixed at some height. Therefore, it can compensate for positional errors in the horizontal direction and has high motion accuracy in the vertical direction.

The Cartesian robot has three translational DOFs along the X, Y, and Z directions. Behavior of the Cartesian robot is mainly linear motion along the X, Y, and Z axes, and rigidity of the Cartesian robot is high [19]. Therefore, the Cartesian robot has higher motion precision and faster movement speed, and is suitable for high-speed and high-precision electronic product assembly. Moreover, the Cartesian robot is automatically controlled and reprogrammable. It can be used in harsh environments and can work for a long time, and is easy to operate and maintain.

2.2.1 Machinery components

Machinery components are devices that perform the work for the robot. They can pick and place objects or perform other operations in space. Machinery components are generally composed of the arm connecting component, wrist joint, rotating joint, base component, and end effector. The function and specific structure of each part will be specifically explained later [20]. The mechanical diagram of the assembly robot is shown in Fig. 2.5.

To increase the working range, a walking mechanism equipped with wheels, crawler, or leg–leg structure can be installed on the base to enable the moving function. The number of independent motion parameters required to determine the position of the assembly robot is called the robot's DOF. The DOF depends on the action required by the job target. Three DOFs are required for two-dimensional plane operations; at least six DOFs are

Figure 2.5
Mechanical diagram of assembly robot: (A) base component, (B) rotating joint, (C) arm connecting component, (D) wrist joint, and (E) end effector.

required to reach any position in space; more DOFs are required for robots that need to avoid obstacles more flexibly [21]. Assembly robots often use revolute pair or sliding pair drive joints to achieve various DOFs.

2.2.1.1 Base component

The base component is used to support the entire operating machine. It is located at the bottom of the operating machine. The base component bears all of the weight of the robot and must have sufficient strength and rigidity. It is usually made of cast iron or cast steel. The base component must be of a certain size to ensure stability of the operating machine and meet the installation of the drive unit and cable [2]. The general frame can be fixed on an automatic guided vehicle (AGV), which will drive the assembly robot as a whole to move with the AGV.

2.2.1.2 Rotating joint

The rotating joint supports the movement of the boom by a swinging motion; it is composed of structures such as a bracket, brake, and waist bracket. The rotating joint is the most loaded motion shaft that has the greatest influence on the motion accuracy of the end effector, so its design precision is high. The shaft of the rotating joint can adopt the support structure of the ordinary bearing, and it has the advantages of a simple structure and convenient installation [2]. Most of the robot's rotating joints are supported by large-diameter cross-roller bearings, which can reduce the height and better support stiffness.

2.2.1.3 Arm connecting component

The function of the arm connecting component is to connect the rotating joint and wrist joint to enable movement of the manipulator in space, including linear and rotary motion. Among them, linear motion can be achieved with a rack and pinion, linear motor, or hydraulic cylinder. Rotary motion can be achieved by directly driving the arm connecting component with the piston rod of the hydraulic cylinder and driving the arm connecting component with a direct current (DC) motor with a harmonic transmission device.

The arm connecting component should generally have at least three DOFs, divided into a boom component and a small arm component. The Programmable Universal Machine for Assembly industrial robot is a six-DOF joint type industrial robot driven by a DC servo motor [22]. The arm connecting component is made of high-strength aluminum alloy materials; each movement is geared. The arm connecting component structure should be designed as a closed and partially reinforced rib structure to increase stiffness and strength. The arm connecting component structure can be divided into telescopic, rotary telescopic, and flexion and extension structures.

The length of the arm connecting component should meet the requirements of the working space. Because the rigidity and strength of the arm connecting component directly affect the overall motion rigidity and compliance movement of the robot, a high-strength, lightweight material should be used as much as possible to reduce weight. A design involving a light arm connecting component structure is a hot topic of research. By optimizing the arm connecting component structure to reduce the weight of the robot's main motion component, it improves the high-speed performance of the assembly robot. Measures are (1) the use of high-strength and high-rigidity lightweight composite materials, such as epoxy-based carbon fiber composite materials; and (2) designing thin-walled structures or honeycomb structures under conditions that meet strength requirements, allowing larger elastic deformation to reduce the weight of an arm connecting component.

Research on assembly robot arm connecting components promotes continuous development in an intelligent direction, so that it can sense and feed back changes in the external environment such as position shift, light and shadow change, and response in time. This requires the installation of a microcomputer, an optical rangefinder, and a vision camera on the robot arm connecting component.

The use of mechatronics technology for robotic arm connecting components improves its repeatability and modular assembly capabilities, and the use of new materials promotes energy conservation and environmental protection. This requires mechanical controllers with sensors, programmable controllers, and hydraulic components.

2.2.1.4 Wrist joint

The wrist joint is located in the middle of the arm connecting component and end effector, including the mechanical interface, drive gear, drive shaft, and wrist joint housing, which determine the attitude of the end effector in space, to change the overall orientation of the operating machine [22]. The wrist joint generally requires two to three DOFs and is composed of three rotary joints. The structure of the wrist joint is compact and the mass is small, and the moving shafts are driven separately. Specific movements of the wrist joint include arm rotation, wrist swing, and hand rotation; most use pneumatic, hydraulic, and rack and pinion transmission.

2.2.1.4.1 Remote center compliance wrist joint

To adapt the assembly robot to a variety of products, assembly tasks, and peripheral equipment, researchers proposed a compliant assembly method. These methods include the passive compliance wrist joint, passive compliance airflow method, active compliance wrist joint, and active compliance workbench. Among them, the passive remote center compliance (RCC) wrist joint application is the most successful and has begun commercial production [23].

A compliance wrist joint is an elastic, compliant device with multiple DOFs, usually mounted on the robot's hand. The RCC wrist joint is designed according to the far-center compliance theory, in which there is a compliant center point in the lower end of the compliance wrist joint. At this point, lateral force is applied, which produces the same lateral displacement as the direction of the force, without accompanying angular displacement; likewise, moment acting on the compliant center produces angular displacement around the compliant center along the direction of the moment without accompanying lateral displacement. In this way, when assembly work is carried out, as long as the compliant center is placed at the end of the part, position error and posture error caused by the positioning of the robot can be better compensated, and the assembly can be smoothly performed.

Three or more elastic rods are used in the RCC wrist joint. The increased number of elastic rods improves the load carrying capacity and reliability, improving the isotropy of the wrist joint's compliance parameters; however, it also increases the stiffness of the RCC wrist joint, complicates the structure, and increases manufacturing costs.

2.2.1.4.2 Passive remote center compliance wrist joint

The passive RCC wrist joint is a purely mechanical device because of its structural characteristics. The passive RCC wrist joint has a certain DOF. It overcomes assembly lag caused by positioning error of the assembly robot through elastic deformation or a slight displacement of components and achieves the effect of compliant assembly. The passive

RCC wrist joint is available in a variety of configurations, including flexure-driven, three-bar linkage, laminated elastic rods, gear mechanisms, and planar linkages.

The passive RCC wrist joint adopts a laminated elastic rod as an elastic body, which can better solve the problems of large size and poor dynamic damping characteristics of the early RCC wrist joint with a metal elastic body. The laminated elastic rod is composed of a rubber sheet and a rigid metal sheet, and the material is lightweight and has good shock-absorbing performance [24]. Moreover, the connection between the laminated elastic rods is a spherical joint, which is favorable for the wrist joint to rotate under the condition of maintaining rigidity, and the self-adaptive ability of the ball joint is used to prove that the workpiece is smoothly inserted into the hole in the chamfering range, and no jamming occurs.

2.2.1.4.3 Active remote center compliance wrist joint

The active RCC wrist joint obtains contact and force information directly through sensors and feeds it back to the arm connecting component or wrist joint of the assembly robot. The compliant assembly is achieved by a slight compliant movement or by the direction and magnitude of the correcting force. Although the passive compliance assembly robot can solve the problem of translational positioning error in most assembly operations and successfully complete the task of inserting the shaft into the hole, the passive RCC wrist joint has difficulty in the case of no chamfering or two-point contact. The active RCC wrist joint with a displacement sensor, attitude sensor, and computer controllers can enable a hole search and posture adjustment. The RCC has a wide range of active compliance wrist joints, including an RCC wrist joint with mechanical obstruction, RCC wrist joint without chamfering, RCC wrist joint with variable structural parameters, and RCC wrist joint with detection.

The RCC wrist joint, which is suitable for use without chamfering, is a structure that can be arbitrarily adjusted according to the stiffness and compliant center [25]. It sets the appropriate stiffness matrix and the position of the compliant center through the control software and changes the force signal into the corresponding position adjustment. The main controller controls the RCC wrist joint to rotate around the compliant center to make the end actuator clamp. The workpiece is in the best insertion position and posture to ensure smooth execution of the assembly. The active RCC wrist joint can enable a hole search without chamfering and adjust the post position after entrance. It can be applied to the assembly work of a shaft and hole with large initial error and has high matching precision [25].

The active RCC wrist joint and passive RCC wrist joint have advantages and disadvantages. The active RCC wrist joint searches for movement and processing information for a long time, and the working time is also long. In addition, the active RCC wrist joint is generally complicated in structure and relatively expensive. However, compared with the passive RCC wrist joint, the active RCC wrist joint can be adapted to large positioning errors and parts without chamfering, and the active RCC wrist joint can use control and sensing technology better. Its various DOFs can be controlled at the same time. Therefore, the active RCC wrist joint has a broader development prospect.

2.2.1.5 End effector

The end effector is located at the end of the wrist joint and is used to grab the workpiece directly during assembly. It is usually driven by a motor and uses a screw or worm gear mechanism to transmit power. It also attaches sensors for control.

The assembly robot performs various assembly operations and is completed by the end effector. To facilitate assembly, it is sometimes necessary to equip different types of assembly tools for the end effector, including two-, three-, or multifinger claws, which may have multiple DOFs and are programmable to perform work [22]. The intelligent gripper can largely reflect the intelligence of the assembly robot. The claw mechanism is equipped with various sensors such as tactile, force, and displacement. Functions such as determining or recognizing the position and shape of the object to be grasped can be achieved. Then, the gripper or the robot can have adaptive and intelligent control.

Depending on the application and structure, end effectors can be divided into mechanical grippers, special tools, and universal hands.

2.2.1.5.1 Mechanical gripper

The mechanical gripper is composed of a variety of structures such as a rotary mechanical gripper, a mobile mechanical gripper, and an internal mechanical gripper. The wedge bar rotary mechanical gripper uses a wedge and lever structure to cause a gripping action and clamping force of the gripper to clamp the workpiece. The elastic force is released by the wedge rearward movement to release the gripper [26]. The rack and pinion parallel link type of moving gripper drives the rack and the sector gear through the electromagnetic drive and drives two connecting rods, a claw, and a gripper to form a parallel four-bar mechanism. The two jaws are driven to translate and clamp, finally releasing the workpiece. The inner-supporting mechanical gripper adopts a four-bar linkage mechanism to transmit the supporting force, and the supporting direction is opposite the outer clamping type of these two modes. The jaws support the workpiece from the inner hole of the workpiece; most of the three grippers are used to position the inner bore after the tension.

2.2.1.5.2 Special tools

Special tools include a vacuum suction device, spray gun, arc welding torch, electromagnetic adsorption device, and spot welding gun, which can help the assembly robot to complete some specific tasks [26].

2.2.1.5.3 Universal hand

In general, the end effector has a specific function, but it can also be designed as a multipurpose structure for a variety of work tasks [26]. For quick and automatic end effector changes, a universal hand with electromagnetic chuck or pneumatically locked adapter can be used.

2.2.2 Sensors

Various environmental factors interfered with the operation of the assembly robot [22]. To ensure the quality of the entire assembly work, it is necessary to install various sensors monitoring each assembly process in real time. Sensors can sense, transmit, process, and display measurement information. They can sense the specified measurements and convert them into available output signals according to certain rules.

There are many types of sensors. According to the measurement object, these can be divided into internal information sensors for detecting the position, angle, speed, force, torque, temperature, and abnormal change inside the equipment, and external information sensors for detecting the external environmental state of the equipment. According to whether the sensor is in contact with the measurement object, it can be divided into contact sensors including tactile, pressure, slide, and noncontact sensors including laser ranging, vision, and ultrasonic ranging sensors.

The force sensor is usually installed in the wrist joint to detect force and is generally employed in precision assembly or force control. The visual sensor can be mounted on the gripper to identify the pose of the object and to compensate for the position of the part or workpiece. Tactile and proximity sensors are typically attached to the fingertips to compensate for positional errors in parts or workpieces and prevent collisions.

Industrial assembly robots require sensors with (1) good stability and high reliability; (2) high accuracy and good repeatability; (3) small mass, small size, and convenient and reliable installation; and (4) strong antiinterference ability [27].

The sensor is intelligent, systematic, miniaturized, digital, and networked, and is the primary link for automatic detection and automatic control [28]. The sensing performance of an assembly robot highly depends on the sensitivity and accuracy of the sensor. Detection and sensing technologies are also gradually developing toward integrating detection sensor devices and multidimensional sensors. The following discussion introduces in detail sensors commonly used in several assembly robots.

2.2.2.1 Position sensor

The role of the position sensor is to measure the position of the target object. It can measure the position of the target object and convert it into a usable output signal. Using position sensors, the assembly robot can measure its position and that of the parts to be assembled when it is working. Commonly used position sensors are resistive position, photoelectric position, magnetic position, and capacitive position sensors [29].

2.2.2.1.1 Resistive position sensor

Resistive position sensors include wire-wound resistance position and slide-wire resistance position sensors. The core of resistive position sensing is a potentiometer that uses a fine tooth screw drive brush as the middle tap. The pitch of the screw and the resistance of the resistance wire determine the resolution of the sensor. Generally, the resistance is small, and the pitch of the screw is small. However, the requirements for resistance and wire diameter conflict with each other. Therefore, the resolution of the resistive position sensor is not high and the linearity is poor [29].

The potentiometer is one of the most common position sensors. It usually includes an electronic component and a vernier sliding on the surface of the component, and a convex piece with multiple angles that can be deformed is arranged on top of the wall surface of the potentiometer fixing body to clamp the electronic component.

The measurement position of the potentiometer is proportional to the output voltage. By measuring the voltage drop of the electrical contact along the resistance bar, the corresponding position information can be obtained. Unlike ordinary sliding varistors, the terminal voltage and terminal resistance of the potentiometer will not change with the change in external position.

Potentiometers have many advantages. They are compact in structure and lightweight. Common potentiometer structures are rotary, linear, or curved. Although the price of potentiometers will increase as accuracy increases, the price of potentiometers is generally lower. Moreover, fine-tuning the resistance bar by laser can effectively improve the linearity of the potentiometer.

Of course, there are many disadvantages to potentiometers. The potentiometer itself is not accurate and it cannot be used as a precision servo control device. The potentiometer is easily to be interfered with by the environment. When there are interfering substances such as dust and sand in the environment, it easily causes wear to the potentiometer. This will cause the resistance to change and greatly reduce the accuracy of the potentiometer. In addition, the potentiometer has poor vibration resistance and is unsuitable for working in high-temperature environments. With the continuous development of material technology, especially conductive plastics, the performance of potentiometers continues to improve, which can maintain a longer service life and better performance parameters.

2.2.2.1.2 Photoelectric position sensor

The photoelectric position sensor is a photoelectric device based on the lateral photoelectric effect. It is relatively sensitive to the position of the light spot on the incident photosensitive surface. When the light beam hits the grating, the photoelectric position sensor will detect the position of the light spot and output a corresponding position signal.

The photoelectric position sensor has the positioning performance of the photodiode array; it can obtain good accuracy after precision installation and has the characteristics of a simple circuit configuration, high resolution, fast response speed, and high sensitivity, and it is widely used [29]. However, similar to the resistive position sensor, the photoelectric position sensor has poor shock resistance, is susceptible to environmental interference, cannot work under extreme temperature conditions, is sensitive to foreign objects, and has no symptoms before its failure.

2.2.2.1.3 Magnetic position sensor

The magnetic position sensor is a sensor that outputs a position signal using changes in magnetic field performance. When the position of the measured object changes, the relative displacement between the permanent magnet and the magnetic detector will cause a change in the magnetic field performance. The change is proportional to the relative displacement. Finally, the change in the magnetic field will be converted as an electrical signal to obtain information about the position of the measured object.

Compared with the photoelectric position sensor, the magnetic position sensor is less susceptible to interference from foreign substances and is more robust. However, because of the hysteresis effect, the magnetic field will change the initial magnetic force of the iron block and the accuracy of the magnetic position sensor is not high [29]. The magnetic position sensor has poor impact resistance and is easily affected by precision machining errors and installation tolerances of various parts. In addition, over time, permanent magnets will absorb more and more debris, causing the sensor to fail.

2.2.2.1.4 Capacitive position sensor

The capacitor is a device that holds charges. It is a component widely used in electronic equipment. It usually consists of two electrode plates that are insulated from and close to each other. The amount of charge that a capacitor can store is affected by the size of the electrode plates, the dielectric constant of the material between the plates, and the overlap and distance between them.

Capacitive position sensors use various types of capacitors as sensing elements and are devices that can measure the distance between electrode plates. When one of the plates moves relative to the other, capacitance along the axis in the circuit will change, and the amount of capacitance change will be output in the form of an electrical signal. The corresponding position information can be obtained through the electrical signal. The effect of the distance between the electrode plates on capacitance is nonlinear, so a measurement circuit with a compensation function must be used to perform nonlinear compensation on the output capacitance.

The capacitive position sensor has a simple structure and low power consumption. However, capacitance is affected by temperature, humidity, surrounding materials, and foreign objects, and the temperature and humidity drift coefficients are large, which means that the performance of the capacitive position sensor is unstable and it is difficult to improve the accuracy, so the application range of the capacitive position sensor is narrow [29].

2.2.2.2 Angle sensor

The angle sensor is a sensor that can convert the measured angle into a usable output signal. The position and velocity of the measured object can be obtained by measuring the angle. Common angle sensors include inclination, magnetic angle, and capacitive angular displacement sensors [27].

2.2.2.2.1 Inclination sensor

The inclination sensor is based on Newton's second law and can be used to measure inclination. An inclination sensor uses the internal acceleration sensor chip to collect the magnitude of the component of gravity acceleration on the sensor's sensitive axis. When the inclination sensor is stationary, there is no acceleration in the side and vertical directions, so only gravity acceleration acts on it [27]. The angle between the vertical axis of gravity and the sensitive axis of the sensor can be obtained by using the component of the acceleration of gravity on the sensitive axis of the sensor [27]. This is the angle of inclination. The control unit will use this measurement principle to process the pulse signal transmitted by the chip accordingly and obtain the magnitude of the inclination, and finally to output it to the network or display screen through the communication unit.

Inclination sensors are often used for system-level measurement. From the working principle, they can be divided into solid pendulum inclination, liquid pendulum inclination, and gas pendulum inclination sensors [27]. The inclination sensor can also be used to measure the amount of inclination change relative to the horizontal plane.

The solid pendulum of the solid pendulum inclination sensor is composed of a pendulum, cycloid, and bracket. The pendulum is affected by gravity and a pendulum pulling force. A force-balanced servo system is widely used in the design. Based on this principle, a strain-type inclination sensor can be obtained. The solid pendulum inclination sensor has a clear pendulum length and a pendulum center. Its principle of action is similar to that of an acceleration sensor. There are many product types in practical application. Among them, the electromagnetic pendulum is widely used, and the measurement accuracy, overload resistance, and measurement range of the product are high.

The glass case of the liquid pendulum tilt sensor is filled with a conductive liquid and is connected to the outside by three platinum electrodes that are equidistant and parallel to each other. When the case is in a horizontal state, the three electrodes are located at the same depth in the conductive liquid. When the case is tilted, the depth of the three electrodes immersed in the liquid will change, causing the strain gauge to change, which will also cause the output electrical signal to change, thereby obtaining the amount of change in the tilt angle. The liquid pendulum inclination sensor system is relatively stable and is widely used in high-precision systems.

The inertial element of the gas pendulum inclination sensor is composed of a closed cavity, a gas, and a hot wire. Gas is subject to buoyancy when heated. Like the sensitive mass of a solid pendulum and liquid pendulum, hot air flow always tries to keep vertical, so it also has the characteristics of a pendulum. When the cavity is subject to acceleration or its plane is inclined relative to the horizontal plane, the resistance value of the hot wire will change owing to the energy exchange between the gas and the hot wire, and the amount of change will have a certain function relationship with the acceleration and angle. The gas has a small mass and a small inertial force and is not susceptible to the impact of vibration. Therefore, the gas pendulum inclination sensor has a strong ability to resist vibration or shock. However, many factors affect the gas motion; thus, the accuracy of the gas pendulum inclination sensor is not high [27].

2.2.2.2.2 Magnetic angle sensor

The magnetic angle sensor is a new type of angle sensor that uses high-performance integrated magnetic-sensitive elements. It works based on magnetic sensitivity. The magnetic circuit generating device is placed at the rotating end of the magnetic angle sensor. When the rotating shaft rotates, the magnetic field direction rotates, which in turn affects the sensitive components fixed on the circuit board. This effect is small, but after the microprocessor performs intelligent signal processing, it becomes a voltage or current signal that can be recognized by electronic equipment and output to the upper computer system.

The magnetic signal has the characteristics of noncontact induction, so the magnetic angle sensor has the advantages of no contact and no friction, high sensitivity, strong reliability, no noise, long service life, and so forth. It is widely used in harsh environments such as frequent mechanical changes [27].

2.2.2.2.3 Capacitive angular displacement sensor

The capacitive angular displacement sensor is a precision measurement sensor based on the noncontact capacitive principle. Generally speaking, a capacitive angular displacement sensor is composed of one or several groups of fan-shaped fixed plates and rotating plates.

To ensure the accuracy and sensitivity of the sensor and avoid indirect changes of the dielectric constant and the shape of the electrode plate owing to changes in environmental temperature and other factors, which adversely affect the performance of the sensor, higher requirements are placed on the material, processing technology, and installation accuracy of the sensor. Among them, designing the capacitive angular displacement sensor as a differential structure is a better solution [27].

Capacitive angular displacement sensors are noninert and nondestructive and have the characteristic of nonfriction common to general noncontact instruments; they also have the advantages of a large signal-to-noise ratio, small zero drift, wide frequency response, and small nonlinearity and are widely used in various fields [27].

2.2.3 Controllers

Controllers are equivalent to the human brain and are used to govern the movement of the operating machinery in the desired sequence, along with a defined position or trajectory. There are many ways to classify controllers. Their structure includes open-loop and closed-loop; modes include program, adaptive, and intelligent controllers.

2.2.3.1 Controller structure

Most industrial robots use two-level computer control; the first level is the main control and the second is the servo control.

The main control level consists of the main control computer, main control software, peripherals, and a teach-in control box. The functions of the main control level are to establish an information channel between the operation and the industrial robot, transfer the work instructions and parameters, feed back the working state, complete various calculations required for the job, and establish an interface with the servo control level [30].

The servo control level consists of a servo control system, including a servo drive and servo controller circuit. Each servo control system corresponds to a certain joint and is used to receive motion command signals such as the position and speed sent by the main control computer to each joint to control the operation of each joint of the operating machine in real time.

2.2.3.2 Controller performance requirements

The role of the controller in the assembly robot is critical. Failure of any part of equipment can affect the failure of the entire assembly operation. Therefore, the performance requirements of the controller are also strict.

2.2.3.2.1 Reliability

The controller consists of communication, data acquisition, gesture recognition, and robot control units. To satisfy the reliability of the control equipment, each part must be in a normal working state [9]. In this way, even in some special cases, if there is a problem in one part, the entire controller will not go out of control, which ensures that the assembly robot is always within the controllable range.

2.2.3.2.2 Accuracy

The correct assembly operation requires the operation of the mechanical end effector to have accurate positioning, and the controller controls the position and posture of the various components of the assembly robot and the operation, or not. Therefore, the accuracy of the controller determines the quality of the assembly operation [9]. The adopted robots can improve the accuracy of the controller to ensure the efficiency of the production of the factory.

2.2.3.2.3 Security

Safety is the bottom line requirement for all tasks. In the event of a safety incident, life and property as well as humans will be greatly harmed. The mechanical structure of the assembly robot's controller is tight and the electrical equipment is complex. Whether it is working or in maintenance, if it cannot be effectively controlled, the assembly robot may harm humans. Therefore, it is important to set the control parameters of the assembly robot and ensure its safety.

Taking the SCARA type robot as an example, the mechanical body in its control structure includes three rotating joints and one moving joint; each joint adopts a stepping motor as a driving device [31]. The multiaxis control card of the stepping motor controls the movement of the stepping motor. It is easy to achieve speed regulation and the maintenance is convenient. The control card of the SCARA controller is composed of an address decoding circuit, a data buffer and a driving circuit, a clock circuit, a state output circuit, and a control logic circuit. Control of speed is achieved by controlling the frequency of the pulse; control of acceleration or deceleration is realized by changing the pulse timing, and control of the position is enabled by controlling the number of pulses.

2.2.4 Actuators

The actuators transmit power and drive the operating machine, including the power unit and the transmission mechanism. The power sources of the common actuators are the pneumatic, hydraulic, and electric drives.

2.2.4.1 Pneumatic drive

The pneumatic robot drives the operating machine with compressed air and has the advantages of convenient air source, quick action, a simple structure, safety and reliability, low price, and no pollution, and it is suitable for an explosion-proof environment. The disadvantage is that the air source pressure of the pneumatic device is only about 6 KPa, so the snatch force is small, generally only a few tens of newtons and the maximum is more than 100 N [32]. Because of the compressibility of the air, it will compress and deform under the load, it is difficult to control the precise position of the cylinder, and the stability of the working speed is poor.

The gas source of the robotic pneumatic drive system includes an air compressor, a gas storage tank, and a pressure-regulating filter. The pneumatic triple joint is composed of a water separation filter, a pressure regulator, and an oil mister. The pneumatic valve includes electromagnetic gas, throttle speed regulating, and deceleration pressure valves. The pneumatic power mechanism mostly uses linear and swing cylinders.

2.2.4.2 Hydraulic drive

The hydraulic system of the hydraulic robot includes a hydraulic actuator and some peripheral devices that drive the hydraulic actuator. The hydraulic actuator is a device that converts the pressure energy of the working oil generated by the hydraulic pump into mechanical energy. The peripheral devices are provided as (1) a hydraulic pump that forms a hydraulic pressure, (2) a conduit for supplying hydraulic oil, (3) a hydraulic control valve that controls the flow of the working oil, and (4) a control loop.

Because the hydraulic pressure is higher than the air pressure, generally about 70 KPa, the hydraulic robot has a large snatch capacity, up to thousands of newtons [32]. The hydraulic robot is compact and sensitive. Owing to inevitable leakage, noise, and low-speed instability in the hydraulic system, and because the power unit is cumbersome and expensive, it is not used much except for some special applications.

2.2.4.3 Electric drive

The electric servo drive system of the electric robot is an actuator that directly or indirectly drives the robot body by a mechanical transmission mechanism using torques and forces generated by various motors to obtain various movements of the robot [33].

There are many types of motors, which offer a variety of options for designing electric robots; electric robots can use a variety of flexible control methods. Electric motors were mostly driven by stepper motors in early days. Later, DC servo drive units were developed. At present, alternating current (AC) servo drive units are also rapidly developing.

Almost all assembly robots use electric power, and high-performance drive motors have been researched internationally to achieve high-speed, high-precision drive performance. High-performance drives include direct-drive motors and brushless AC motors. Among them, direct-drive motors have been successfully used in assembly robots and have been widely employed in various countries. The direct-drive motor is the product of the development of servo technology. The assembly robot application directly drives the motor.

2.2.4.3.1 Higher precision and repeatability

The direct-drive motor is a low-speed, high-torque output directly connected to the load without a speed reduction mechanism. Without additional mechanical components, the axial and radial mechanical runout values of the equipment are minimized, and the operating and measurement accuracy of the equipment are maximized. The accuracy of a precision direct-drive torque motor can typically be as high as 1 arc second [32]. Its high positioning accuracy, high response speed, and other features better guarantee and improve the accuracy of the equipment, simplifying the equipment structure and saving costs.

2.2.4.3.2 Higher energy efficiency

The use of direct-drive equipment reduces intermediate drive components and unnecessary wear, greatly increasing energy efficiency. Experience has shown that direct-drive motors can increase energy efficiency by approximately 40% compared with conventional servo motors, saving 60% in energy consumption [32].

2.2.4.3.3 Compact design

The direct-drive motor eliminates the concept of lifting the transmission torque through the traditional mechanical structure, and the motor is directly connected to the mechanical equipment without the need for a reduction gear, which saves a lot of space by eliminating the complicated transmission structure [33]. In this way, the entire equipment structure is compact, reliable, and easy to control. For device manufacturers, the overall device design is more optimized; for the end user, the device requires less installation space.

2.2.4.3.4 High dynamic response

The traditional mechanical transmission components limit the start—stop speed and adjustment time of the machine. The direct-drive motor has a low-speed and high-torque output; it does not require elastic coupling components and can match the load moment of inertia to 50 to 1000 times the rotor moment of inertia [32]. Therefore, the drive equipment of the direct-drive motor has better real-time performance and the response speed of the equipment is higher, which greatly shortens the time required for starting and stopping operations.

2.2.4.3.5 Longer life

Gears, belts, and other components in conventional mechanical transmission equipment are subject to wear and require regular maintenance and overhauling. In direct-drive equipment, these transmission components are not present and there is almost no wear, so the maintenance time is greatly reduced, the life cycle of the entire equipment is prolonged, and the overall cost of use is also reduced [33].

2.3 Arm dynamics of rail transit assembly robots

The robotic arm dynamics of the assembly robot are mainly used to study the relationship between the arm movement variable and force. The robotic arm is a complex dynamic system with multiple inputs and outputs. The mechanical dynamics of the assembly robot are studied. Through the simulation, control, and optimization design of the arm, analysis of control among the various structures under the high dynamic response demand is achieved.

In previous studies of arm dynamics, Newton—Euler, Kane, Lagrange, and Gauss methods have been applied based on different principles [31]. The Lagrange method is representative of analytical mechanics. The absolute coordinate is used to describe the large displacement motion of the system. Based on the perspective of system energy, the finite-dimensional dynamic model can be obtained. The derivation process is relatively simple. The method analyzes the dynamics of the robotic arm of the assembly robot.

2.3.1 Arm postures

An articulated robotic arm of an assembly robot is generally composed of several booms connected in series. In 1956, Hartenberg and Denavit proposed the Denavit—Hartenberg (D-H) method for the mathematical modeling of robotic arms [34]. The adjacent two arms of the robot arm are connected in series by a rotating or moving pair, their rotating axes may not be parallel, and the respective arms may not be in the same plane.

2.3.1.1 Denavit—Hartenberg coordinate transformation

Establishing the mechanical arm linkage coordinate system by the D-H method is achieved in this way: (1) a corresponding coordinate system is established on each joint to describe the coordinates of the points on each respective arm; (2) the coordinate transformation is performed; and (3) the coordinates of the end of the robot arm are obtained using the overall coordinate system.

In three-dimensional space, assuming that the length of each arm is C^2, the torsion angle of each arm is (s, m), the distance between the two arms is $dis[m]$, and the angle between the two arms is $w(s, m)$. The coordinate transformation can be completed between two adjacent arms by two rotations and two translations. The specific steps are presented next [34]:

(a) Rotate the coordinate system $f(n) = g(n) + h(n)$ of the first arm by $f(n)$ around the $g(n)$ axis so that the $h(n)$ axis of the first coordinate system is coplanar with the r_b axis.
(b) Translate l_b along the l_a axis of the coordinate system $O_iX_iY_iZ_i$ such that the X_i axis of the first coordinate system is collinear with the X_{i+1} axis.

(c) Translate d_i along the X_{i+1} axis of the coordinate system $O_{i+1}X_{i+1}Y_{i+1}Z_{i+1}$ so that the coordinate origins of the two-arm coordinate system coincide.
(d) Rotate α_i around the X_i axis of the coordinate system $O_iX_iY_iZ_i$ so that the Z_i axis of the first coordinate system is on the same line as the Z_{i+1} axis.

Each transformation is based on a new coordinate system, assuming that $C\theta_i$ represents $\cos\theta_i$, $S\theta_i$ represents $\sin\theta_i$, $C\alpha_{i+1}$ represents $\cos\alpha_{i+1}$, and $S\alpha_{i+1}$ represents $\sin\alpha_{i+1}$. The coordinate transformation matrix is presented as [34]:

$$^{n}T_{n+1} = A_{n+1} = Rot(Z_i, \theta_{i+1}) \times Trans(0,0,l_{i+1}) \times Trans(d_{i+1},0,0) \times Rot(X_{i+1}, \alpha_{i+1})$$

$$= \begin{bmatrix} C\theta_{i+1} & -S\theta_{i+1}C\alpha_{i+1} & S\theta_{i+1}S\alpha_{i+1} & d_{i+1}C\theta_{i+1} \\ S\theta_{i+1} & -C\theta_{i+1}C\alpha_{i+1} & C\theta_{i+1}S\alpha_{i+1} & d_{i+1}S\theta_{i+1} \\ 0 & S\alpha_{i+1} & C\alpha_{i+1} & l_{i+1} \\ 0 & 0 & 0 & 1 \end{bmatrix} \tag{2.1}$$

The final coordinate transformation matrix is finally obtained as [34]:

$$T_n = {}^{0}T_1 {}^{1}T_2 {}^{2}T_3 \ldots {}^{n-1}T_{n1} = A_1A_2A_3\ldots A_n \tag{2.2}$$

2.3.1.2 Lagrange equation

After the coordinate transformation is performed by the D-H coordinate method, the Lagrange function can be established; that is, total kinetic energy K and total potential energy P of the system are made [35]:

$$L(q,\dot{q}) = K(q,\dot{q}) - P(q,\dot{q}) \tag{2.3}$$

The Lagrange equation for assembling a robotic arm can be expressed as [35]:

$$F_i = \frac{d}{dt}\bullet\frac{\partial L}{\partial \dot{q}_i} - \frac{\partial L}{\partial q_i}, i = 1,2,3,\ldots,n \tag{2.4}$$

where F_i represents the generalized force acting at the i_{th} joint; q_i represents the generalized coordinates of potential energy or kinetic energy; and n represents the number of links.

Substituting Eq. (2.3) into Eq. (2.4), the simplification is available as [35]:

$$F_i = \frac{d}{dt}\bullet\frac{\partial K}{\partial \dot{q}_i} - \frac{\partial K}{\partial q_i} + \frac{\partial P}{\partial q_i}, i = 1,2,3,\ldots,n \tag{2.5}$$

2.3.1.3 Kinetic energy of the connecting rod

For any point on the link i, ${}^0Q_i$ represents the transformation matrix of the link to the base coordinate, ${}^ir$ represents the position matrix of the point in the link coordinate system $O_iX_iY_iZ_i$, and its position vector is ${}^0r = {}^0Q_i{}^ir$. Assuming q is a joint variable, position vector ${}^0r$ at any point on the link is derived from time t to obtain its speed [36]:

$$v_i = \frac{d}{dt}\left({}^0Q_i{}^ir\right) = \left(\sum_{j}^{i}\frac{\partial^0 Q_i}{\partial q_j}\dot{q}_i\right){}^ir \tag{2.6}$$

Let the kinetic energy of any mass point dm on the connecting rod i be dE_{ki}; the kinetic energy of the connecting rod i can be presented as [36]:

$$K_i = \int dK_i = \int \frac{1}{2}v_i^2 dm = \frac{1}{2}Trace\left[\sum_{j=1}^{i}\sum_{k=1}^{i}\frac{\partial^0 Q_i}{\partial q_j}I_i\left(\frac{\partial^0 Q_i}{\partial q_k}\right)^T\dot{q}_j\dot{q}_k\right] \tag{2.7}$$

where I_i denotes the pseudoinertia matrix of the i points of the link and the bar centroid coordinates are denoted by $\overline{x}, \overline{y}$, and $\overline{z}$; then, I_i is presented as [36]:

$$I_i = \begin{bmatrix} \dfrac{-I_{ixx}+I_{iyy}+I_{izz}}{2} & I_{ixy} & I_{ixz} & m_i\overline{x}_i \\ I_{ixy} & \dfrac{I_{ixx}-I_{iyy}+I_{izz}}{2} & I_{iyz} & m_i\overline{y}_i \\ I_{ixz} & I_{iyz} & \dfrac{I_{ixx}+I_{iyy}-I_{izz}}{2} & m_i\overline{z}_i \\ m_i\overline{x}_i & m_i\overline{y}_i & m_i\overline{z}_i & m_i \end{bmatrix} \tag{2.8}$$

2.3.1.4 System dynamics calculation equation

Let the mass of the connecting rod i be m_i and the gravity acceleration matrix be g^T. The potential energy P_i of the connecting rod i can be expressed as [37]:

$$P_i = -\int g^T \cdot {}^0Q_i \cdot {}^irdm = -m_i g^{T0}Q_i{}^ir \tag{2.9}$$

Combining Eqs. (2.3)–(2.9), assuming q is the joint coordinate, τ_i is the force or torque on each joint, D is the symmetric positive fixed inertia matrix of the robot arm, C is the centrifugal force and the Coriolis force term, and G is the gravity term, the torque that can be applied to the rotating joint of the connecting rod i is presented as [37]:

$$\tau_i = D(q)\ddot{q} + C(q,\dot{q})\dot{q} + G(q) \tag{2.10}$$

2.3.2 Forward dynamics

Robot dynamics analysis includes forward and inverse dynamics. Forward dynamics are the driving force or driving torque of robot joints known to find the displacement, velocity, and acceleration of the manipulator; the inverse dynamics are the opposite. The positive problem of the robot is related to the simulation, and the inverse problem is usually used for the real-time control of the robot.

There are generally two ways to implement forward dynamics algorithms. The first is to list the dynamic equations of the rigid body system, find the coefficient matrix of the equation, obtain the analytical solution of the acceleration, and then solve the joint acceleration according to the given joint torque. The second is to use the recursive method to calculate the acceleration of all rigid bodies calculated. This section introduces the first method, which is to solve the inertia matrix by the combined inertia method and obtain the analytical solution of the acceleration.

In the analysis of the forward dynamics, it can be known from the general equation of dynamics in Eq. (2.10) that after the coefficient matrices D, C, G are obtained, the joint angular acceleration $\ddot{q}$ can be obtained from the joint torque τ_i, and the positive solution to the dynamic problem is completed.

In the combined inertia method, the diagonals above $D(q)$ (that is, the elements above the diagonal) are calculated in the order $d_{nn}, d_{n-1,n}, \ldots, d_{1n}; d_{n-1,n-1}, d_{n-2,n-1}, \ldots, d_{1,n-1}; \ldots; d_{22}, d_{12}; d_{11}$.

Let N_B denote the number of rigid bodies; then, the kinetic energy of the robot arm can be expressed as [36]:

$$T = \frac{1}{2} \sum_{k=1}^{N_B} \widehat{v}_k^T \widehat{I}_K \widehat{v}_k \tag{2.11}$$

Let the set of joints between the rigid body and the base be $\kappa(k)$; then, the speed of assembling the robot arm is presented as [36]:

$$\widehat{v}_k = \sum_{i \in \kappa(k)} \widehat{S}_i \dot{q}_i \tag{2.12}$$

By sorting these two equations, it can be launched as [36]:

$$T = \frac{1}{2} \sum_{i=1}^{N_B} \sum_{j=1}^{N_B} \sum_{k \in v(i) \cap v(j)} \widehat{\dot{q}}_i^T \widehat{S}_i^T \widehat{I}_k \widehat{S}_j \widehat{\dot{q}}_j \tag{2.13}$$

The kinetic energy of the robotic arm system of the assembly robot is expressed by inertia matrix D; then, there is [36]:

$$T = \frac{1}{2}\dot{q}^T D\dot{q} = \frac{1}{2}\sum_{i=1}^{N_B}\sum_{j=1}^{N_B}\dot{q}_i^T D_{ij}\dot{q}_j \tag{2.14}$$

2.3.3 Inverse dynamics

The inverse dynamics of a robot need to know the displacement, velocity, and acceleration of each joint. They are used to calculate the driving torque or driving force required for each joint.

For the inverse dynamics algorithm, the most widely used and computationally efficient algorithm is the Newton-Eulerian. To make the form more concise, this section will use rotation to represent the Newton-Eulerian algorithm. The Newton-Eulerian recursive dynamics algorithm consists of two parts; the first part is the outward recursion of kinematics and the second is the inward recursion of the mechanical quantity.

The outward recursive equation for kinematics is presented as [35]:

$$\begin{cases} \widehat{v}_0 = 0 \\ \widehat{a}_0 = -\widehat{a}_g \\ \widehat{v}_{Ji} = \widehat{S}_i\dot{q}_i \\ \widehat{c}_{Ji} = \dot{\widehat{S}}_i\dot{q}_i \\ \widehat{v}_i = {}^i\widehat{X}_{\lambda(i)}\widehat{v}_{\lambda(i)} + \widehat{v}_{Ji} \\ \widehat{a}_i = {}^i\widehat{X}_{\lambda(i)}\widehat{a}_{\lambda(i)} + \widehat{S}_i\ddot{q}_i + \widehat{c}_{Ji} + \widehat{v}_i \times \widehat{v}_{Ji} \end{cases} \tag{2.15}$$

The inward recursive equation for the amount of mechanics is presented as [35]:

$$\begin{cases} \widehat{f}_i^B = \widehat{I}_i\widehat{a}_i + \widehat{v}_i \times {}^*\widehat{I}_i\widehat{v}_i \\ \widehat{f}_i = \widehat{f}_i^B - {}^i\widehat{X}_0^*\widehat{f}_i^x + \sum_{j\in\mu(i)} {}^i\widehat{X}_j^*\widehat{f}_j \\ \tau_i = \widehat{S}_i^T\widehat{f}_i \end{cases} \tag{2.16}$$

2.3.4 Arm trajectory planning

With the continuous development of assembly robots, they are constantly applied in various fields and have a huge value and role. The performance of assembly robots is also constantly improved to meet increasing needs, and trajectory planning has an important part. Trajectory planning helps the assembly robot to run smoothly and quickly according to the desired trajectory, reducing vibration and shock and improving work efficiency.

Trajectory planning of the robot arm belongs to bottom layer planning in robot planning. It is a trajectory generation method that considers task requirements, robot kinematics, and robot dynamics. The parameters describing the movement trajectory of the robot arm include the displacement, velocity, and acceleration of the robot arm. The trajectory of the robot arm refers to the position, velocity, or acceleration curve of the joints of the arm as time changes. Through trajectory planning, motion trajectory from the initial to the target state can be obtained according to the assembly task requirements of the robot [38].

With continuous research on the trajectory planning of the robot arm of assembly robots, scholars have proposed many trajectory planning algorithms. From the perspective of planning space, trajectory planning generally includes joint space trajectory and Cartesian space trajectory planning [39]. Advantages and disadvantages of joint space trajectory planning and Cartesian space trajectory planning are shown in Table 2.2.

Trajectory planning in joint space expresses the joint variable with a time function and a second-order time derivative, and then describes the motion trajectory of the robot arm of the assembly robot as a function of the joint angle [39]. The joint space trajectory planning method easily satisfies the kinematics and dynamics constraints of each joint, and calculation is simple. Because there is no continuous correspondence between joint space and Cartesian coordinate space, the singular phenomenon of the mechanism does not occur. However, in joint variable space, the joint trajectory and the trajectory of the robot arm are nonlinear. The planned trajectory is a controlled variable that directly adopts motion time. It is impossible to determine the position of each member and hand in the motion accurately, so that the end trajectory is often not intuitive enough.

For operations in which instantaneous variation in the trajectory and attitude requires spraying, arc welding, cutting, and so on, trajectory planning must be performed in

Table 2.2: Advantages and disadvantages of joint space trajectory planning and Cartesian space trajectory planning.

Trajectory planning methods	Advantages	Disadvantages
Joint space trajectory planning	The calculation is simple and does not cause the singularity of the mechanism	Unable to determine the position of each member and hand in motion accurately; the end track is not intuitive enough
Cartesian space trajectory planning	More intuitive, the planned trajectory has a high degree of compliance	A large amount of calculation and long control time

Cartesian coordinate space. Trajectory planning in Cartesian space is a function of the pose, velocity, and acceleration at the end of the robot arm as time [39]. The Cartesian space trajectory planning method is relatively straightforward, and the planned trajectory has a high degree of conformity. Because all available control algorithms are based on joint coordinates, when the intermediate point of the trajectory is found in Cartesian operation space, the planned result needs to be mapped to joint space first, and then the joint variable value is obtained. However, in the mapping process, the kinematics inverse solution is required, and the amount of calculation will be greatly increased, so that the control time is greatly extended. Cartesian space trajectory planning often has a special configuration.

Whether planning in the joint or Cartesian coordinate system, the planned trajectory must be continuous and smooth, and the target trajectory between the points must be generated to ensure the continuity of position, velocity, and acceleration [39].

According to the advantages and disadvantages of joint space trajectory planning and Cartesian space trajectory planning, for point-to-point planning to achieve a pick and place operation, only the motion of the initial state and the target state is described, which is generally performed in joint space. Continuous trajectory planning (that is, not only the starting and the ending points of the target task but also the intermediate path point between the two points is required) is generally performed in Cartesian space [39].

In practical applications, not only task requirements of movement need to be considered, but also factors such as time, energy consumption, and stability. Therefore, more and more scholars have begun to study the optimal trajectory planning scheme under each working condition. Because optimal trajectory planning can use many targets as evaluation indicators, most studies focus on optimizing time, energy, and impact. There are still some comprehensive considerations of these evaluation indicators to obtain the optimal solution.

From the perspective of planning time, a time-optimal algorithm is a research hot topic of current trajectory planning. a time-optimal algorithm solves the movement trajectory of the robot arm with the shortest time to assemble it under condition of satisfying various constraints.

In addition to the time-optimal algorithm, the energy-optimal algorithm is an optimal algorithm for solving the trajectory of the robotic arm of the assembly robot. Although energy-optimal algorithms are not as popular as time-optimal algorithms, many scholars have studied energy-optimal algorithms and used the minimum energy consumed by the system as an index of evaluation to optimize the motion trajectory of the assembly robot.

According to this description, trajectory planning algorithms for assembly robot arms are shown in Fig. 2.6.

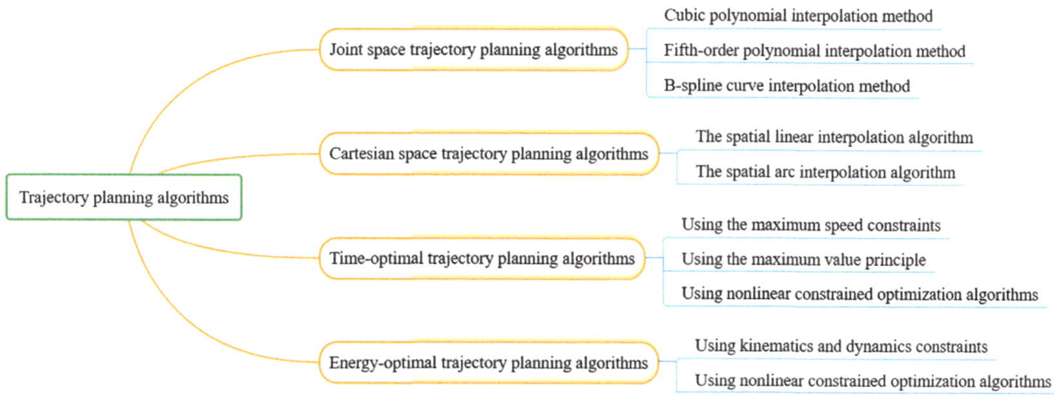

Figure 2.6
Trajectory planning algorithms.

2.3.4.1 Joint space trajectory planning algorithm

Scholars have conducted a lot of research on joint space trajectory planning algorithms to solve the problem of trajectory planning of assembly robot manipulators. To plan the trajectory in joint space, the start and end poses of the robot need to be known. When interpolating the joint, the direction of the initial point object, the moving direction of the lifting point object, the position of the object at the lower point, and the speed requirement of the stopping point object should be satisfied. The displacement, velocity, and acceleration must meet the continuity requirements.

Under constraints, different types of joint interpolation functions can be selected to generate different trajectories. Commonly used joint space trajectory planning algorithms include cubic polynomial interpolation, fifth-order polynomial interpolation, and B-spline curve interpolation [40]. The first two interpolation methods have a small amount of calculation and can obtain a continuous speed curve; B-spline curve interpolation has the advantages of both derivative continuity and local support and is more widely used.

2.3.4.1.1 Cubic polynomial interpolation method

Therefore, a description of the trajectory of the robot arm of the assembly robot can be expressed by a smooth interpolation function of the joint angle of the starting point and the joint angle of the ending point.

In the cubic polynomial interpolation of the trajectory of a robot arm of the assembly robot, usually the initial and final poses of the robot arm and the initial set of joint angles are known conditions, and it is necessary to solve a set of corresponding joint angles by inverse kinematics. Therefore, a description of the trajectory of the robot arm of the assembly robot can be expressed by a smooth interpolation function $\theta(t)$ of the joint angle of the starting point and the joint angle of the ending point.

The trajectory function $\theta(t)$ should satisfy at least four constraints [41]:

(a) Starting point:

$$\begin{cases} \theta(0) = \theta_0 \\ \theta(t_f) = \theta_f \end{cases} \tag{2.17}$$

(b) Starting speed:

$$\begin{cases} \dot{\theta}(0) = 0 \\ \dot{\theta}(t_f) = 0 \end{cases} \tag{2.18}$$

A unique cubic polynomial can be determined according to the boundary conditions of Eqs. (2.17) and (2.18) [41]:

$$\theta(t) = a_0 + a_1 t_f + a_2 t_f^2 + a_3 t_f^3 \tag{2.19}$$

Then, the joint velocity and acceleration on the trajectory of the robot arm can be obtained as [41]:

$$\begin{cases} \dot{\theta}(t) = a_1 + 2a_2 t + 3a_3 t^2 \\ \ddot{\theta}(t_f) = 2a_2 + 6a_3 t \end{cases} \tag{2.20}$$

Substituting the corresponding constraints, the four linear equations for the coefficients a_1, a_2, a_3 are shown as [42]:

$$\begin{cases} \theta_0 = a_0 \\ \theta_f = a_0 + a_1 t_f + a_2 t_f^2 + a_3 t_f^3 \\ 0 = a_1 \\ 0 = a_1 + 2a_2 t_f + 3a_3 t_f^2 \end{cases} \tag{2.21}$$

Solving these linear equations, coefficients a_1, a_2, a_3 of the cubic polynomial are provided as [42]:

$$\begin{cases} a_0 = \theta_0 \\ a_1 = 0 \\ a_2 = \dfrac{3}{t_f^2}(\theta_f - \theta_0) \\ a_3 = -\dfrac{3}{t_f^3}(\theta_f - \theta_0) \end{cases} \tag{2.22}$$

Cubic polynomial interpolation is suitable only for solving the case in which the end point speed is zero. However, the assembly robot often needs to pass some specific points in the motion and does not stay in actual operation. In this case, cubic polynomial interpolation is not suitable. If several intermediate points are given in the task, the entire trajectory can be divided into several segments using these intermediate points as separation points, and three interpolations are respectively performed on each segment, but the motion trajectory thus solved may not be smooth enough.

2.3.4.1.2 Fifth-order polynomial interpolation method

Trajectory planning of the robot in joint space requires different types of joint interpolation functions to be selected and designed for different constraints. When the starting point and the target position, velocity, and acceleration are known conditions, fifth-order polynomial interpolation can be used.

The value of $\theta(t)$ at time $t_0 = 0$ is starting joint angle θ_0, and the value at end time t_f is ending joint angle θ_f; then, trajectory function $\theta(t)$ should satisfy the following constraints [41]:

(a) Starting point:

$$\begin{cases} \theta(0) = \theta_0 \\ \theta(t_f) = \theta_f \end{cases} \tag{2.23}$$

(b) Starting speed:

$$\begin{cases} \dot{\theta}(0) = \dot{\theta}_0 \\ \dot{\theta}(t_f) = \dot{\theta}_f \end{cases} \tag{2.24}$$

(c) Starting acceleration:

$$\begin{cases} \ddot{\theta}(0) = \ddot{\theta}_0 \\ \ddot{\theta}(t_f) = \ddot{\theta}_f \end{cases} \tag{2.25}$$

The first and second derivative functions of trajectory function $\theta(t)$ are respectively used as time functions of the joint velocity and acceleration, and the expressions are shown as [43]:

$$\theta(t) = a_0 + a_1 t + a_2 t^2 + a_3 t^3 + a_4 t^4 + a_5 t^5 \tag{2.26}$$

$$\dot{\theta}(t) = a_1 + 2a_2 t + 3a_3 t^2 + 4a_4 t^3 + 5a_5 t^4 \tag{2.27}$$

$$\ddot{\theta}(t) = 2a_2 + 6a_3 t + 12a_4 t^2 + 20a_5 t^3 \tag{2.28}$$

According to the constraint, the coefficients of the polynomial can satisfy six constraint equations [43]:

$$\begin{cases} \theta_0 = a_0 \\ \theta_f = a_0 + a_1 t_f + a_2 t_f^2 + a_3 t_f^3 + a_4 t_f^4 + a_5 t_f^5 \\ \dot{\theta}_0 = a_1 \\ \dot{\theta}_f = a_1 + 2a_2 t_f + 3a_3 t_f^2 + 4a_4 t_f^3 + 5a_5 t_f^4 \\ \ddot{\theta}_0 = 2a_2 \\ \ddot{\theta}_f = 2a_2 + 6a_3 t_f + 12a_4 t_f^2 + 20a_5 t_f^3 \end{cases} \tag{2.29}$$

Solve them to obtain [43]:

$$\begin{cases} a_0 = \theta_0 \\ a_1 = \dot{\theta}_0 \\ a_2 = \dfrac{\ddot{\theta}}{2} \\ a_3 = \dfrac{20\theta_f - 20\theta_0 - (8\dot{\theta}_f + 12\dot{\theta}_0)t_f - (3\ddot{\theta}_0 - \ddot{\theta}_f)t_f^2}{2t_f^3} \\ a_4 = \dfrac{-30\theta_f + 30\theta_0 + (14\dot{\theta}_f + 16\dot{\theta}_0)t_f + (3\ddot{\theta}_0 - 2\ddot{\theta}_f)t_f^2}{2t_f^4} \\ a_5 = \dfrac{12\theta_f - 12\theta_0 - (6\dot{\theta}_f + 6\dot{\theta}_0)t_f - (\ddot{\theta}_0 - \ddot{\theta}_f)t_f^2}{2t_f^5} \end{cases} \tag{2.30}$$

Compared with the cubic polynomial interpolation algorithm, the fifth-order polynomial interpolation algorithm has a large amount of computation because of the increased constraints. However, this method can ensure that the joint angular displacement, velocity, and acceleration of each point are continuous, the movement of the robot arm of the assembly robot is steady, and the resulting motion trajectory is smooth.

2.3.4.1.3 B-Spline curve interpolation method

Let $d_i(i = 0, 1, \ldots, n)$ be the control vertex and $B_{\mathrm{i},k}(u)(i = 0, 1, \ldots, n)$ be the k-norm B-spline basis function. The equation of the B-spline curve is presented as [44]:

$$P(u) = \sum_{i=0}^{n} d_i B_{i,k}(u) \tag{2.31}$$

The B-spline basis function has many definitions. This section uses the de Boor–Cox recursion equation. This recursive method is easily implemented on a computer and reflects the properties of the B-spline basis function. The definition is presented as [44]:

$$\begin{cases} B_{i,0}(0) = \begin{cases} 1 & u \in [u_i, u_{i+1}) \\ 0 & \text{else} \end{cases} \\ B_{i,k}(u) = \dfrac{u - u_i}{u_{i+k} - u_i} B_{i,k-1}(u) + \dfrac{u_{i+k+1} - u}{u_{i+k+1} - u_{i+1}} B_{i+1,k-1}(u) \end{cases} \tag{2.32}$$

In this equation, i in $B_{i,k}(u)$ represents the serial number and k represents the number of times.

Eqs. (2.31) and (2.32) show that when $k = 3,\ i = 0, 1, 2, 3$, the basic functions of the cubic uniform B-spline curve are [44]:

$$\begin{cases} B_{0,3}(u) = \dfrac{1}{6}(-u^3 + 3u^2 - 3u + 1) \\ B_{1,3}(u) = \dfrac{1}{6}(3u^3 - 6u^2 + 4) \\ B_{2,3}(u) = \dfrac{1}{6}(-3u^3 + 3u^2 + 3u + 1) \\ B_{3,3}(u) = \dfrac{1}{6}u^3 \end{cases} \tag{2.33}$$

Eq. (2.33) is converted into a matrix, which is presented as [44]:

$$P_i(u) = \sum_{l=0}^{3} d_{i+1} B_{l,k}(u) = \frac{1}{6}\left[u^3 u^2 u 1\right] \begin{bmatrix} -1 & 3 & -3 & 1 \\ 3 & -6 & 3 & 0 \\ -3 & 0 & 3 & 0 \\ 1 & 4 & 1 & 0 \end{bmatrix} \begin{bmatrix} d_i \\ d_{i+1} \\ d_{i+2} \\ d_{i+3} \end{bmatrix}, u \in [0.1] \tag{2.34}$$

The equation for sorting Eq. (2.34) is presented as [44]:

$$P_{\mathrm{i}}(u) = R_0 + R_1 u + R_2 u^2 + R_3 u^3$$

$$\begin{cases} R_0 = \dfrac{1}{6}(d_i + 4d_{i+1} + 3d_{i+2}) \\ R_1 = \dfrac{1}{6}(-3d_i + 3d_{i+2}) \\ R_2 = \dfrac{1}{6}(3d_i - 6d_{i+1} + 3d_{i+2}) \\ R_3 = \dfrac{1}{6}(-d_i + 3d_{i+1} - 3d_{i+2} + d_{i+3}) \end{cases} \tag{2.35}$$

When taking the cubic uniform B-spline function to perform trajectory planning, it is necessary to preset the value point, and then, according to Eq. (2.35), the vertices are controlled by the inverse of the type value and the curve equation of the trajectory is obtained.

After the curve equation of the trajectory is obtained, it is necessary first to obtain parameter period increment Δu in the interpolation process, and then to solve period parameter $u_{j+1} = u_j + \Delta u_j$ at the next moment according to Δu. Finally, the coordinates of the interpolation point at the next moment are obtained according to Eq. (2.35).

The interpolation can be performed by the acceleration–constant–deceleration method to obtain parameter period increment Δu. Let a be the acceleration, ΔL_j the period step in the interpolation process, T the interpolation period, and $j = 0, 1, 2, \ldots, (\text{int})\frac{v}{a}$. Then, the specific calculation equation is presented as [45]:

$$\Delta L_{\text{j}} = \begin{cases} (0.5 + j)aT^2 & \text{Acceleration section} \\ vT & \text{Uniform speed section} \\ -(0.5 + j)aT^2 & \text{Deceleration section} \end{cases} \tag{2.36}$$

$$\Delta u_j = \frac{\Delta L_j}{\sqrt{\dot{x}_j^2 + \dot{y}_j^2 + \dot{z}_j^2}} \tag{2.37}$$

$$\sqrt{\dot{x}_j^2 + \dot{y}_j^2 + \dot{z}_j^2} = \sqrt{a_4 u_j^4 + a_3 u_j^3 + a_2 u_j^2 + a_1 u_j}$$

$$\begin{cases} a_4 = 9\left(R_{3x}^2 + R_{3y}^2 + R_{3z}^2\right) \\ a_3 = 12(R_{3x}R_{2x} + R_{3y}R_{2y} + R_{3z}R_{2z}) \\ a_2 = 4\left(\left(R_{2x}^2 + R_{2y}^2 + R_{2y}^2\right) + 6(R_{3x}R_{1x} + R_{3y}R_{1y} + R_{3z}R_{1z})\right) \\ a_1 = 4(R_{2x}R_{1x} + R_{2y}R_{1y} + R_{2z}R_{1z}) \\ a_0 = \left(R_{1x}^2 + R_{1y}^2 + R_{1z}^2\right) \end{cases} \tag{2.38}$$

where R_{2y} is the component of the y coordinate direction of R_2 in Eq. (2.35) and the other quantities are the same principle.

Find Δu_j according to Eq. (2.37), and then find the coordinates of the interpolation point at the next moment by Eq. (2.35), which is presented as [45]:

$$P_{ij+1} = P_i(u_j + \Delta u_j) \tag{2.39}$$

2.3.4.2 Cartesian space trajectory planning algorithm

When planning the trajectory in joint space, it can be guaranteed that the end effector may reach the given intermediate point and target point, but in Cartesian coordinate space, the trajectory is often more complicated and sometimes is required. The end effector walks through a specific trajectory shape such as a regular curve, a straight line, or an arc. Trajectory planning in Cartesian space is to plan the time function directly according to the position and attitude of the end effector in the Cartesian coordinates. The trajectory shape is a regular curve. The intermediate point obtained by using the interpolation algorithm in the Cartesian coordinate system is the Cartesian coordinate value instead of the joint angle. Because the control system realizes the motion control of the robot through the joint angle of the robot, it is also necessary to use inverse kinematics to transform into the corresponding joint angle.

Compared with the joint space trajectory planning algorithm, there is less research on the Cartesian space trajectory planning algorithm. However, in some cases, the robot's end trajectory is more stringent and requires moving along a given trajectory. Trajectory planning in Cartesian space is necessary for the task in this case, so it is necessary and practical to conduct in-depth research on Cartesian space trajectory planning. Research on Cartesian space trajectory planning mainly focuses on spatial linear interpolation and spatial arc interpolation algorithms.

2.3.4.2.1 Spatial linear interpolation algorithm

The spatial linear interpolation algorithm refers to the method of solving the posture of the starting point and the end point of the linear motion of the mechanical arm, and then solving the positional interpolation of the linear interpolation point.

Let the starting point coordinate be $P_1(x_1, y_1, z_1)$, the ending point coordinate be $P_2(x_2, y_2, z_2)$, and the moving speed of the assembly robot arm be v. The movement time of the entire trajectory can be obtained as [46]:

$$T = \frac{\sqrt{(x_2 - x_1)^2 + (y_2 - y_1)^2 + (z_2 - z_1)^2}}{v} \tag{2.40}$$

If the interpolation period is Δt, the number of interpolations needs to be $N = T/\Delta t$, and N is an integer. The position increments of the x axis, y axis, and z axis can be determined as [46]:

$$\begin{cases} \Delta x = \dfrac{x_2 - x_1}{N} \\ \Delta y = \dfrac{y_2 - y_1}{N} \\ \Delta z = \dfrac{z_2 - z_1}{N} \end{cases} \tag{2.41}$$

Then, the coordinates of the interpolation point are shown as [46]:

$$\begin{cases} x_{i+1} = x_i + \Delta x \\ y_{i+1} = y_i + \Delta y \\ z_{i+1} = z_i + \Delta z \end{cases} \tag{2.42}$$

In this equation, i is an integer and $i \in [0, N]$. The position coordinates of each interpolation point can be found according to the equation. Then, the inverse kinematics of each interpolation point can be obtained; that is, the joint values of the robot arm can be obtained.

2.3.4.2.2 Spatial arc interpolation algorithm

The spatial arc interpolation algorithm is a method of calculating the coordinates of the intermediate trajectory in this determined plane, assuming that the trajectory of the robot arm approximates an arc; it determines the plane of the arc based on three points that are not on the same line. The arc in any plane in three-dimensional space is the space arc. Space arc interpolation can transform the three-dimensional problem into a two-dimensional one and find the plane of the arc [47]. Interpolation of the space arc interpolation algorithm is performed in the plane where the arc is located. Therefore, it is necessary first to convert the coordinates of the Cartesian coordinate system into the newly established space coordinate system, and then to convert it to the original Cartesian coordinate system after interpolation is completed.

It is assumed that the space arc passes from the starting position $p_1(x_1, y_1, z_1)$ through the intermediate position $p_2(x_2, y_2, z_2)$ to the end position $p_3(x_3, y_3, z_3)$, which determines a plane L. Let plane M be perpendicular to p_1p_2 and pass through the midpoint of p_1p_2; the other plane N is perpendicular to p_2p_3 and passes through the midpoint of p_2p_3. The three planes L, M, N intersect at one point; that is, the center is $p_0(x_0, y_0, z_0)$ and the radius of the circle is r. This can be shown as [48]:

$$r = \sqrt{(x_i - x_0)^2 + (y_i - y_0)^2 + (z_i - z_0)^2}, i = 1, 2, 3 \tag{2.43}$$

In the new coordinate system, the origin coordinates are $Q_0(0, 0, 0)$, the z'_0 axis is perpendicular to plane L, and the direction cosine is $\{a'_x, a'_y, a'_z\}^T$; the p_0p_1 is axis x'_0 and the direction cosine is $\{n'_x, n'_y, n'_z\}^T$. According to the right-hand rule, axis y'_0 is obtained and the direction cosine is $\{o'_x, o'_y, o'_z\}^T$. Thus, the transformation matrix of the new coordinate system can be obtained as [48]:

$$A = \begin{bmatrix} n'_x & o'_x & a'_x & p'_x \\ n'_y & o'_y & a'_y & p'_y \\ n'_z & o'_z & a'_z & p'_z \\ 0 & 0 & 0 & 1 \end{bmatrix} = \begin{bmatrix} R' & P \\ 0 & 1 \end{bmatrix} \tag{2.44}$$

The coordinates in the new coordinate system are presented as [48]:

$$[noa1]^T = A^{-1}[xyz1]^T \tag{2.45}$$

The arc in the space is turned into an arc of the L plane, the trajectory planning of the arc is performed in the L plane, and the angle from p_1 to p_3 is θ_3, presented as [49]:

$$\theta_3 = \begin{cases} \arctan 2(o_3, n_3) & o_3 > 0 \\ \pi & o_3 = 0 \\ 2\pi + \arctan 2(o_3, n_3) & o_3 < 0 \end{cases} \tag{2.46}$$

Interpolation points of the space arc can be solved as the equation [49]:

$$\begin{aligned} &\theta = \lambda\theta_3 \\ &\begin{cases} n = r \times \cos\theta \\ o = r \times \sin\theta \\ a = 0 \end{cases} \end{aligned} \tag{2.47}$$

Finally, the coordinate sequence in the original Cartesian coordinate system can be found as [49]:

$$[XYZ]^T = A[X'Y'Z']^T \tag{2.48}$$

Through these steps, the interpolation calculation of the space arc can be completed.

2.3.4.3 Time-optimal trajectory planning algorithm

The optimal trajectory plan for the robot arm of the assembly robot is to find the trajectory with the shortest time for the robot to complete the same trajectory under the conditions for which the task is satisfied. Using the time-optimal trajectory planning algorithm, the assembly robot can complete more tasks at the same time and the work efficiency is higher, which helps to improve profits. The time-optimal trajectory planning algorithm is one of the earliest and most popular optimal trajectory planning algorithms [50].

To solve time-optimal trajectory planning, there are three methods: considering the constraints of maximum speed and maximum acceleration, select the optimal shape of the joint speed curve and the joint acceleration curve; use the principle of maximum value to solve the problem; and under the conditions of kinematic or dynamic constraints, use various nonlinear constraint optimization algorithms to solve the problem [51]. Scholars have developed many methods for trajectory planning with the shortest time based on robot kinematics [52]. In addition, they have used neural networks or genetic algorithms to study this problem.

Early on, Lin et al. proposed a trajectory planning method for the optimal time after considering various kinematic constraints of the robot, and used a flexible polyhedron search algorithm to solve it [53]. To displace each set of joints, through cubic spline curve fitting, the high-order polynomial curve was used to achieve an effective connection at the transition point, and the time-optimal trajectory was obtained. However, this algorithm is a local search algorithm, and the performance of the algorithm is closely related to selection of the initial conditions. Tondu et al. used the same constraint to connect key points in joint space with a fold line with a smooth transition, which achieved a simplified effect through the transition [54]. Bazaz et al. pointed out that in the process of time-optimal trajectory planning, based on the speed and acceleration constraints, the cubic spline curve is the simplest polynomial curve form connecting key points in joint space, according to which they proposed an algorithm [55].

However, this method does not consider the continuity of the acceleration at the joint of key points in joint space, which may cause vibration during the movement of the robot. Later, Bazaz et al. summarized the previous scheme and proposed that the cubic curve segment with a smooth transition could be used as the connection curve of key points of joint space. The new algorithm designed according to this has achieved certain effects. *Choi* et al. proposed a method for solving trajectory planning problems using only kinematics and evolutionary strategies, and obtained some optimization solutions [56]. Rong et al. proposed a time-optimal 3−5−3 polynomial interpolation trajectory planning algorithm based on the particle swarm optimization (PSO) algorithm and the kinematic constraint of the manipulator [57]. The PSO algorithm is easy to use and the parameters are adjustable. The advantage of the polynomial algorithm in time directly involves the optimization target space search, which solves the shortcomings of polynomial interpolation trajectory planning with a high order and no convex hull property. The experiment proves that the algorithm has a shorter running time and better fluency. However, application of this algorithm to other trajectories besides polynomial remains to be verified [58].

Although these studies proposed their own optimal trajectory planning algorithms for specific situations or under certain preconditions, each has defects. There is no general algorithm to optimize the trajectory planning of all robot arms. The time-optimal 3−5−3 polynomial interpolation trajectory planning algorithm based on the PSO algorithm is introduced next in detail.

2.3.4.3.1 Origin of particle swarm optimization algorithm

The PSO algorithm was originally inspired by the results of artificial life research by Eberhart and Kennedy in 1995, who proposed a population intelligence-based evolutionary computation technique when simulating the migratory and clustering behavior of foraging birds [59].

In the initial state, each bird in the flock was flying in random directions. However, over time, these birds gradually aggregated into small communities through self-organization and flew in the same direction at the same speed, and then several small communities aggregated into large communities. The overall behavior of the synchronized flight of flocks is based on the local perception of each bird's surroundings; there is no centralized controller. *Eberhart* and *Kennedy* learned from the foraging behavior of social animals such as birds and found that they can greatly speed the search for food by sharing information about the locations of food among group members during activities such as foraging [59]. Usually, the benefits of group search are greater than the loss caused by competition for resources among group members.

Based on this factual experience, the following rules can be drawn: When the entire group is searching for a certain target, for one individual, the individual in the current optimal position of the group and the optimal position that it has reached is often referenced to adjust the next search. *Eberhart* and *Kennedy* modified the model of this simulated group interaction and designed it as a general method to solve optimization problems; they called it the PSO algorithm [59].

It has been widely used in functional optimization, neural network training, fuzzy system control, pattern recognition, signal processing, robotics, and other fields. PSO researchers have studied particle population organization, cooperation modes, and algorithm parameters. They proposed cluster analysis PSO, fuzzy PSO, PSO with Gaussian variation, collaborative PSO, PSO with selection, multistage PSO, and so on.

2.3.4.3.2 Construction of 3—5—3 spline interpolation function

Generally, k-degree polynomials with $k-1$-order derivative continuity at the interpolation points are used to represent the spline curve.

Suppose $h_{j1}(t), h_{j2}(t), h_{j3}(t)$ represent the third-order polynomial trajectory of the first segment of the j-segment joint, the fifth-order polynomial trajectory of the second segment, and the third-order polynomial trajectory of the third segment; $a_{j1i}, a_{j2i}, a_{j3i}$ are the i-th coefficients of the interpolation function of the first, second, and third segments of the j-th joint trajectory, and n represents the number of joints, $j = 1, 2, \ldots, n$. General equations of the 3—5—3 spline polynomial are presented as [57]:

$$h_{j1}(t) = a_{j13}t^3 + a_{j12}t^2 + a_{j11}t^1 + a_{j10} \tag{2.49}$$

$$h_{j2}(t) = a_{j25}t^5 + a_{j24}t^4 + a_{j23}t^3 + a_{j22}t^2 + a_{j21}t^1 + a_{j20} \tag{2.50}$$

$$h_{j3}(t) = a_{j33}t^3 + a_{j32}t^2 + a_{j31}t^1 + a_{j30} \tag{2.51}$$

where t_1, t_2, t_3 represent the time of the 3-segment polynomial interpolation of the j-th joint. Based on these conditions, the relational expressions (2.52)—(2.54) to solve unknown coefficient a_{ji} and the interpolation point are obtained [57]:

$$A=\begin{bmatrix} t_1^3 & t_1^2 & t_1 & 1 & 0 & 0 & 0 & 0 & 0 & -1 & 0 & 0 & 0 & 0 \\ 3t_1^2 & 2t_1 & 1 & 0 & 0 & 0 & 0 & 0 & -1 & 0 & 0 & 0 & 0 & 0 \\ 6t_1 & 2 & 0 & 0 & 0 & 0 & 0 & -2 & 0 & 0 & 0 & 0 & 0 & 0 \\ 0 & 0 & 0 & 0 & t_2^5 & t_2^4 & t_2^3 & t_2^2 & t_2 & 1 & 0 & 0 & 0 & -1 \\ 0 & 0 & 0 & 0 & 5t_2^4 & 4t_2^3 & 3t_2^2 & 2t_2 & 1 & 0 & 0 & 0 & -1 & 0 \\ 0 & 0 & 0 & 0 & 20t_2^3 & 12t_2^2 & 6t_2 & 2 & 0 & 0 & 0 & -2 & 0 & 0 \\ 0 & 0 & 0 & 0 & 0 & 0 & 0 & 0 & 0 & 0 & t_3^3 & t_3^2 & t_3 & 1 \\ 0 & 0 & 0 & 0 & 0 & 0 & 0 & 0 & 0 & 0 & 3t_3^2 & 2t_3 & 1 & 0 \\ 0 & 0 & 0 & 0 & 0 & 0 & 0 & 0 & 0 & 0 & 6t_3 & 2 & 0 & 0 \\ 0 & 0 & 0 & 1 & 0 & 0 & 0 & 0 & 0 & 0 & 0 & 0 & 0 & 0 \\ 0 & 0 & 1 & 0 & 0 & 0 & 0 & 0 & 0 & 0 & 0 & 0 & 0 & 0 \\ 0 & 1 & 0 & 0 & 0 & 0 & 0 & 0 & 0 & 0 & 0 & 0 & 0 & 0 \\ 0 & 0 & 0 & 0 & 0 & 0 & 0 & 0 & 0 & 0 & 0 & 0 & 0 & 1 \\ 0 & 0 & 0 & 0 & 0 & 0 & 0 & 0 & 0 & 1 & 0 & 0 & 0 & 0 \end{bmatrix} \tag{2.52}$$

$$b=\begin{bmatrix}0 & 0 & 0 & 0 & 0 & 0 & X_{j3} & 0 & 0 & X_{j0} & 0 & 0 & X_{j2} & X_{j1}\end{bmatrix}^T \tag{2.53}$$

$$a=A^{-1}\cdot b=\left[a_{j13}a_{j12}a_{j11}a_{j10}a_{j25}a_{j24}a_{j23}a_{j22}a_{j21}a_{j20}a_{j33}a_{j32}a_{j31}a_{j30}\right]^T \tag{2.54}$$

2.3.4.3.3 Solution process based on particle swarm optimization algorithm

Let V_{j1}, V_{j2}, V_{j3} be the speeds of each polynomial in the j-th joint with time; the shortest motion time of all joints satisfying kinematic constraints is the optimization goal. The objective function is presented as [57]:

$$\begin{aligned} & \left. f(t) = \min \sum_{j=0}^{n} (t_{j1}+t_{j2}+t_{j3} \right) \\ & s.t. \max\{|V_{j1}|\} \le V_{\max} \\ & \max\{|V_{j2}|\} \le V_{\max} \\ & \max\{|V_{j3}|\} \le V_{\max} \end{aligned} \tag{2.55}$$

Because 3−5−3 polynomial interpolation does not have the nature of the traditional optimization method, its time optimization can adopt only the intelligent method, and so the PSO algorithm is used. Specific solution steps are [57]:

(a) Initialize the flying speed and position of each particle at each iteration.
(b) Optimize each joint of the robot arm separately to obtain the optimized objective function of each joint.

(c) The switch control of two fitness functions is adopted to keep the joint motion speed converge within the constraint conditions as soon as possible.
(d) After satisfying the kinematic constraints, perform a time-optimal optimization iteration in the search space to be optimized.
(e) Update the position and flying speed of each particle.
(f) If the termination conditions are not met, continue to optimize. Otherwise, the algorithm ends.
(g) Complete time optimization of all joints and take the maximum value of each joint's time.

According to these seven steps, the PSO-based 3–5–3 polynomial interpolation algorithm is used to determine the optimal time of each interpolation, and time-optimal 3–5–3 polynomial interpolation trajectory planning based on the PSO is completed.

2.3.4.4 Energy-optimal trajectory planning algorithm

Time-optimal trajectory planning can make industrial production more efficient, so it occupies an important part of assembly robot trajectory planning. However, some assembly robots, such as underwater and aerial robots, cannot carry many energy storage devices when working in a specific environment and have high energy requirements [50]. Optimal energy trajectory planning has a huge role. The energy-optimal trajectory of the assembly robot refers to the trajectory of a robot that consumes the least amount of energy in completing that trajectory. On the one hand, energy-optimal trajectory planning tries to find the smoothest trajectory to reduce energy loss between joints. On the other hand, it optimizes the whole power system to achieve the best energy distribution. Research on this aspect started relatively recently [60].

In 1970, the concept of energy-optimal trajectory planning was proposed by scholars, but after 20 years, scholars began to conduct comprehensive research [39]. In 1996, *Hirakawa* et al. discussed the trajectory generation problem of redundant robots and introduced the variational method and B-spline curve to optimize energy consumed by the robot system, so that energy consumption was in the same mission mode [61]. This method reduced the inherent inconvenience of using the variational method; for example, design of the forward vector is difficult and the calculation time is long. However, this algorithm performs better only on redundant robots and has greater limitations. In 2002, Garg et al. focused on two connecting robots and two cooperating robots [62]. Combined with the genetic algorithm and adaptive simulated annealing algorithm, the optimization target of the articulated robot was set to the minimum torque. Martin et al. used the B-spline curve for energy-optimal trajectory planning with no excessive constraints and integrated the square of each joint torque to obtain the energy function. Several parameters were optimized to obtain the energy-optimal trajectory [63]. This algorithm largely guarantees the working performance of the robot and has good practicability and flexibility.

Similar to time-optimal trajectory planning, energy-optimal trajectory planning uses kinematics and dynamics as constraints and many optimization algorithms such as neural networks, simulated annealing algorithms, genetic algorithms, and improved chaos optimization algorithms to arrive at the planned trajectory results quickly and efficiently [60]. However, during actual movement of the assembly robot, the joints may undergo a flexible change different from rigid body motion, and the solution result may not be optimal. Moreover, actual working conditions are complicated, and these algorithms cannot achieve global optimization. According to this point of view, there is currently no general optimization algorithm for the energy-optimal trajectory planning of assembly robots.

With the continuous development of the modern manufacturing industry, considering only optimal energy trajectory planning is not suitable for many production conditions. Researchers have gradually begun to consider time, impact, and energy factors; carry out trajectory planning; and obtain a comprehensive optimal track.

2.4 Arm inverse dynamic application of rail transit assembly robots

Inverse dynamic calculation of the robotic arm of the intelligent robot is an important problem in motion planning and control of the robot. However, owing to the complexity of the motion of the robot arm, the inverse dynamic problem is an overdetermined nonlinear equation-solving process. Research has provided two methods for solving inverse dynamic problems: algebraic analysis and the neural network [64].

Algebraic analytic methods can solve general inverse dynamic problems, but the difficulty of solving them may be greatly improved when there are more constraints. The neural network method has great advantages in solving nonlinear problems. Through continuous training, the accuracy of the neural network can be improved further. Therefore, when solving the inverse dynamic problem, the neural network is greatly superior and has been widely used.

The artificial neural network (ANN) is a kind of model that imitates the structure of human brain neurons from the perspective of information processing. It is composed of different connections and forms of different networks. It has been a hot topic in artificial intelligence optimization since the 1980s [65].

The basic processing unit of the ANN is a complex network of a large number of neurons connected to each other, and is highly nonlinear. The main structural unit of the ANN is the input, synthesis processing, and output of signals. Each node represents a special function, which is weighted by the connection of neurons to complete complex logical operations and the realization of nonlinear relationships.

The composition of the neural network requires three basic elements. The is the neuron, the basic processing unit of the neural network. Its function is to weigh several inputs and then perform nonlinear processing and output. The second basic element is the network topology [66], which represents the structure of the network and the way neural units are connected. According to the way it is connected, the network can be divided into a feedback and nonfeedback network. The third basic element is the network training algorithm, in which initial values and thresholds of the connected neurons are determined. These initial values and thresholds can be adjusted according to training modes to meet the performance requirements of the task.

In the ANN, the processing power of the model is changed with different network connection weights. The ANN has four basic characteristics [66]:

(a) Nonlinear. The ANN is connected to form a network with a large number of neurons, completing the input, integrating processing, and outputting the signal, and each parallel network interacts with each other, forming a nonlinear mathematical model.
(b) Nonlimiting. The nature of an ANN depends on the interaction of individual neurons, not just the characteristics of any one of them. Its mathematical model is not a simple superposition of individual functions.
(c) Fault tolerance. The ANN has the function of memorizing information, which is stored in the threshold of neurons. This greatly improves the storage capacity and fault tolerance of the network.
(d) Adaptability. The ANN has strong adaptive and self-learning capabilities. It can be trained by continuously processing information. The structure and threshold of the network will also change constantly to adapt to the changing environment.

With the unremitting research of researchers, the ANN is constantly developing, and its applications in daily life are increasingly extensive. The ANN has been applied in medicine, intelligent robot, biology, automatic control and other fields. The ANN solves many practical problems and has great potential for development.

The basis of neural networks lies in neurons. The structure of neurons is simple and easy to understand. However, neural networks are complex network systems formed by a large number of connected neurons whose function is subtle. After many years of development, neural networks have dozens of different models, including radial basis function, backpropagation (BP), convolutional neural, and recurrent neural network.

For a multijoint assembly robot system, joint coordinates can be used as system generalized coordinates, based on Eq. (2.10), and the uncertainty dynamic force or torque N such as disturbance and nonlinear damping is considered to establish the rigid body dynamics of the system. The equation is presented as [37]:

$$\tau = f(q, \dot{q}, \ddot{q}) + N \tag{2.56}$$

Because the uncertainty of the disturbance is uncertain, it can only be compensated by feedback and is generally not considered. Control torque τ can be regarded as a function uniquely determined by joint coordinates and their derivatives, $q, \dot{q}, \ddot{q}$. The dynamics model is the mapping of $\tau \rightarrow (q, \dot{q}, \ddot{q})$, whereas the inverse model is the mapping relationship of $(q, \dot{q}, \ddot{q}) \rightarrow \tau$. According to research, multilayer feed-forward neural networks such as the BP neural network can approximate any nonlinear mapping with precision, which can reflect a static mapping relationship well, and the parallel structure makes it capable of implementing such large-scale complex operations quickly [67].

Establishment of the dynamics inverse model of the feed-forward network is a process of continuous learning according to actual experimental data using the actual motion experiment as a training sample. Input of the actual dynamic system is taken as the output sample of the network, and the output results of the displacement, velocity, acceleration, and so on of the actual dynamic system are taken as input samples of the network. Training of the network fits the training samples using the selected network structure, and the use of the model is equivalent to predicting the behavior of the object based on the fitting result.

The reachable state space of the dynamic system is represented by X, the sample data of a certain point x in state space is recorded as $(\tau \rightarrow \ddot{q})|x$, the state set of the training sample is X_p, $X_p \subset X$ can be seen, any element x_i in X_p is a discrete point in X space, and the envelope X_c of all sample points constitutes a subspace on X. For any state point x in reachable state space, define the error of the network output at x as ξ_x.

Accuracy of the network at the sample point is the highest; it depends on the tolerance of the training. Because of the inertia of the actual dynamic system and the continuity of the network transfer function, the state point error closer to the sample point is generally smaller [68].

Assume that the average error of the sample points is represented by ξ, and d_p and d_q represent the distance to the two sample points, respectively. K is a positive coefficient that increases as the nonlinearity of the object itself increases. Farther sample points with smaller effects should be ignored, whereas the influence of the two sample points closest to the state point on the state point should be considered. According to the experimental data, the empirical equation for the error expectation of the network at state point x is presented as [69]:

$$E(\xi_x) = K\sqrt{d_p d_q}\xi \tag{2.57}$$

As shown in Fig. 2.7, the process of using the ANN for inverse dynamics calculation can be roughly summarized in three steps: (1) train the ANN with training samples composed of actual data, (2) use the force or torque of each joint to be calculated as the input to the ANN inverse dynamics model, and (3) output the joint coordinates of the assembly robot.

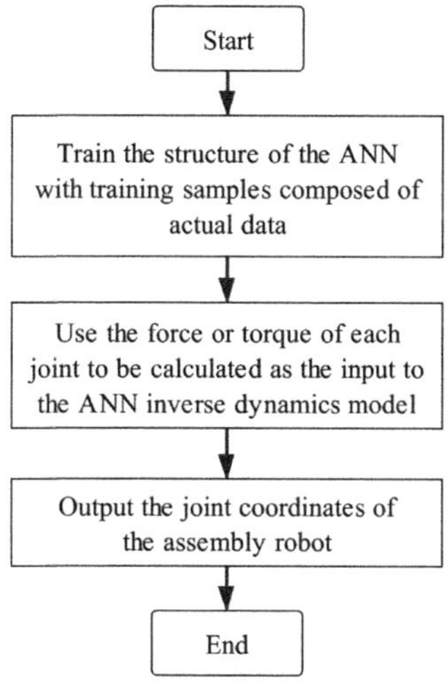

Figure 2.7
Process of using the artificial neural network (ANN) for inverse dynamics calculation.

2.5 Conclusion and outlook

This chapter first introduces the development progress and key technologies of assembly robots. Then, it introduces the main components of assembly robots: machinery components, sensors, controllers, and actuators. Arm dynamics is a research hot topic of assembly robots. This chapter introduces mathematical models of forward and inverse dynamics and discusses various trajectory planning algorithms in joint and Cartesian space. Finally, the ANN algorithm is introduced to perform an inverse dynamics optimization calculation of the robot arm.

Assembly robots have great application and market value because of their unique advantages. The third-generation assembly robot (the intelligent assembly robot) is still in the research and development stage, and there are still many shortcomings. Researchers across the world are working hard to improve the technical level of assembly robots and develop them in the direction of "high speed, high precision, heavy load, light-weight and intelligence" [70]. The research content of intelligent assembly robots in the future mainly focuses on six aspects:

(a) New material research and structural optimization design

High-quality materials are sought with higher strength and lower self-weight, and a lightweight design for the operating mechanism; at the same time, with the continuous development of modular mechanisms, assembly robots have gradually become integrated, which not only can improve their load level but can save resources and reduce environmental pollution.

(b) Multisensor information fusion technology

With the emergence of various new types of sensors, sensor-type intelligent assembly robots are appearing in people's field of vision. Because of the complexity of the assembly work environment and the operation process, a single sensor cannot capture more effective information, and there is no way to ensure the accuracy of system decisions. Multisensor information fusion technology is used to capture and process environmental information in time, which greatly improves the speed and reliability of system decision-making.

Commonly used information fusion algorithms are embedded constraint, evidence combination, and ANN methods. In the research of multisensor information fusion technology for assembly robots, it is necessary to select an appropriate information fusion algorithm according to actual requirements [9].

(c) Assembly robot control technology

The assembly robot control system is gradually developing in the direction of easy control, modularization, and openness. In addition to a variety of sensor input positions, force, acceleration, and other information, force control and feedback control, such as network control, offline programming, and other control technologies based on personal computers are also research hot topics.

(d) Assembly robot autonomous technology

A major feature of intelligent assembly machine robots is their strong learning and self-adaptation capabilities. Using a variety of sensors, information fusion, expert systems, and other artificial intelligence means, the assembly robot can continuously adapt to changes in the environment, make dynamic responses to trajectory planning, and independently carry out related operations to achieve work autonomy.

(e) Multirobot collaboration technology

With the continuous development of assembly production lines, collaboration of the same robotic arms and the cooperation of multiple robots have been applied. Multiple assembly robots need to be equipped with visual and tactile sensors during work to sense in advance physical contact that may occur and plan the route in time to avoid collisions [71].

Compared with a single robot, in the face of more complicated work, the multirobot collaboration system can use multiple simple robots to work together to avoid the complicated structure and control system of the single robot, improve work efficiency, and work flexibility. The multirobot collaboration system can autonomously change the collaboration relationship according to changes in the environment and cooperate to fulfill specified task requirements.

(f) Assembly robot human–robot interaction technology

Assembly robot human–robot interaction technology studies the interaction between workers and assembly robots. Through remote control and monitoring technology, multimedia, and virtual reality technology, a large-scale simulation environment and robot remote control system are established to realize automatic offline programming and virtual remote operation of the assembly robot.

The assembly robot human–robot interaction system generally includes a control box and control system. The control system is composed of a human–robot interaction interface, various function buttons, and a controller, and can control the direction and speed of movement of the robot, greatly improving the intuitiveness and flexibility of assembly operations.

References

[1] M. Hillman, 2 Rehabilitation Robotics from Past to Present–A Historical Perspective. Advances in Rehabilitation Robotics, Springer, 2004, pp. 25–44.

[2] T. Owen, Assembly with Robots, Springer Science & Business Media, 2012.

[3] M.J. Dunne, Industrial Robots: Present and Future. Robotics and Artificial Intelligence, Springer, 1984, pp. 667–678.

[4] S. Lewis-Workman, D. Brod, Measuring the neighborhood benefits of rail transit accessibility, Transp. Res. Rec. 1576 (1997) 147–153.

[5] M. Ghobakhloo, The future of manufacturing industry: a strategic roadmap toward Industry 4.0, J. Manuf. Technol. Manag. 29 (2018) 910–936.

[6] G. Boothroyd, Assembly Automation and Product Design, CRC Press, 2005.

[7] T.M. Wang, Y. Tao, Research status and industrialization development strategy of Chinese industrial robot, J. Mech. Eng. 50 (1) (2014).

[8] H. Hoffmann, Perception through visuomotor anticipation in a mobile robot, Neural Netw. 20 (2007) 22–33.

[9] T. Brogårdh, Present and future robot control development–an industrial perspective, Annu. Rev. Control 31 (2007) 69–79.

[10] C. Xu, J. Wang, G. Cao, et al., Design of a new current sensing device for joint torque force control of the precision assembly robot, in: Fifth World Congress on Intelligent Control and Automation (IEEE Cat No 04EX788), vol. 5, 2004, pp. 4609–4613.

[11] R. Taylor, P. Jensen, L. Whitcomb, et al., A steady-hand robotic system for microsurgical augmentation, Int. J. Robot. Res. 18 (1999) 1201–1210.

[12] Y. Yamada, S. Nagamatsu, Y. Sato, Development of multi-arm robots for automobile assembly, in: Proceedings of 1995 IEEE International Conference on Robotics and Automation, vol. 3, 1995, pp. 2224–2229.

[13] C. Gong, J. Yuan, J. Ni, Nongeometric error identification and compensation for robotic system by inverse calibration, Int. J. Mach. Tool Manufact. 40 (2000) 2119–2137.

[14] E. Tuci, R. Groß, V. Trianni, et al., Cooperation through self-assembly in multi-robot systems, ACM Trans. Autonom. Adapt. Syst. 1 (2006) 115–150.

[15] B. Horn, B. Klaus, P. Horn, Robot Vision, MIT Press, 1986.

[16] M. Hu, Z. Lin, X. Lai, et al., Simulation and analysis of assembly processes considering compliant, non-ideal parts and tooling variations, Int. J. Mach. Tools Manufact. 41 (2001) 2233–2243.

[17] F. Roy, Rigid Body Dynamics Algorithms, Springer, 2014.
[18] F. Pierrot, V. Nabat, Olivier Company, et al., Optimal design of a 4-DOF parallel manipulator: from academia to industry, IEEE Trans. Robot. 25 (2009) 213–224.
[19] H.S. Kim, Kinematic calibration of a Cartesian parallel manipulator, Int. J. Contr. Autom. Syst. 3 (2005) 453–460.
[20] T. Sasaki, T. Miyata, K. Kawashima, Development of remote control system of construction machinery using pneumatic robot arm, in: 2004 IEEE/RSJ International Conference on Intelligent Robots and Systems (IROS)(IEEE Cat No 04CH37566), vol. 1, 2004, pp. 748–753.
[21] C.M. Gosselin, E. Lavoie, On the kinematic design of spherical three-degree-of-freedom parallel manipulators, Int. J. Robot. Res. 12 (1993) 394–402.
[22] M.E. Rosheim, Robot Evolution: The Development of Anthrobotics, John Wiley & Sons, 1994.
[23] S. Joo, F. Miyazaki, Development of variable RCC and its application, in: Proceedings 1998 IEEE/RSJ International Conference on Intelligent Robots and Systems Innovations in Theory, Practice and Applications (Cat No 98CH36190), vol. 2, 1998, pp. 1326–1332.
[24] Y. Xu, R.P. Paul, A robot compliant wrist system for automated assembly, in: Proceedings, IEEE International Conference on Robotics and Automation, 1990, pp. 1750–1755.
[25] M.R. Cutkosky, P.K. Wright, Active control of a compliant wrist in manufacturing tasks, J. Eng. Ind. 108 (1986) 36–43.
[26] C.C. Nguyen, F.J. Pooran, Dynamic analysis of a 6 DOF CKCM robot end-effector for dual-arm telerobot systems, Robot. Autonom. Syst. 5 (1989) 377–394.
[27] S. Ruocco, Robot Sensors and Transducers, Springer Science & Business Media, 2013.
[28] N. Wang, N. Zhang, M. Wang, Wireless sensors in agriculture and food industry—recent development and future perspective, Comput. Electron. Agric. 50 (2006) 1–14.
[29] J. Borenstein, H.R. Everett, L. Feng, et al., Mobile robot positioning: sensors and techniques, J. Robot. Syst. 14 (1997) 231–249.
[30] S. Hutchinson, G.D. Hager, P.I. Corke, A tutorial on visual servo control, IEEE Trans. Robot. Autom. 12 (1996) 651–670.
[31] J. Wang, T. Cheng, Structural design and algorithms for trajectory planning of the SCARA robot, J. Hefei Univ. Technol. (Nat. Sci.) 31 (2008) 10126–11028.
[32] H.S.A. Da Silva, Design for Manufacture of Brushless Permanent Magnet Synchronous Servomotors, University of East London, 2014.
[33] R.J. Wai, C.M. Lin, C.F. Hsu, Adaptive fuzzy sliding-mode control for electrical servo drive, Fuzzy Sets Syst. 143 (2004) 295–310.
[34] R.S. Hartenberg, J. Denavit, A kinematic notation for lower pair mechanisms based on matrices, J. Appl. Mech. 77 (1955) 215–221.
[35] N. Orlandea, M.A. Chace, D. Albert Calahan, A sparsity-oriented approach to the dynamic analysis and design of mechanical systems—part 1, J. Eng. Ind. 99 (1977) 773–779.
[36] R. Featherstone, D. Orin, Robot dynamics: equations and algorithms, in: Proceedings 2000 ICRA Millennium Conference IEEE International Conference on Robotics and Automation Symposia Proceedings (Cat No 00CH37065), vol. 1, 2000, pp. 826–834.
[37] T. Geng, Dynamics and trajectory planning of a planar flipping robot, Mech. Res. Commun. 32 (2005) 636–644.
[38] K. Shin, N. McKay, A dynamic programming approach to trajectory planning of robotic manipulators, IEEE Trans. Automat. Control 31 (1986) 491–500.
[39] K. Shengzheng, H. Wu, L. Yao, et al., Coordinated workspace analysis and trajectory planning of redundant dual-arm robot, in: 2016 13th International Conference on Ubiquitous Robots and Ambient Intelligence, URAI, 2016, pp. 178–183.
[40] F.E. Kydland, E.C. Prescott, Time to build and aggregate fluctuations, Econometrica 50 (1982) 1345–1370.

[41] C. Gosselin, S. Foucault, Dynamic point-to-point trajectory planning of a two-DOF cable-suspended parallel robot, IEEE Trans. Robot. 30 (2014) 728–736.
[42] R. Kelly, V.S. Davila, J.A.L. Perez, Control of Robot Manipulators in Joint Space, Springer Science & Business Media, 2006.
[43] N. Sun, Y. Fang, Y. Zhang, et al., A novel kinematic coupling-based trajectory planning method for overhead cranes, IEEE ASME Trans. Mechatron. 17 (2011) 166–173.
[44] D.F. Rogers, N.R. Fog, Constrained B-spline curve and surface fitting, Comput. Aided Des. 21 (1989) 641–648.
[45] L.M. Capisani, A. Ferrara, Trajectory planning and second-order sliding mode motion/interaction control for robot manipulators in unknown environments, IEEE Trans. Ind. Electron. 59 (2011) 3189–3198.
[46] L. Guo, P.S. Ge, M. Yue, et al., Lane changing trajectory planning and tracking controller design for intelligent vehicle running on curved road, Math. Probl. Eng. (2014) 1–9.
[47] O.J. Blanchard, C.M. Kahn, The solution of linear difference models under rational expectations, Econometrica 48 (1980) 1305–1311.
[48] T. Wang, Trajectory planning of articulated robotic arm in the Cartesian space, J. Funct. Mater. Devices 21 (2015) 118–123.
[49] W. Lin, W. Jiang, Trajectory planning of industrial robot in Cartesian space, Mech. Eng. Autom. 10 (2014) 141–143.
[50] S. Behzadipour, A. Khajepour, Time-optimal trajectory planning in cable-based manipulators, IEEE Trans. Robot. 22 (2006) 559–563.
[51] A.M. Zalzala, A.S. Morris, A distributed on-line trajectory generator for intelligent sensory-based manipulators, Robotica 9 (1991) 145–155.
[52] T.Y. Wu, T.J. Yeh, B.H. Hsu, Trajectory planning of a one-legged robot performing a stable hop, Int. J. Robot. Res. 30 (2011) 1072–1091.
[53] C. Lin, P. Chang, J. Luh, Formulation and optimization of cubic polynomial joint trajectories for industrial robots, IEEE Trans. Automat. Control 28 (1983) 1066–1074.
[54] B. Tondu, H. El-Zorkany, Identification of a trajectory generator model for the PUMA-560 robot, J. Robot. Syst. 11 (1994) 77–90.
[55] S.A. Bazaz, B. Tondu, Minimum time on-line joint trajectory generator based on low order spline method for industrial manipulators, Robot. Autonom. Syst. 29 (1999) 257–268.
[56] T. Choi, C. Park, H. Do, et al., Trajectory correction based on shape peculiarity in direct teaching manipulator, Int. J. Control Autom. Syst. 11 (2013) 1009–1017.
[57] F.U. Rong, J.U. Hehua, Time-optimal trajectory planning algorithm for manipulator based on PSO, Inf. Control 40 (2011) 802–808.
[58] Y. Shi, R. Eberhart, A modified particle swarm optimizer, in: 1998 IEEE International Conference on Evolutionary Computation Proceedings IEEE World Congress on Computational Intelligence (Cat No 98TH8360), 1998, pp. 69–73.
[59] J. Kennedy, R. Eberhart, Particle swarm optimization, in: Proceedings of the IEEE International Conference on Neural Networks, vol. 4, 1995, pp. 1942–1948.
[60] J. Gregory, A. Olivares, E. Staffetti, Energy-optimal trajectory planning for robot manipulators with holonomic constraints, Syst. Contr. Lett. 61 (2012) 279–291.
[61] A.R. Hirakawa, A. Kawamura, Trajectory generation for redundant manipulators under optimization of consumed electrical energy, in: IAS'96 Conference Record of the 1996 IEEE Industry Applications Conference Thirty-First IAS Annual Meeting, vol. 3, 1996, pp. 1626–1632.
[62] D.P. Garg, M. Kumar, Camera calibration and sensor fusion in an automated flexible manufacturing multi-robot work cell, in: Proceedings of the 2002 American Control Conference (IEEE Cat No CH37301), vol. 6, 2002, pp. 4934–4939.
[63] B.J. Martin, J.E. Bobrow, Minimum-effort motions for open-chain manipulators with task-dependent end-effector constraints, Int. J. Robot Res. 18 (1999) 213–224.

[64] W.Q.D. Do, D.C.H. Yang, Inverse dynamic analysis and simulation of a platform type of robot, J. Robot. Syst. 5 (1988) 209–227.
[65] Y. Li, B. Li, J. Ruan, et al., Research of mammal bionic quadruped robots: a review, in: 2011 IEEE 5th International Conference on Robotics, Automation and Mechatronics (RAM), 2011, pp. 166–171.
[66] K.L. Hsu, H.V. Gupta, S. Sorooshian, Artificial neural network modeling of the rainfall-runoff process, Water Resour. Res. 31 (1995) 2517–2530.
[67] G. Cybenko, Approximation by superpositions of a sigmoidal function, Math. Control Signals Syst. 2 (1989) 303–314.
[68] H.C. Nho, P. Meckl, Intelligent feedforward control and payload estimation for a two-link robotic manipulator, IEEE/ASME Trans. Mechatron. 8 (2003) 277–282.
[69] N. Luan, Study on robot inverse dynamics modeling with neural network, Electric Mach. Control 12 (2008) 174–178.
[70] S. Pruthi, Wireless robotics: a history, an overview, and the need for standardization, Wirel. Pers. Commun. 64 (2012) 597–609.
[71] H. Schiøler, T.S. Toftegaard, Wireless communication in mobile robotics a case for standardization, Wirel. Pers. Commun. 64 (2012) 583–596.

CHAPTER 3

Rail transit collaborative robot systems

3.1 Overview of collaborative robots

3.1.1 Definition of collaborative robots

Collaborative robots can work and interact directly with human workers in a shared workspace. Since the security of human–robot collaboration has been fully considered at the time of design, the collaborative robots are inherently secure. In the event of a collision, pinching, and other accidents or unexpected faults, collaborative robots can handle these situations well to ensure the safety of human workers, workspace, and the robots themselves.

Collaborative robots usually refer to the robotic arms and their control systems. However, collaborative robots are designed to collaborate with human workers on workpieces in the shared workspace, and should be equipped with different kinds of end effectors for different work tasks to satisfy the operation of different workpieces. Therefore, a safe and efficient collaborative robot system should include collaborative robots, end effectors, fixtures, workpieces, and accessories. When it comes to the safety of collaborative robots, it is not enough to consider only the safety of the robotic arms, and the overall safety of the collaborative robot system should also be paid attention to.

With more sensors installed, collaborative robots can effectively perceive the working environments, the position, and speed of human workers, and then work safely in the human–robot coexistence working environments, which are unstructured and complex. Collaborative robots usually have the characteristics of small structures, low power, low running speeds, low noises, and convenient installation. They do not have to work within the security fences and can effectively reduce initial construction investment and space occupancy [1]. Compared with traditional industrial robots, which must be programmed by professional programmers, the collaborative robots usually have the function of teaching programming, and they can be programmed by ordinary workers. The programming methods of collaborative robots can effectively reduce the time consumption of programming and debugging. The applications of collaborative robots can effectively reduce risk levels by helping human workers to perform those jobs that will be harmful [1].

Although collaborative robots have many advantages mentioned earlier, they have some losses in accuracy and efficiency due to the emphasis on safety design. To reduce the damage

Robot Systems for Rail Transit Applications. https://doi.org/10.1016/B978-0-12-822968-2.00003-6

caused by an accidental collision that may occur in the process of collaboration, the speed and mass of collaborative robots are limited to a certain range to reduce their inertia and energy during a collision. Therefore, collaborative robots generally operate at a slower speed, with lower power and load [1]. Due to the small self-weight, the stiffness of collaborative robots is usually lower than that of the traditional industrial robots, and the repetitive positioning accuracy of collaborative robots is usually also lower than that of the traditional industrial robots. The loss of stiffness can also cause problems in collision detection.

3.1.2 Development progress

3.1.2.1 Background on the development of collaborative robots

Traditional industrial robots are usually confined to physical fences to perform independent, repetitive, and high-intensity work [2]. Since these robots usually have very powerful driving systems and complex end effectors and are not equipped with sensors for detecting workers, if they are not confined within the fences, they are very likely to cause unnecessary harm to human workers. Since the work performed by traditional industrial robots has been highly specialized, these robots are often programmed to autonomously work in fixed programs to perform repetitive and arduous tasks. For this reason, traditional industrial robots usually do not need to cooperate with human workers, and they only need routine maintenance. During maintenance, traditional industrial robots are also in a power-off state and locked in a safe position, which means that human–robot interaction (HRI) is still not required.

However, traditional industrial robots lack the necessary flexibility and adaptability. For example, on the assembly line, there are still some assembly operations that cannot be fully automated for the time being. However, human work is more prone to errors when product complexity and reliability requirements are higher. Collaborative robots are designed to take full advantage of human flexibility and adaptability to solve complex tasks that cannot be fully automated. In the collaborative work, the collaborative robots can perform tasks that are dangerous or burdensome to human workers, who use their experience, flexibility, and ability to cope with emergencies to deal with imprecise problems that require creativity and imagination [3]. It is worth noting that virtual fences will replace physical fences in the collaborative work environment, allowing the collaborative robots to interact closely with human workers. The collaboration between collaborative robots and human workers can effectively reduce the risk of injury to human workers, and make full use of the advantages of robots in terms of accuracy, power, and repetition to improve production efficiency and product quality.

Traditional industrial robots are often difficult to meet the production occasions where there are many kinds of products, few production batches, and highly flexible requirements. In the case of small parts assembly work and small space operations, which are difficult to achieve fully automated, and where flexibility is required, traditional industrial robots are also difficult to apply. In this case, the cooperative operation between the collaborative robots and human workers can meet these production requirements, improve manufacturing efficiency, manufacturing quality, and effectively improve the comfort and safety of human workers [4].

3.1.2.2 Early collaborative robots

Early collaborative robots are very different from the collaborative robots on the market today in terms of definition, shape, function, task, and security. Collaborative robots originated in the 1990s. In 1995, General Motors proposed a project to develop sufficiently safe robotic equipment that would assist human workers in assembly operations on assembly lines. In 1996, Edward et al. of Northwestern University first proposed the concept of collaborative robots and subsequently applied for patents [5]. Early collaborative robots were mainly intelligent assist devices designed to solve the problem of combining robotics with human assistive devices. The power to operate these devices was completely exerted by the operator to ensure the safety, and there was usually no additional power source. Collaborative robots developed later were gradually equipped with power sources. Early collaborative robots supported and guided workpieces primarily through software-defined virtual boot surfaces. They also provided amplification of the workforce to reduce the labor intensity of human work.

3.1.2.3 Current collaborative robots

3.1.2.3.1 Multirobot collaboration

Before the collaborative robots for HRI are developed, the cooperation of collaborative robots mainly referred to the interaction and cooperative movement between two robots. The two robots shown in Fig. 3.1 can cooperate with each other to carry out task assignments. The work done by the two-arm robots is still considered to be multirobot collaboration. The collaboration at this stage tended to be more "multirobot collaboration" than "human–robot collaboration." For example, the Slim Dual Arm (SDA) series and the Slim Individual Arm (SIA) series of lightweight robots launched by the robot division of the YASKAWA company, although they are almost identical in structure and appearance to the current mainstream robots, they can only collaborate with each other [6]. It is difficult for them to achieve collaboration with human workers.

Figure 3.1
Multirobot collaboration.

Nevertheless, due to the limited capability of a single robot, in some areas with special needs, multiple robots are required to complete tasks together. Compared with a single robot, multirobot collaboration has the characteristics of high efficiency, stability, and rapidity, which make up for the deficiency of a single robot in performance. Therefore, multirobot collaboration system technology has become a hot research direction [7]. The key technologies that may be involved in the research of multirobot collaborative control include path planning, formation control, environmental perception, obstacle avoidance, and task allocation.

3.1.2.3.2 Human–robot collaboration

At present, the collaborative robot products produced by many robot manufacturers in the world have redefined the meaning of collaborative robots. The collaborative robot has realized the transformation from "intelligent auxiliary lifting equipment" to the current collaborative robot. Compared to earlier collaborative robots, these collaborative robots are often equipped with multidegree-of-freedom robotic arm structures similar to traditional industrial robots. They have their power and are equipped with end effectors for specific tasks, but they are not restricted to work in the fences. This type of collaborative robot can collaborate with human workers. Moreover, new collaborative robots are typically smaller in size and power than traditional industrial robots. Because of active or passive safety precautions, they are safe enough for human workers when working with them. The first collaborative robot that fits the current meaning is the UR5 collaborative robot introduced by Universal Robots (UR) in 2008. UR10 and UR3 collaborative robots were subsequently launched in 2012 and 2015, respectively. Since then, the collaborative robot market has been opened, and traditional industrial robot companies have expanded their development goals to this new field. The human worker can safely access and work with a running collaborative robot as shown in Fig. 3.2.

Significant differences between traditional industrial robots and collaborative robots are summarized in Table 3.1.

Figure 3.2
Human–robot collaboration.

Table 3.1: Comparison of traditional industrial robots and collaborative robots.

	Traditional industrial robots	Collaborative robots
Collaboration	Work for human	Work with human
Manufacturing mode	Single variety, mass production	Multivariety, small and medium batch production
Safety features	Inadequate security and need to be isolated	High security, can work with human workers safely
Task type	Repetitive, high-intensity	Complex and flexible
Sensor	Equipped with fewer sensors	Equipped with more sensors
Programming	Require advanced programming skills and long debugging time	Programming is usually easy and debugging time is shorter

3.1.3 Application fields

Collaborative robots have been widely used, especially in those fields that require HRI.

3.1.3.1 Traditional production line

Collaborative robots are mainly used in production lines. It can be used for auxiliary assembly and assembly quality inspection on an assembly line. For example, Volkswagen integrated UR collaborative robots into the large-scale production line of its engine manufacturing plant in 2013 [8]. Trelleborg Sealing Solutions used 42 UR collaborative robots to tend Computer Numerical Control (CNC) machines [1]. This allows one operator to be responsible for eight CNC machines at the same time. A Greek factory uses UR5 collaborative robots for product packaging [1]. The robot picks up three bottles from the production line at the same time and places them in the packaging device. This can effectively increase work efficiencies and free human workers from repetitive tasks. Audi has set up a lot of robots in its factory called PART4you, which can pick up the workpieces and pass them to human workers, thereby eliminating the harms that repeated actions may cause to human workers [1].

Makrini et al. proposed a collaborative robot architecture that combines gesture recognition, face recognition, visual inspection, and human-like robot behavior module to assist and guide human workers in the assembly process [9]. The system was verified on the Baxter collaborative robot from Rethink Robotics. In the experiment, the robot assisted human workers to assemble a box and realized quality control through visual inspection. The collaborative robot system with guidance and assistance is particularly suitable for the assembly of complex products consisting of many different components.

3.1.3.2 Aviation manufacturing and maintenance

It is worth mentioning that there are also applications of collaborative robots in the field of aviation manufacturing. For example, Fokker Aerostructures used 7 Degrees of Freedom

(DOFs) LBR iiwa (LBR stands for Lightweight Robot and iiwa for intelligent industrial work assistant) collaborative robot launched by KUKA with Automatic Drilling Unit (ADU) to achieve automatic hole drilling [10]. The collaborative robot can be mounted on an Automatic Guided Vehicle (AGV) or fixed guide rail on the ground to achieve a greater range of work. The company also used the 6-DOF UR10 collaborative robot launched by UR to carry out the coating of the landing gear bushing [11]. The operators only need to load the mixed sealant into the end effectors of the collaborative robots, and the rest of the work can be done by the collaborative robots. The pressure and speed of glue application are determined by the robots, and the operators only need to complete the quality inspection and abnormal situation handling. It can be said that the applications of collaborative robots effectively reduce the labor intensity of human workers, improve work efficiency, and obtain higher work quality.

Collaborative robots have also been used in aircraft maintenance. The French Air-Cobot project, for example, aims to improve preflight maintenance of aircraft and provide traceability of inspections that have been carried out [12]. To achieve these functions, a collaborative mobile robot has been built. Although such mobile robots usually cannot work in close proximity with human workers, they can safely travel and work in environments where human workers are present [2]. The robot can navigate by itself near the aircraft and conduct nondestructive tests with some sensors to detect possible aircraft faults. All the electronic equipment of the collaborative robot is carried by a mobile platform, which is driven by four wheels and can run at a maximum speed of 2 m per second. Obstacle detection buffers are set in front and rear of the platform.

When they are compressed, the platform will stop automatically. The collaborative robot is equipped with navigation sensors and nondestructive test sensors. The former includes two stereo cameras, two laser range finders, a Global Positioning System (GPS) receiver and an Inertial Measurement Unit (IMU), while the latter includes a Pan-Tilt-Zoom (PTZ) camera and a three-dimensional (3D) scanner. Of course, the robot supports human–robot collaboration while supporting autonomous navigation and nondestructive testing. After the detection process is over, the robot gives the diagnosis and asks the human operator to verify or reject it.

3.1.3.3 Scientific research

Collaborative robots can also provide new experimental platforms for universities and research institutes. For example, Jubien et al. studied the dynamic identification of KUKA's lightweight robot LWR4+ [13]. Since information about the dynamics parameters of the robot was not given by KUKA, it is necessary to build a dynamics model for model-based control and simulation. In this study, actual dynamic parameters of LWR4+ were identified, as well as weight, inertia, and stiffness parameters. The identification process is accomplished by using an offline identification method which is based on the Inverse Dynamic Identification Model and linear Least Squares technique (IDIM-LS). The results show that the identified parameters are similar to the parameters given by the manufacturer. Ju et al. reconstructed the dynamics model of the Baxter collaborative robot and conducted experimental verification [14]. In this study, the Denavit-Hartenberg (D-H) representation was used to establish the dynamics model, which was then implemented in the Robotics Toolbox of MATLAB. It should be noted that the D-H representations are widely used to describe the kinematic model of a robot. Many comparisons between simulation and experimental results confirmed the validity of the dynamic model.

3.1.3.4 Entertainment

Some researchers have also studied the application expansion of collaborative robots, such as the application of collaborative robots for entertainment purposes. For example, Kirschner et al. proposed and evaluated an example of a collaborative robot for game application, in which a collaborative robot worked with a human partner on the tangram puzzle [15]. In this study, Asea Brown Boveri (ABB)'s dual-arm collaborative robot YuMi was used to conduct research on tangram recognition, grasping, game sequence, and emergency stop during a collision, etc. Finally, the game was successfully completed through human–robot collaboration. AlAttar et al. proposed a collaborative robot that could compete with human opponents against air hockey [16]. The Panda collaborative robot developed by Franka Emika GmbH was used in this study. The architecture consisted of two layers including a tactical layer and an active layer. The former was used to realize high-speed perception, motion prediction, and strategy planning, while the latter was used to realize the execution of robot motion through optimal control. During the experiment, the collaborative robot demonstrated robust game performance and it was able to adapt to a variety of situations.

3.1.3.5 Other fields

There are also many applications of collaborative robots. For example, automobile manufacturing plant, Computer, Communication, Consumer Electronics (3C) industry manufacturing plant, medical auxiliary operation, packaging, loading and unloading, parts assembly, and parts testing, etc. [1,17,18].

3.1.4 Key technologies

3.1.4.1 Standards and specifications for collaborative robots

ISO 10218 and ISO 15066 are current standards and specifications for collaborative robot design. ISO 10218–1:2011, entitled "Robots and robotic devices - Safety requirements for industrial robots - Part 1: Robots," specifies requirements for inherent safety design, protective measures and use the information for industrial robots [19]. ISO 10218–2:2011, entitled "Robots and robotic devices - Safety requirements for industrial robots - Part 2:

Robot systems and integration," specifies safety requirements for industrial robots, industrial robot integration and industrial robot cells in terms of design, manufacturing, installation, operation, maintenance, decommissioning, etc. [20].

When the ISO 10218 was issued, the development of collaborative robots in the market was not mature, but with the increasing popularity of the collaborative robots, new technical specifications are urgently needed. The title of ISO/TS 15066:2016 is "Robots and robotic devices - Collaborative robots" [21]. It is based on ISO 10218-1 and ISO 10218-2 and specifies the safety requirements for collaborative industrial robot systems and work environments. It also complements the requirements for collaborative industrial robot operations given in ISO 10218-1 and ISO 10218-2.

3.1.4.2 Types of collaborative operations

Four types of collaborative operations and corresponding security measures are defined for collaborative robots. They are safety-rated monitored stop mode, hand guiding mode, speed and separation monitoring mode, and power and force limiting mode.

3.1.4.2.1 Safety-rated monitored stop

In this type of collaboration, if a worker invades a collaborative workspace, the robotic system immediately triggers a stop action. And after the worker leaves the collaborative space, the robot system will resume work. It should be noted that after triggering a safety-rated monitored stop mode, the robot is only in a shutdown state and has not been disconnected from the power supply. After that, the robot will work in a noncollaborative manner like a traditional industrial robot. Although such an approach ensures that robotic systems and human workers do not move at the same time in the collaborative area, this is a security measure at the expense of efficiency.

To achieve this type of collaboration, associated safety sensor devices are required for detecting the human workers in the collaborative area. The detection of human invasion is mainly achieved by installing light curtains, laser scanners, cameras, floor sensors, and other sensors. With these devices, it is possible to detect whether there is a worker in the collaboration area in real time and trigger the start or stop action of the robot system accordingly. For example, Mitsubishi used Aspect Automation's floor mat sensor to transform its traditional industrial robot, making the robot more secure and friendly [2]. The sensor consists of two parts, an inner ring and an outer ring. When a human worker enters the outer ring, the robot will slow down its running speed, and when the human worker continues to approach the robot and enters the inner ring, the robot will immediately stop moving.

This method is applied in the case of manual loading and unloading of the end effectors, routine inspection of the robot, and maintenance of the fault. In fact, for current collaborative robots, this method usually needs to be combined with other methods to achieve better collaboration. For example, the safety-rated monitored stop must be triggered before the hand guiding mode of the collaborative robot begins. Collaborative robots running in speed and separation monitoring mode will also be triggered to safety-rated monitored stop mode when the minimum safety distance is exceeded.

3.1.4.2.2 Hand guiding

In the hand guiding mode, the human worker uses a manual guiding device to drag and guide the robot to move along the desired trajectory. The manual guiding device is usually located on or near the end effector, using force/torque sensors to obtain the operator's intention and transmitting it to the robot controller. When determining the installation position of the manual guiding device, the convenience of the operator to observe the movement of the robot should be fully considered and other safety hazards should not be introduced.

Before the operator is ready to begin the manual boot operation, the robot should have been triggered to the safety-rated monitored stop mode mentioned earlier. Subsequently, when the operator prepares to start guiding the robot work in the hand guiding manner, the safety-rated monitored stop mode will be released. After the hand guiding procedure is over, once the manual guiding device is released by the operator, the robot will trigger the safety-rated monitored stop mode again. After the operator leaves the collaborative space, the robot will resume movement and follow the trajectory and actions recorded during manual guidance.

This mode needs to be carried out in situations where robotic systems are only used to amplify human forces, or where robots frequently switch tasks. In the case where a collaborative robot itself cannot achieve the precise positioning of the workpiece, the robot can also enter the hand guiding mode and request the assistance of a human worker. Subsequently, the robot's end effector is manually operated by the human worker to perform relatively accurate positioning and clamping of the workpiece. When the robot ends the hand guiding mode and the human worker exits the danger zone, the collaborative robot will resume autonomous operation.

3.1.4.2.3 Speed and separation monitoring

In this mode, the operator and the robot can move and work simultaneously in the collaborative space. When simultaneously moving in hand guiding mode, the robot usually moves passively without active power. In this mode, the robot independently performs active movement and works under certain restrictions. The limitation is that the distance between the operator and any part of the robot must be under the minimum protective safety distance at any time. The minimum protective safety distance is the minimum permitted distance between the operator and any moving part of the robot. This distance is not a constant, but a variable that decreases as the speed of the robot slows.

When the minimum protective safety distance is broken and cannot be guaranteed, the robot should stop immediately. To ensure this, when the operator approaches the robot, the running speed of the robot should decrease, thereby correspondingly reducing the minimum safety distance. The robot will enter a protective stop state when it is close enough to a human worker and perform related safety functions, such as shutting down a potentially dangerous tool. Another way to ensure that the minimum protective safety distance is not violated is for the robot to continue the task by performing an alternative motion path that does not exceed the minimum safety distance. When the worker leaves the vicinity of the robot and the distance meets the requirements of safety, the robot will resume motion autonomously. Compared with the safety-rated monitored stop mode, this mode has better human–robot collaboration efficiency.

The minimum protective safety distance should have a certain value at any time, so the calculation of its value is crucial. The calculation of the minimum protective safety distance requires several parameters, including data on the motion status of the human operator, such as the operator's position and speed, which usually need to be monitored by a laser device or a visual system. The calculation of the minimum protective safety distance S at time t_0 can be done by Ref. [22]:

$$S(t_0) \geq \left(\int_{\tau=t_0}^{\tau=t_0+T_R+T_S} v_H(\tau)d\tau \right) + \left(\int_{\tau=t_0}^{\tau=t_0+T_R} v_R(\tau)d\tau \right) + \left(\int_{\tau=t_0+t_R}^{\tau=t_0+T_R+T_S} v_S(\tau)d\tau \right) + (C + Z_S + Z_R) \tag{3.1}$$

where t_0 is the current time. v_H is the speed at which the human worker approaches the robot, v_R is the speed at which the robot approaches the human worker, and v_S is the speed of the robot during braking. T_R is the time it takes for the robot to respond to the appearance of a human worker, and T_S is the time it takes for the robot to stop safely. Different running speeds or different loads will lead to different braking times of the robot, so T_S is a function of the robot's running speed and load. C is the safety margin reserved for the intrusion distance. Z_S is the distance caused by the position uncertainty of the human worker, and Z_R is the distance caused by the position uncertainty of the robot.

A tactile floor is used in the SAPARO project for safety monitoring in a human–robot collaboration environment [1]. The hardware part of the system is mainly composed of a hard-safety sensor and a soft-safety component. The former refers to a tactile floor with high spatial resolution for workspace monitoring, while the latter refers to a projection system that can show the boundaries of safety zones to human workers by projecting different colors onto the tactile floor. The software part of the system is mainly used to dynamically calculate the safety zones around the robot based on the movement of the robot and human workers, such as the velocities and joint angles of the robot, and the position and velocities of the human workers. The calculated safety zones will then be projected on the tactile floor by the projection system to alert human workers. The colors of the projection area include red, orange, and green, which represent the critical zone, the warning zone, and the safe zone, respectively. Once a human worker enters the warning or critical zone, the robot will slow down or even stop. Compared with the traditional detection method that can only form a static safety zone, the proposed methods can determine the safety zones dynamically and therefore can protect the robot.

3.1.4.2.4 Power and force limiting

In power and force limiting mode, the worker is allowed to have physical contact with the robot. Therefore, robots must be specially designed to meet the needs of inertia, shape, materials, control methods, etc. For example, robots should have low mass and inertia, rounded edges and corners, smooth surfaces, deformable materials, etc. The robot can read the forces in the joints and determine if a collision has occurred by distinguishing between the forces involved, such as friction, drag, and impact. Once a collision occurs, the robot will immediately stop or run in the opposite direction.

Direct physical contact involves a variety of situations, not all of which are accidental, but still require effective limits on power and force. These contacts can be classified into the following three types: (a) foreseeable contact within the collaborative mission plan, (b) accidental contact caused by the operator's failure to follow the operating procedures, and (c) accidental contact caused by the failure of the robot system.

The contact between the robot and its operator can be divided into two types, namely quasistatic contact and transient contact. Among them, quasistatic contact usually occurs in the case of clamping and squeezing, when a certain part of the human body is limited between the movable part of the robot and other fixed objects. The duration of such contact is usually longer until it is alleviated. Transient contact usually occurs in an accidental collision, when the trajectory of a part of the human body collides with the trajectory of the robot arm. However, unlike the case of quasistatic contact, the human body is only subjected to force in one direction during transient contact. When a collision occurs, the operator's active protection and precautionary awareness cause him to retreat. Therefore, the action time of the transient contact is much shorter than that of the quasistatic contact. Considering that the response strategy of the robot to the collision may include the design of an active retreat, the action time of transient contact will be further shortened. The action time also depends on the relative speed at the time of the collision, as well as the inertia of the collision part.

In the design of the robot, it is also necessary to consider the tolerance of the human body to the pain at the time of a collision to limit the relevant operational characteristics of the robot. Some studies summarize the threshold limits that can be tolerated in different parts of the body and gives these specific thresholds. These thresholds are generally considered to be the limits that the human can tolerate. When the pressure and force on the human body exceed these values, the human around is very likely to be injured. The relationship between the threshold value for quasistatic contact and the threshold value for transient contact is shown as follows [21]:

$$P_T = 2p_{QS} \tag{3.2}$$

$$F_T = 2F_{QS} \tag{3.3}$$

where P_{QS} is the maximum permissible pressure for quasistatic contact, P_T is the maximum permissible pressure for transient contact, F_{QS} is the maximum permissible force for quasistatic contact, and F_T is the maximum permissible force for transient contact.

Therefore, in the power and force limiting mode, the most important safety protection mechanism is that the robot cannot exceed the operating parameters under the above threshold limits. These limits can be applied to structural weight, joint force or torque, speed, momentum, power, and stroke. To ensure safety under all conditions, safety issues in the worst case must be considered during the design process.

Power and force limiting modes are widely used in the fields where human–robot collaboration is very frequent, or where additional time consumption caused by safety-rated monitored stop mode cannot be accepted.

Different from the previous three types of cooperation, power and force limiting mode has inherent security, which fundamentally limits the degree of harm that robots may cause to human beings. Traditional industrial robots can also realize the other three types of collaboration mentioned earlier after proper transformation and upgrading, but it is usually difficult to realize the collaboration in the mode of power and force limiting mode. This is because traditional industrial robots typically have large dimensions, weight, power, and speed, which are in contradiction with the limitations mentioned earlier regarding the operational parameters of the collaborative robot.

For this reason, robots under the power and force limiting mode can be truly called collaborative robots. They can work safely around human workers without the need for additional safety devices.

The aforementioned four modes can also be used in hybrid ways. Shin et al. proposed a method in which the robot runs at the allowable maximum velocity [23]. The method is a combination of the third and fourth collaboration modes, which can improve production efficiency without affecting the safety of human–robot collaboration. The allowable maximum velocity is the maximum velocity at which the peak pressure and force will not exceed the thresholds when a collision occurs between the robot and a human worker. The allowable maximum velocity is calculated based on a layered collision model, which can predict the peak pressure and force when the collision occurs. The results of simulation experiments showed that, compared with the general method, this safe velocity method can effectively reduce the travel time of the robot and ensure the safety of collision.

3.1.4.3 Robot operating system

Although the controllers and control systems used by the collaborative robots in the current market are different, the collaborative robots launched by some manufacturers are based on the Robot Operating System (ROS) which is widely used in the development of robots. For example, the workstation or control system of the Baxter collaborative robot launched by Rethink Robotics is based on ROS architecture to achieve seamless cross-platform integration [9]. The ROS can be used to optimize and improve the existing collaborative robots. Mokaram et al. proposed an application programming interface (API) for the KUKA LBR iiwa collaborative robot [24]. The interface is implemented based on the ROS, which can meet the needs for the rapid development and devices integration of the robot without destroying the original safety features.

The ROS is a robot secondary operating system architecture specially developed for robot software development. The ROS originated from the artificial intelligence laboratory of Stanford University. The lab was working with Willow Garage on the PR2 robot, which was running under an earlier version of the ROS. In 2010, Willow Garage officially released the ROS system framework, which is a free and open source software framework for robotics. Since then, the ROS has been loved by many robot developers and enterprises, and has developed rapidly in just a few years. Although the ROS is not really an operating system, it provides operating-system-like services, including hardware abstraction, low-level device control, implementation of common functions, interprocess messaging, package management, etc. The main mission of ROS is to provide an open source, standardized programming framework for robots, so libraries and tools are provided to help robot software developers create programs. To meet various development needs of robots, the ROS also includes 3D reconstruction, dynamic simulation, real-time scene prevention, and physical control.

In terms of programming languages, as robots are a system highly related to hardware, and different developers have different professional backgrounds, the ROS supports the development of multiple programming languages. Currently, the ROS supports Python, C++, Java, Octave, LISP, and other programming languages. Developers can flexibly choose the programming language they are familiar with to complete their work, or use different languages in different modules, and then integrate them to complete the task.

In terms of module communication, the ROS uses distributed network communication. The ROS uses the mature transmission control protocol (TCP) communication mode. It expands and packages the functions on this basis to realize the loose coupling of point-to-point connections between different modules. This approach makes communication between nodes easy. Instead of spending time and energy to build communication protocols and methods between different functions, developers only need to focus on the realization of specific module functions.

3.1.4.4 Programming method

The programming of robots can usually be divided into online programming and offline programming.

The online programming of the collaborative robots includes the drag teaching method and somatosensory teaching method. The drag teaching method is human–robot friendly and easy to operate. In this way, the collaborative robot can learn the task it will perform. During programming, the operator drags the end effector to follow the expected trajectory, and all the joint motion parameters are recorded by its controller and stored in the internal memory. In the following works, it only needs to execute the parameters and procedures stored by the user in the teaching process to realize the automatic repetition of the job. Since industrial robots usually have good repetition accuracy, the accuracy of reaching any position in space can be guaranteed. However, the traditional drag teaching method usually requires the use of sensors, such as force/torque sensors, which will lead to expensive costs. Researchers have also studied sensorless hand guiding methods.

For example, Lee et al. used torque control based on the robot dynamics model, friction model, and motor current to determine the operator's intention to move the end effector [25]. Some researchers used the IMU to program robots through somatosensory teaching. In this way, the noncontact synchronous movement of the robot and human can be realized. This method is very practical in the service industry where the precision requirement is not high. The use of virtual reality and augmented reality technology can achieve direct and intuitive natural interaction between humans and robots. The aforementioned two online programming methods allow the debugging and correction of the actual movement of the robot, but online programming also has some insurmountable disadvantages. For example, when a robot needs to track a specific irregular curve or perform a complicated task, online programming will be difficult to implement due to too many factors to be considered.

Offline programming means that the operator builds the virtual scene of the robot working environment in the programming software and completes the program design of the robot according to the production requirements. Offline programming typically allows a program to be simulated and tested. It can also predict collisions, or simply define posture. The advantages of offline programming include the ability to (1) reduce robot downtime; (2) optimize the program; and (3) program complex tasks. Although the offline programming method is usually adopted by traditional industrial robots, collaborative robots can also be programmed by this method if the teaching method cannot meet the task requirements.

To ensure the simplicity of collaborative robot programming, many collaborative robot manufacturers have developed software systems that make the programming process very simple. For example, the UR robot can be programmed by a tablet [1]. This process is user-friendly and simple but can implement very complex functions. The collaborative robot YuMi released by ABB can also be programmed by the YuMi app running on a tablet [1]. The programming of the YuMi can be achieved by moving the robotic arms and the end effectors along the trajectories, and at the same time recording the trajectory points and the postures of the end effectors.

Collaborative robots with integrated vision systems can learn and adjust tasks through observation. These robots can complete the task through autonomous learning. For example, a collaborative robot can use a visual system to recognize the position, size, and shape of a workpiece, and carry out autonomous job programming and motion. In this case, the workpiece can be processed with high precision without being accurately positioned in advance. For the collaborative robots that perform tasks through fixed programs without autonomous learning capabilities, the workpieces need to be accurately positioned to ensure that the robots can work properly.

3.1.4.5 Collision detection

When collaborative robots are in the same collaborative space as human workers, safety should be considered first, and the most common safety problem is the damage caused by the collision between robots and human workers. Whenever human workers or robots in motion come into contact with each other, there is a risk of injury [26]. Therefore, to prevent the collision from causing harm to human workers, the collaborative robot must have the function of collision detection.

At present, collision detection solutions are mainly divided into two categories. At present, the mainstream solutions are to install sensors outside the robot, such as wrapping a sensitive artificial robot skin sensor on the surface of the robot skin, adding vision sensors [27], and installing joint torque sensors. The joint torque sensor is the most common solution. Accurate joint torque can be obtained through torque sensors in each joint, and collision detection can be realized by observing the change of moment at the joint. However, attention should be paid to distinguish the moments of accidental collision and normal contact, to realize the recognition of interference and collision without affecting the normal work of the collaborative robot. Although the project of installing the joint torque sensors can ensure the safety of human–robot collaboration to a certain extent, it has the following two problems. On the one hand, sensor data acquisition and analysis and processing will increase the complexity of the control system, which may have the problem of poor real-time performance. On the other hand, the installation of the joint torque sensors will further increase the complexity of joint structure, the cost of manufacturing, and the stiffness of the joint.

Another method of collision detection is implemented by algorithms without the need for additional external sensors. On the one hand, this method can reduce the cost of the newly manufactured collaborative robot. On the other hand, for the traditional industrial robots without joint torque sensors, the method can realize the collaborative transformation of them. For example, by measuring and monitoring the current of the motor at each joint, a peak current can be detected to identify the collision, then the robotic arms can be stopped safely [2]. The UR5 introduced by Universal Robots can obtain key joint force information by detecting motor current changes, to ensure safety without torque sensors. As a result, production cost has been reduced and market competitiveness has been improved.

An observer of the moment outside the joint can be constructed based on the dynamic model to detect and judge the collision according to the observed value. When a running robot collides with a person or an object in the workspace, the observed value of the external torques will suddenly change and be significantly different from the observed values in normal operation, so the occurrence of the collision can be judged. Lee et al. proposed a method to observe the external moment generated by collision based on the generalized momentum observer and friction moment model [28]. This method only used the encoder of the robot without using additional sensors. By using a 6-DOF robot, the experimental results showed that collision detection is reliable without additional sensors. Tian et al. proposed a method to estimate the joint impact moment and contact force of the end effector of the collaborative robot [29]. This method designed a collision moment observer, which can identify the joint friction parameters by the friction torque model. This method can track the torque quickly and accurately with consideration of the friction factor and system model error.

By analyzing the observed values and changes of the external torque of each joint, the location of the collision can also be calculated. After the collision detection, different collision response strategies can be implemented to realize the safety function of collision protection of collaborative robots and ensure the safety of human–robot collaboration. For example, after a collision is detected, the robot can be stopped and driven into reverse motion to buffer.

Although collaborative robots may have the ability to detect the collision and can be stopped safely after the collision, it is still best to avoid a collision for safety and efficiency. When the collision may be caused by the arm or head of a human worker, the study on the prediction of the collision and the timely avoidance strategy is particularly important. Dumonteil et al. studied how to use a single Kinect sensor to monitor the workspace, and used a reactive planner to achieve real-time avoidance trajectory selection [30]. The experimental results show that when an obstacle is detected, the planner will try to find an alternative path that allows the robot to avoid the obstacle and reach its destination. If the search for the alternative path fails, the robot will stop and wait for the obstacle to disappear.

This section summarizes the collaborative robots and collaborative robot systems, and introduces their development process, key technologies, and core issues. The knowledge of collaborative robots covered in this chapter can basically be represented by the system block diagram shown in Fig. 3.3. The system block diagram mainly includes human workers, collaborative robots, workpiece or task assignments, HRI devices and methods, collaborative workspaces, and links between various parts.

In the collaborative space, human workers and collaborative robots work together on workpieces or complete tasks, and communicate with each other through HRI devices and methods. To further ensure the security of collaboration, sensors are also deployed in the collaborative space in some scenarios to achieve a common perception of human workers and collaborative robots, and to provide effective reference data for collision detection and security protection. Thus, between the human worker and the collaborative robot, a virtual fence is formed, which replaces the traditional physical entity fence.

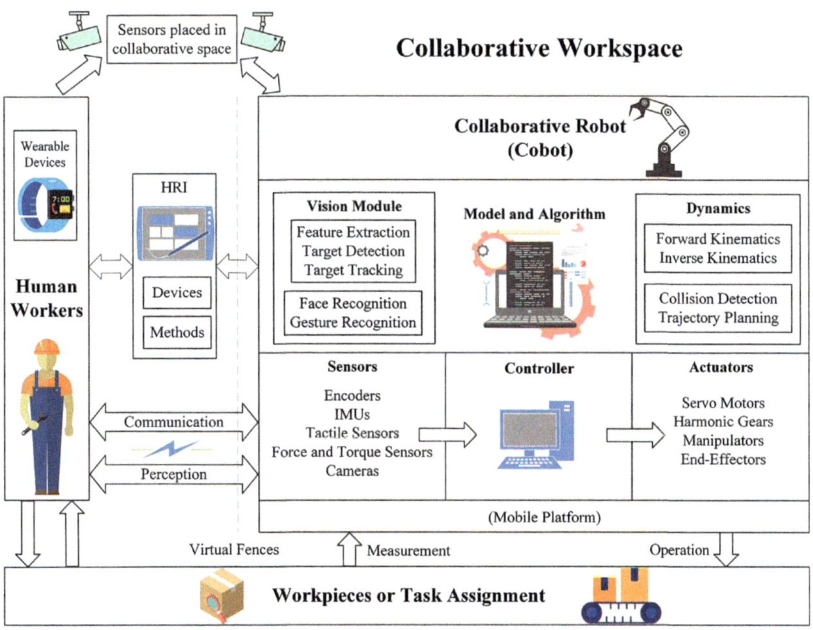

Figure 3.3
The block diagram of collaborative robot system.

In this chapter, collaborative robots are roughly divided into two levels: the hardware level and the algorithm (software) level. The former mainly includes sensors, controllers, and actuators; the latter mainly includes kinematics and dynamics algorithms of the manipulator, collision detection algorithms, computer vision algorithms, etc. Under the control and coordination of the software level, the collaborative robot uses the sensors to sense the environment and the workpieces, and then issues commands through the controller to control the actuators to operate the workpiece. This process constitutes a closed-loop control system.

In some applications, human workers need to wear wearable devices or hold HRI devices (such as indoor positioning devices and smartphones) to achieve control of the collaborative robot and an understanding of its state. The programming of the collaborative robot can be done using a teaching device or other handheld terminal.

A mobile platform is placed in the lower part of the block diagram. This means that the usual collaborative robots can be installed on mobile platforms (such as AGVs or rails) for a wider range of more flexible operations.

3.2 Main components of rail transit collaborative robots

The overall structure of collaborative robots tends to be miniaturized and lightweight. Its main purpose is to reduce its inertia to improve its safety in the event of a collision. Currently, collaborative robots on the market widely use aluminum, magnesium, carbon fiber, and other materials to reduce weight. Some collaborative robots wrap plastic on the surface [1]. In case of accidental contact or collision, the plastic layer can also absorb contact force to a large extent, which plays the role of shock absorption and protection.

The collaborative robots usually have a more rounded geometry. They have almost no clamping points, and most of the time they have internalized driving motors, sensors, and cables, thanks to their hollow joints.

To achieve flexible interaction with human workers, current collaborative robots usually have six or seven DOFs. Among them, the 6-DOF arm can reach any position and any direction within its working range [31]. However, a robot with six DOFs does not have redundant DOFs, and its ability to avoid obstacles is not as good as a robot with redundant DOFs. The robot with redundant DOFs usually has seven DOFs and has good compliance. But the disadvantage of the robot with redundant DOFs is that its reverse kinematic analysis is more complicated. The human arm is usually considered to have seven DOFs. Therefore, some robots have adopted the design of seven DOFs to further improve the flexibility of robots and the flexibility of interaction. The 7-DOF arm can realize the transformation of other configurations on the premise of the same position of the end effector. In other words, the 7-DOF joint robot has more advantages in obstacle avoidance, singularity overcoming,

Table 3.2: Several mainstream collaborative robots and manufacturers.

Manufacturer	Collaborative robots
ABB	YuMi
Rethink	Baxter
	Sawyer
Universal Robots	UR3
	UR5
	UR10
KUKA	LBR IIWA 7
	LBR IIWA 14
FANUC	CR-35IA

flexibility, and fault tolerance. Therefore, the robot with redundant DOFs is very suitable for tasks such as precision assembly and working in tight working spaces. These characteristics are very important for the robot's active avoidance in the speed and separation monitoring mode and power and force limiting mode mentioned earlier.

To achieve the flexibility of the manipulator, some collaborative robots adopt a Series Elastic Actuator (SEA) [32,33]. SEA mainly includes motor, gear box, and spring. In the SEA, the motor and the joint are connected by a spring, instead of directly connecting the motor output shaft to the joint through a gear box as in the traditional method. Because of this, the torque that drives the joint comes from the torsion of the spring, not directly from the motor (or gearbox). The SEA collaborative robots can achieve the same elasticity as humans, so the impact resistance of collaborative robots will be improved and the risk of collision will be reduced. The SEA was assembled by the Baxter collaborative robot from Rethink Robotics. The spring in the SEA can also be used as a force/torque sensor, that is, by measuring the torsion of the spring to obtain force information and use it for feedback to control the output. Another structure that can be used for collaborative robots is the Variable Stiffness Actuator (VSA). In the VSA, flexibility or stiffness can be changed by changing mechanical parameters in the joint [34]. In the event of a collision, the VSA can make the collaborative robot safer because its stiffness and collision inertia can be reduced.

Table 3.2shows some of the current mainstream collaborative robots and manufacturers [35].

3.2.1 Sensors

Sensors are necessary equipment for collaborative robots to sense their state and the external environment. The sensors frequently used by collaborative robots mainly include motion sensors for obtaining the angles and speeds of joints, force sensors for measuring joint torques and external forces, range sensors, and vision sensors for observing the working environment and workpieces. These sensors are shown in Fig. 3.4.

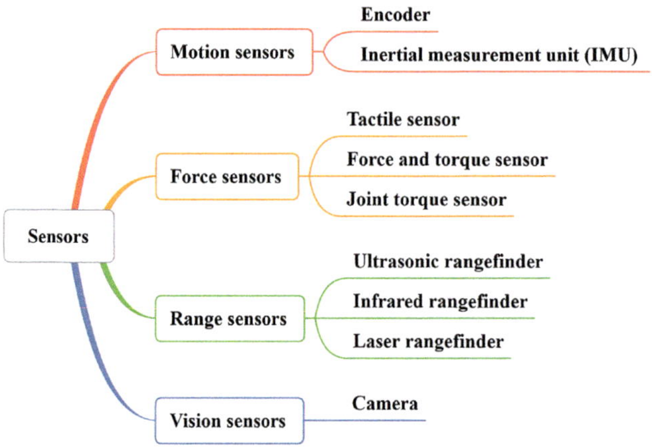

Figure 3.4
Sensors for collaborative robots.

3.2.1.1 Motion sensors

The collaborative robot is a closed-loop control system, and its controller needs to know the angular displacement, angular velocity, and angular acceleration of each joint at all times, to calculate the state of the manipulator and carry out feedback control. It is also a necessary condition for a collaborative robot to realize trajectory planning and obstacle avoidance. Therefore, motion sensors installed at each joint are crucial.

A *Encoder*

3.2.1.2 Optical encoder

The rotary encoder is one of the most used motion sensors in a robot. It can be used to measure angular displacement, angular velocity, and angular acceleration. Because of the noncontact sensor, its service life is relatively long. The measuring principle of the rotary encoder is to convert the rotation of the input shaft into an optical pulse signal, which is then converted into an electrical pulse signal. The controller calculates the rotation parameters of the input shaft through the analysis of the number, frequency, and phase of the electrical pulse signal. Optical rotary encoder usually consists of a light source, grating, photosensitive element, etc. Grating made of disk shape, according to a certain rule on the circumference of the groove. The shaft of the grating is connected to the input shaft and rotates with the input shaft. The light-emitting diode is usually used as a light source of an optical encoder, and a phototransistor or a photodiode is usually used as a photosensitive element. When working, the light source and photosensitive elements are arranged symmetrically on both sides of the rotating grating. When the input shaft turns the grating, it indirectly passes the light through the through-hole or is shielded. Thus, the photosensitive element receives or does not receive light intermittently and converts this state into an electrical signal.

There are two types of optical rotary encoders: incremental encoders and absolute encoders. The incremental encoder measures the relative rotation of the input shaft. In the simplest case, its output signal is only one pulse signal. The code disk of the incremental encoder has only one track with equally spaced slots along the circumference of the track. The resolution of an incremental encoder is determined by the number of these slots. Only one pair of the light source and the photosensitive element are needed for one track. As the input shaft turns, the output produces a series of electrical impulses. The angle of rotation of the input shaft can be determined by counting the pulse signal.

Nevertheless, the simplest case mentioned earlier is not very powerful because it cannot obtain the rotation direction and absolute position of the input shaft. To obtain the direction of rotation of the input shaft, two pairs of light sources and photosensitive elements are usually provided. Because of the arrangement of these two pairs, one always goes through a slot first, and the other will go through the same slot after a certain interval. This produces two pulses, one of which lags or leads the other, where the angle signal is usually used. The controller can determine the rotation direction by comparing the lead and lag relations between the two signals, and the displacement parameters can be obtained by counting the pulses of any one signal. To solve the problem of the unknowable absolute position, another slot is usually placed inside the track on the code disk, which is used as a reference point. Therefore, another pair of light sources and photosensitive elements must be added. When a reference position is needed, the controller drives the input shaft to rotate at a low speed until the reference point is reached or the limit of travel is reached. If no reference point is found when the limit is reached, drive the input shaft to turn in reverse and continue to look for it.

The absolute encoder is used to measure the absolute position of the input shaft. Its output signal corresponds to the absolute angular position of the input shaft, and the signal is output in the form of a binary numerical signal. The number of bits of the numerical signal determines the resolution of the rotary encoder. Unlike the incremental rotary encoder, the absolute encoder has multiple tracks with different radius. There are grooves or notches on the circumference of a track, depending on its location and coding convention. Accordingly, the same number of light sources and photosensitive elements as the number of tracks shall be provided. The combination of light states all of the photosensitive elements are uniquely determined at a given position. The combination changes as the input shaft rotates, and the parallel signal is then converted into a serial signal output, which is the binary numeric signal mentioned earlier (with the same number of bits as the number of tracks). The absolute position of the input axis can be determined by the controller through the judgment of the signal value, and the rotation direction of the input shaft can be determined by the judgment of the numerical change. The collaborative robot equipped with the absolute encoders for position feedback does not need to move to the original reference position to find a reference for positioning when it starts [2].

3.2.1.3 Magnetic encoder

Unlike optical encoders, magnetic encoders use changes in magnetic field intensity to measure position. On the code plate of the magnetic encoder, a series of magnetic poles are usually distributed on the circumference. When the input shaft rotates, the alternating N and S poles rotate accordingly. The sensor of the magnetic encoder is a hall component or reluctance component. The output voltage of the hall component is proportional to the strength of the magnetic field, while the reluctance component senses the change of reluctance caused by the change of the magnetic field. In general, magnetic encoders are more stable than optical encoders.

Harmonic gears are used by more collaborative robots to reduce joint size and weight, thus reducing joint stiffness. The lack of stiffness can lead to the deviation between the actual position and the command position, so two high-resolution rotary encoders are needed. Encoders on the input and output sides of the joint can help the control system to detect the rotation deviation caused by the lack of stiffness in real time, which is conducive to the improvement of control accuracy.

3.2.1.3.1 Inertial measurement unit (IMU)

The IMU is a sensor that measures triaxial acceleration and triaxial angular velocity. The IMU consists of an accelerometer, which can output linear acceleration signals on three axes in space, and a gyroscope, which can output angular velocity signals on three axes in space. The attitude of the sensor in space can be calculated based on the six outputs, which is of great application value when attitude control needs to be considered. Since the IMU cannot give an angle in the horizontal direction, the IMU is often used in combination with a triaxial magnetometer for further attitude determination.

The data measured by the inertial sensors usually have certain errors. One type of error is offset error, which means that accelerometers and gyroscopes have nonzero outputs even when they are stationary. In the process of using the IMU, displacement data is usually needed. Integrating the output of the accelerometer twice can get linear displacement, and integrating the output of the gyroscope once can get angular displacement. However, as time goes by, errors will accumulate so that great errors will be produced. The background noise in the working environment will also interfere with the measurement results. Therefore, in actual use, the data output by the IMU must be filtered and corrected, which can usually be achieved by using a Kalman filter (KF). In fact, in many application fields, it is difficult to obtain accurate results using the data output by the IMU alone, so the IMU usually needs to perform data fusion with other sensors to obtain better results. The KF can be also used for data fusion [36].

The IMU techniques can be used in the somatosensory programming of collaborative robots. The IMU can also be used in the end effectors of collaborative robots to observe their attitude. Since the collaborative robot is usually fixed on its base or placed on the slide track, the position, speed and acceleration of the collaborative robot at any time can be determined by using motion sensors. Therefore, the application of IMU in the self-localization of collaborative robots is not very common. At present, IMUs are widely used in AGVs, Unmanned Aerial Vehicles (UAVs) and other robots that need to determine their attitude and position in space [37,38].

Nevertheless, in the field of collaborative robot research, many scholars have researched the application of the IMU [36,39]. These studies focus on the positioning of human workers who collaborate with collaborative robots. The positioning of human workers is a prerequisite for achieving safe human–robot collaboration. On the one hand, the vision sensors, ranging sensors, etc. that will be introduced later can be used to achieve this purpose. On the other hand, the attitude and position of the operator can be captured by using the IMU. The former method usually only records the overall position and direction of the operator, while the motion capture system based on IMUs can measure the movement of the operator's whole body, which is more conducive to human–robot collaboration. In some visual systems (such as SLAM), the IMU is also used for data complementation and fusion [37].

Corrales et al. developed a hybrid tracking system for positioning operators in the robot workplace [36]. This system includes two parts, an inertial motion capture system based on IMUs and an ultrawideband (UWB) radio positioning system. The former is used for the high-precision recording of the movement of the operator's limbs, while the latter is used to obtain the operator's global position in the workspace. The fusion algorithm based on KF is used to combine the two parts of data. In the experiment, the operator picked up an object located outside the robot's operating space (or extreme travel) and passed it to the robot. This proves the effectiveness and accuracy of the hybrid tracking system.

Meziane et al. proposed an operator avoidance algorithm using instrumented safety helmet [39]. The safety helmet combines the IMU module and Received Signal Strength Indication (RSSI) technology. The former gives the posture of the operator's head, while the latter uses WiFi signal strength to measure the global position. Similarly, data fusion is still implemented using KF. Tactile devices, such as vibrating motors, are integrated into the helmet to alert the operator to possible danger. To distinguish the activities of the workers in the workspace, a posture database needs to be created to define the state of the workers. Combined with the algorithm for collision avoidance, the system can effectively determine the posture and position of the operator and realize collision prevention and avoidance.

The data obtained from the instrumented safety helmet can be used for the position and posture recognition of workers. The data is then transmitted wirelessly to the server and the robot's controller for safe path planning.

The earlier research on operator position and posture detection based on IMU are also particularly applicable to the current collaborative needs of traditional industrial robots. This is due to the fact that current collaborative robots are generally small in power (there are exceptions, such as the world's largest six-axis collaborative robot, the CR-35iA, launched by FANUC in 2015, has a payload of 35 kg and can perform high-load work), while the power and force of traditional industrial robots are usually relatively large. When the human–robot collaboration transformation of traditional industrial robots is needed, the aforementioned methods can be adopted to realize the transformation and optimization of the traditional industrial robot.

3.2.1.4 Force sensors

3.2.1.4.1 Tactile sensor

The tactile sensor is a sensor for judging whether contact occurs with the outside world or measuring the characteristics of the contacted object. It obtains environmental information through direct contact. On the one hand, the tactile sensor installed on the gripping device of the end effector can obtain the contact force distribution between the gripping device and the gripped object, and can also be used to distinguish the workpiece from the operator, thereby preventing the operator from being injured. On the other hand, collaborative robots operating in power and force limiting mode can use artificial skin composed of tactile sensors to improve the safety of human–robot collaboration. These sensors are arranged on the outer layer of the robotic arm and can sense the occurrence of collisions in almost any part.

According to different physical principles, tactile sensors can be divided into the following categories:

(a) **Mechanics-based sensor:** This is the simplest tactile sensor. Its sensitive element is a mechanical microswitch, which is used to convert the presence or absence of force into two states, on and off.
(b) **Resistance-based sensor:** The sensitive element of this sensor is a conductive elastic object whose resistance changes with external force. When an external force acts on the elastic object, the external force can be determined by measuring the resistance.
(c) **Capacitance-based sensor:** The sensitive element of this sensor is a capacitor. Its capacitance can change with the change of external force. Generally, the application of force changes the capacitance value by changing the distance or the effective area between the capacitor plates.
(d) **Magnetic-based sensor:** This sensor reflects the action of external forces by measuring changes in magnetic flux density, magnetic coupling, or inductance.
(e) **Optics-based sensor:** This sensor reflects the action of external forces by detecting changes in the intensity of light. There are two ways for external forces to change the intensity of light. One is to change the light intensity by moving obstacles in the light

path, and the other is to change the light intensity by using the photoelastic effect of transparent materials.

(f) **Piezoelectric sensor:** This sensor uses the piezoelectric effect to reflect the action of external forces.

(g) **Strain-based sensor:** This sensor determines the action of external forces by strain. The sensitive element is a strain gauge whose resistance value changes with the change of its geometry under the action of external force.

In fact, for robots with medium and high loads, due to their large power and long braking distances, it is not enough to rely on tactile sensors to detect collisions alone. Fraunhofer IFF developed a hybrid safety sensor system and demonstrated it on an ABB “IRB 4600” robot, which has a load capacity of 60 kg [1]. The hybrid system uses locally distributed capacitive sensors and tactile sensors to achieve proximity detection, so it can detect the presence of human workers in advance, and stop or slow down the robot’s motion before a collision occurs. The detection range of the system for human hand proximity is about 10–15 cm, and longer for other larger body parts.

3.2.1.4.2 Force/torque sensor

Unlike the tactile sensors, which measure the distribution of forces in only a small defined area, force/torque sensors are commonly used to measure the resultant forces and torques applied to an object. The force/torque sensor is usually installed between the end effector and the end of the robotic arm, so it is also called a wrist sensor. The force/torque sensor is usually six-dimensional and can be used to measure forces in three directions and moments in three directions on an end effector. The force/torque sensor usually estimates the force and moment by measuring the strain. The force and torque acting on the structure first causes the deformation of the relevant mechanical structure, and then the deformation will cause the resistance of the strain gauge to change. Changes in resistance can be converted to changes in output voltage through a Wheatstone bridge. The installation of this sensor is very helpful for the realization of programming by manual guidance.

3.2.1.4.3 Joint torque sensor

The joint torque sensor is used to measure the resultant torque on a robot joint. Like the force/torque sensor, the joint torque sensor usually calculates the magnitude and direction of the torque by measuring the strain. The currently used torque sensors include potentiometer torque sensors, magnetoelectric torque sensors, and infrared torque sensors. In collaborative robots, joint torque sensors are the most commonly used solutions for motion control and collision detection. Accurate joint torque can be obtained by installing torque sensors on each joint of a robot, and collision detection of the robot can be achieved by observing changes in these torques and calculating external forces. It should be noted that when a collision occurs, the torque measured by the joint torque sensor includes not only the external torque generated by the collision, but also the effects of inertial force, centripetal force, Coriolis force, and gravity. Therefore, the specific detection needs to use the observer to extract and restore the collision torque from the original joint torque information. Combined with the robot's dynamic model, the joint torques can also be used to estimate the forces on the end effector of the robot.

3.2.1.5 Range sensors

Range sensors are used to detect objects in the surrounding environment without contact. Because of this characteristic, such sensors have a relatively long service life compared with contact sensors. These sensors can be classified into two types: active sensors and passive sensors. The former mainly includes ultrasonic range finder and laser range finder, which can measure the distance by actively sending signals and listening to the signal reflection. The latter mainly refers to various types of cameras, which use various wavelengths of light reflected by ambient objects to detect. Combined with specific algorithms, the data obtained by these sensors can not only be used to obtain the position of the target object, but also can be used to distinguish the shape, size and other information of various objects. Based on the earlier information, the map of the robot working environment can be constructed. In the application field of collaborative robots, range sensors are mainly used for the detection of obstacles in collaborative space and collision prevention.

3.2.1.5.1 Ultrasonic range finder

The ultrasonic range finder is used to measure the distance between the sensor and its object. When it works, it first emits a sound wave at a specific ultrasonic frequency. After a certain period of time, it receives the sound wave reflected by objects in the detection area. The distance from the measured object can be calculated by recording the time between sending and receiving and combining the real-time sound velocity under the current conditions (such as local temperature). The coverage shape of an ultrasonic range finder is usually a cone with the ultrasonic generator as its apex. But as the distance increases, the end of the cone begins to shrink, that is, the expansion rate begins to decay. To expand the detection area, multiple ultrasonic range finders with different angles can be set up. It should be noted that the shape, size, direction, and other factors of the measured object can all influence the measurement of the ultrasonic range finder.

3.2.1.5.2 Infrared range finder

The infrared range finder is a distance measuring device with infrared as the medium. It mainly includes an infrared signal transmitter and an infrared receiver. When the infrared range finder works, the transmitter first transmits an infrared light signal at a specific frequency. After the light signal is reflected by an obstacle, the infrared receiver receives the infrared light signal of this frequency and converts it into an electrical signal. By processing the received electrical parameters related to the intensity of the infrared signal, the distance of the obstacle can be calculated. Generally, the infrared range finder can provide an analog voltage output according to the distance of the obstacle. As the distance of the obstacle changes, the voltage of the sensor changes accordingly. However, the relationship between the output voltage and the distance is usually nonlinear. The analog output voltage needs to be converted into a digital signal by an analog to digital converter (ADC) before it can be processed by the computer system. The infrared range finder has the advantages of high measurement accuracy, fast reflection speed, and strong directivity. But infrared range finder is also very susceptible to environmental factors.

3.2.1.5.3 Laser range finder

The laser range finder can also be used to measure the distance between the sensor and the object being measured. Unlike ultrasonic range finder, the laser range finder calculates distances by sending a laser and calculating the time difference between sending and receiving (the speed of light is known). Another type of laser range finder measures the phase shifts of light sent and received. A laser range finder can measure only one target at a time, while a laser scanner can scan all directions in the environment. Through high-speed sampling and scanning of the environment, the laser scanner will generate the coordinates of a series of points in polar coordinates. The polar distances are obtained by the laser range finder and the polar angles are obtained by the rotary encoder connected to the rotating part. A two-dimensional environment map can be constructed using these sampling points, and if another DOF is added to the scanner, a three-dimensional environment can be sampled to build a three-dimensional map. These maps are important for robot path planning, obstacle detection, and avoidance.

3.2.1.6 Vision sensors

3.2.1.6.1 Camera

The ultrasonic range finder and laser range finder are active sensors, and passive sensors can be used as well. In a collaborative robot system, operators and workpieces position detection can be realized by the machine vision algorithm. The necessary sensors for machine vision are cameras. To detect the position of the operators and the occurrence of a collision, the camera can be placed in a fixed position in the collaborative workspace or a specific position on the top of the collaborative robot. To locate the workpieces and prevent the end effectors from harming the operators, the camera can also be installed at the end of a collaborative robot's mechanical arm. For example, Sawyer, a collaborative robot from Rethink Robotics, has a powerful embedded vision system. The system mainly includes a wide-angle camera equipped on the top, and a camera embedded on the wrist [2]. Among them, the top camera is used to obtain a wide field of view, and the camera on the wrist combined with the robot positioning system can achieve real-time dynamic positioning of the target. Al Attar et al. used a collaborative robot in their research for air hockey games, and fixed a camera on the top of a hockey table through a wooden frame for air hockey tracking [16].

The camera used for monocular vision is usually only used for two-dimensional (2D) imaging of the working environment of a robot. Of course, a single camera can also be used to implement 3D information in some cases, such as taking pictures of the same object in different poses, or using an image to estimate its depth information in combination with known geometric parameters of the object. For the construction of a 3D scene, binocular vision, or multiocular vision technology is usually needed to obtain the depth information of the object. After the acquisition of the environment image, the detection, measurement, and tracking of the objects in the field of view need to be realized through algorithms. These contents will be discussed later in this chapter. The correction for the camera is essential before any further machine vision algorithms can be performed. Correction of camera usually includes two aspects: correction of lens distortion; transformation of camera plane coordinate to collaborative robot reference coordinate system.

3.2.1.6.2 Kinect sensor

Kinect is a red-green-blue-deep (RGB-D) sensor. It mainly includes a red-green-blue (RGB) camera, an infrared emitter, an infrared camera, and a multiarray microphone. It can use the RGB camera to obtain color images and the infrared camera to obtain corresponding depth images. Compared with other traditional 3D cameras, such as stereo cameras, and Time of Flight (ToF) cameras, Kinect has wider applicability and lower cost. Teke et al. combined the depth image obtained by the Microsoft Kinect v2 camera with the nearest-point approach to improve the motion performance of the collaborative robot and the efficiency of HRI [40]. In the research of Cherubini et al., a Microsoft Kinect v2 camera installed on the torso of the collaborative robot BAZAR was used to detect humans in the environment. Kinect can be used to collect information about human skeletons and hand gestures, which can help vision-based intention recognition [41].

3.2.2 Controllers

3.2.2.1 Control system

The control system is used to control the issuing of instructions and control the robot to achieve the mission goals. At present, although there are many manufacturers of collaborative robots, the control system developed by each manufacturer is designed for a single robot, and universal applicability is low.

In terms of control methods, most collaborative robots currently retain the control methods of traditional industrial robots, that is, special controllers are set outside the robot itself. However, some new collaborative robot products have abandoned this method. They have got rid of the dependence on external controllers and internalized the controllers inside the robot body. All the buttons and display screens are installed on the robot itself, which is flexible, simple, and intuitive.

3.2.2.2 Human–robot interaction

HRI is one of the important technologies of collaborative robots. Its main purpose is to realize the exchange of command, status, and other information between the operator and the collaborative robot, to realize mutual understanding and identification. Common interaction methods include speech and gesture interaction, as well as interaction directly through the control device. In situations where multiple workers need to collaborate with robots, it is necessary for the robots to identify and differentiate these different workers. To determine the identities of these workers, the collaborative robot can adopt the method of face recognition and confirm the workers through the visual equipment. It is also possible to assign tasks based on the work experience and capabilities of the specific workers identified. This method can also be used to determine the height of the worker. When handing something to the worker, the robotic arm can be controlled to take the corresponding height. To further improve the coordination of human–robot collaboration, collaborative robots can be equipped with a "human-like behavior system," which may include an eye gaze system, a head movement system, and a text display system.

HRI methods commonly used in collaborative robots are presented as follows.

3.2.2.2.1 Speech human–robot interaction

The effective distance of speech interaction is not very long because of the high noise level in the environment where the collaborative robot works. Barattini et al.'s experiments in 2012 showed that speech interaction is usually only available within 3 m [42]. Although it is possible to extend the interaction distance by wearing microphones for workers, doing so will cause some problems. For example, the use of additional equipment will increase costs and reduce convenience. When workers move between different work areas or tasks, interactions with different robots cannot be quickly switched. It is impossible for a person who is not specifically in charge of the collaborative robot to control the robot because the person is not wearing the relevant equipment.

3.2.2.2.2 Gesture human–robot interaction

Compared with the speech interaction, gesture communication can achieve a longer communication distance, allowing human workers to give orders to collaborative robots from beyond the sound control range of 3 m [9,42]. The operator provides instructions to the robot by performing the agreed gestures, and the robot can read and understand the predefined gestures through the camera, using real-time machine vision. For the reasons mentioned earlier, workers are also generally not allowed to wear special tags or wearable tags when interacting with gestures [42]. In the case of gesture interaction, since the worker may be holding a tool or device with one hand, it is necessary to define single limb action [42]. Despite the rapid development of HRI, there is currently no basic set of gestures for collaborative robotics, and no corresponding standard has been proposed [42]. Given the diversity of collaborative robots and their applications, gesture sets must be easy to learn to reduce the cognitive burden of workers, and must have a certain degree of reusability to meet the broad application of collaborative robots produced by different manufacturers. Therefore, it is necessary to develop gesture recognition algorithms with real-time and accuracy.

3.2.2.2.3 Display equipment

Some robots have display screens on top of their bases, such as the Sawyer and Baxter collaborative robot from Rethink Robotics. When the operator configures the parameters, the information is displayed on the screen. When the robot starts to work, a pattern of eyes will be displayed on the screen. These eyes are always watching the current working position and show different expressions according to whether the current work is performed well. According to the design of the program, the display can also display information such as prompts, which can be used to inform human workers of the specific status of the robot or to guide the human workers to work. This humanized interactive mode can realize real-time monitoring of the working status of the collaborative robot intuitively, and can also add some fun to the boring industrial production.

3.2.2.2.4 Remote control equipment

Collaborative robots may be equipped with simple remote control devices for special purposes. JACO2, for example, is equipped with a remote control, which has a 3-DOF joystick and seven buttons. In the application of some collaborative robots, to facilitate the operation of personnel, the collaboration system uses smartphones to complete the integrated control of the robot and display the working process information. Operators wear smartphones on their arms to monitor the real-time running status of the robot.

3.2.3 Actuators

The actuators of a collaborative robot consist of end effectors and robotic arms. The end effectors are devices that interact directly with workpieces and are highly application-specific. The robotic arm acts as a transmitter of force and motion, and is usually used to directly control the position and attitude of the end effector in space.

3.2.3.1 Driving devices

The power of collaborative robot joints usually comes from electricity, which requires a suitable motor for energy conversion. The motors currently used in the joint of collaborative robots usually include servo motors and direct drive motors.

3.2.3.1.1 Servo motor

The servo motor is the motor used for servo system. The so-called servo system is a control system that acts according to the instructions. It can compare the actual state of the system with the corresponding state of the instructions, and use the comparison result for further control. Servo motors include direct current motors (with brush and brushless) and alternating current motors (synchronous and asynchronous). The function of the servo motor is to convert the control signal of the controller into the rotational angular displacement or angular velocity of the motor output shaft. Servo motor is used to drive the joints. Generally, servo motors used in robots should have the following characteristics: fast response speed, high starting torque, wide speed range, etc. When used in a collaborative robot, the servo motor should have the characteristics of small size, lightweight, and hollow structure to achieve safe human–robot collaboration.

3.2.3.1.2 Direct drive motor

Unlike traditional motors, direct drive motors can be directly connected to the load, which means that they do not need other transmission mechanisms among them, such as gears and belts. In other words, the direct drive motor drives the load directly. Direct drive motors are divided into rotary direct drive motors and linear direct drive motors, and the former is widely used in collaborative robots. Rotary direct drive motors are often referred to as torque motors because they can generate high torque at low speeds (even at rest). Since the mechanical transmission mechanism can be canceled, the transmission efficiency of the driving device will be improved, and the mechanical noise and vibration can be reduced. This results in longer service life, less maintenance, higher accuracy, and higher reliability. Direct drive motors are often more compact than traditional motors and more suitable for the lightweight design of collaborative robots. And many rotary direct drive motors are provided with a center hole, which allows cables and pneumatic lines for driving end effectors to pass through the center of the motor. This design eliminates the need to attach these lines to the surface of the robotic arm, thus avoiding possible dangers.

3.2.3.1.3 Harmonic gear

Harmonic gear is a kind of gear transmission mechanism. Harmonic gears have the following characteristics: small backlash, small size, light weight, high reduction ratio, high accuracy, high efficiency, low noise, high load-carrying capacity, smooth movement, etc. Harmonic gear is simple in structure and easy to install. The input and output shafts of the harmonic gear are coaxial, so the harmonic gear is particularly suitable for being mounted on a collaborative robot. The use of harmonic gears can meet the requirements of collaborative robots for light weight, miniaturization, and round appearance.

Although it has all the advantages mentioned earlier and is widely used in robots, harmonic gear uses controllable deformation of flexible components to transmit motion and power. The transmission process is flexible, and the flexibility of joints will bring additional DOF. Therefore, higher requirements are put forward for the establishment of an accurate dynamic model and the control of the robotic arm.

3.2.3.2 End effector

To interact with the environment and perform specific tasks, the collaborative robot is equipped with tools called end effectors, which are usually installed at the end of the collaborative robot's robotic arm. Depending on the task assignment, the end effector changes accordingly. Fig. 3.5 shows the classification of the end effectors according to the power source. The end effectors can be divided into the following types: electric end effectors, pneumatic end effectors, and hydraulic end effectors.

The structure of the end effector is usually a gripper or task-specific tool, and its shape and material should match the robotic arm to improve its own safety. To reduce the load of the robotic arm and improve the positioning accuracy, the structure of the end effector should be as simple as possible and the weight should be as low as possible.

As an important part of the collaborative robot system, the end effector directly interacts with the workpiece and operator, so its security is also very important. To improve security, it is essential to install numerous sensors on the end effectors to sense the environment. For example, capacitance sensors can be installed to detect the operator in advance; cameras can be installed to detect the working environment; tactile sensors can be used to feel external forces and distinguish the human workers from the workpieces. The end effectors should be restricted to prevent forces or pressures acting on the operator's body parts from exceeding the permissible value in the event of a clipping incident. In case of an accident, the end effector should be able to judge its next action according to the specific conditions. If the end effector is clamping the workpiece at the time of the accident, it should remain in this state to prevent the falling workpiece from hitting the operator (which is also important in case of accidental loss of power). If it pinches the body part of the operator, it should stop applying force immediately.

The six-dimensional force/torque sensor installed between the end of the robotic arm and the end effector can realize the towing and teaching function, enabling the operator to easily and conveniently drag the robot and the end effector at different speeds. Safeea et al. proposed an accurate hand guiding method at the end effector level. This method can

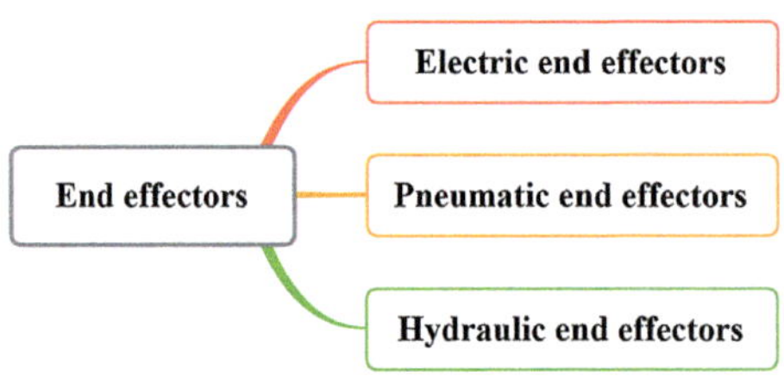

Figure 3.5
Classification of end effectors.

use force feedback to compensate for the weight and inertia of the end effector. Experimental results show that this method can achieve precise hand guiding of the collaborative robot [43].

To improve the ability of interaction and collaboration, it is also possible to consider installing HRI equipment on end effectors, such as monitors, indicators or control buttons (which may include emergency stop button). In some application scenarios, smartphones are used to realize integrated control and display of the work process, which can help the operators to further understand the working situation of the end effector of the collaborative robot.

Although it is hoped that the versatility of the end effector can be improved, the consequences of doing so will inevitably increase the complexity of the end effector, thereby reducing its reliability and accuracy. The alternative method is to design and develop a series of universal end effectors that can be quickly replaced.

3.3 Visual perceptions of rail transit collaborative robots

Since collaborative robots need to work with human workers in a shared operating space, and sometimes even operate the same workpiece together, the collaborative robot's awareness of the environment, the position, size, shape of the workpiece, and the operator's position are critical. The way to realize these perceptions is to use machine vision, which is to use the camera to sense the environment in real-time and then process the image through algorithms to realize the perception of the environment. Face recognition, gesture recognition, and other HRI methods aforementioned all involve visual perception technology.

3.3.1 Feature extraction algorithms

3.3.1.1 Image features and feature extraction

Features are the corresponding essential characteristics of a class of objects that distinguish them from other class objects and their collection. As for the image, each image has its features that can make it different from other images. Image features can be divided into global features and local features. The global features describe the overall information of the image. Common global features include color features, shape features, texture features, and spatial relationship features. These features have their methods of description and extraction. The local features describe the information of the local area of the image, representing some special points or special local areas on the image, such as corner points, edges, spots, etc. In contrast, global features are easy to calculate and express but are easily disturbed by illumination, rotation, and noise. The local features are usually invariable to the rotation, scaling, brightness change, etc., and stable to a certain extent to perspective change, affine transformation, noise, etc.

There are many ways to classify features due to the variety of features. The global features and local features mentioned earlier are classified according to the region size of the extracted features. Features can be classified according to their manifestation. For example, features can be divided into point features, line features, and regional features.

The function of image feature extraction is to extract the aforementioned features from the image through a computer algorithm, that is, to determine whether each pixel on the image belongs to an image feature. Feature extraction is the premise of image analysis and image recognition. It is the most effective way to simplify the expression of high-dimensional image data. Computers do not recognize images, only binary digits. For a computer to understand an image and truly have computer vision, it is necessary to extract useful data or information from the image. Such data or information is usually not represented or described in the form of images but exists in the form of values, vectors, symbols, etc. The aforementioned process is feature extraction, and the result of feature extraction is usually called a feature description or feature vector.

In the process of target classification, target detection, and target tracking, a collaborative robot system needs to apply feature extraction technology. In other words, feature extraction is the first condition for further image processing in many cases. Therefore, the introduction of feature extraction methods and algorithms is conducive to the further understanding of the visual perception of collaborative robots. Based on the extraction of environmental features, collaborative robots have a better understanding of the environment and have a basic guarantee of the interaction with human workers. For example, the nondestructive testing function of the robot developed in the Air-Cobot project uses Speeded Up Robust Features (SURF) to inspect aircraft surface devices that have two possible states [44].

3.3.1.2 Basic concept of image

Before further introducing the methods and algorithms of machine vision, it is necessary to introduce the basic knowledge of images. It mainly includes graphics and images, digital images, pixels, binary images, gray and gray images, color space and color images, image graying, gray gradient, gray histogram, Gaussian blur.

3.3.1.2.1 Images and graphics

Graphics generally refer to computer-drawn pictures, such as straight lines, circles, arcs, arbitrary curves, and charts. Graphics usually only save definitions and feature points. An image is an actual scene captured by an input device or any image stored digitally. An image is made up of a number of arranged pixels.

3.3.1.2.2 Digital image

The digital image is a 2D image with a finite number of pixels. The digital images are images digitized from analog images, with the pixels as the basic element, which can be stored and processed by digital computers.

3.3.1.2.3 Pixel

In digital images, a pixel is a physical point in a raster image. Pixels are generally considered the smallest part of a digital image.

3.3.1.2.4 Color space and color images

Colors in nature, colors perceived by humans and colors used by computers can be described through color space. A color space is a model system that represents various colors in a coordinate system composed of several (usually three or four) color components. Each point in the color space represents a color. Common color spaces include RGB color space and hue-saturation-brightness (HSB) color space. The RGB color space is mainly used for the computer system, which uses different proportions of red, green, and blue to represent the color of a point in the color space.

In practice, the images collected by the machine vision acquisition system are mostly color images. Each pixel of a color image is usually represented by RGB three components. If a component is represented as an 8-bit binary number (1 byte), the value of each component is between 0 and 255. Each pixel is thus represented by a 24-bit binary number, which allows for a theoretical variety of colors, perhaps 16 million different colors. Of course, if more numbers are used to represent colors, there will be more colors.

Generally, image display is carried out in RGB space, while image processing is mostly carried out in HSL or HIS space. Therefore, color images can be converted to the HSL space before they can be processed. Once the image processing is complete, the results need to be converted to the RGB space for display.

3.3.1.2.5 Binary image

Each pixel of the binary image has only two possible colors, either black or white. Usually, when converting other images into binary images, a threshold value needs to be selected first. When the value of a pixel in the original image is greater than this threshold, the pixel is set to white, and if the value of a pixel is less than this threshold, the pixel is set to black. When all pixels of the whole image are processed, a binary image is formed.

3.3.1.2.6 Grayscale and gray image

In digital images, grayscale means that the valuc of each pixel represents only the intensity information of the light. Such images typically display only the darkest black to the brightest white. In other words, the image contains only black, white, and gray colors, in which gray has multiple levels. In grayscale images, the value of each pixel is related to the number of bits of data used to represent the pixel. The value of the gray image is usually represented by 8 bits, that is, the combination of eight binary numbers represents the pixel value of a pixel. Therefore, the value range of pixels is 0–255

(0b00000000-0b11111111, "0b" means the following number is in binary format), with a total of 256 grayscale levels. If a 16-bit number is used to represent the pixel value of a pixel, the value range will be 0–65,535, with a total of 65,536 grayscale levels.

3.3.1.2.7 Image graying

Image graying refers to the conversion of color images containing red, green and blue components into gray images containing only one component. The purpose is usually to prevent the color that is susceptible to light affecting the extraction of key information in the image. The commonly used image graying methods are component method, maximum value method, mean value method, and weighted mean value method [45].

(a) Component method

The component method reassigns one of the RGB three color components of each color pixel to the pixel, thus obtaining the gray image. Its mathematical expression is shown as follows [45]:

$$Gray_1(i,j) = R(i,j) \tag{3.4}$$
$$Gray_2(i,j) = G(i,j) \tag{3.5}$$
$$Gray_3(i,j) = B(i,j) \tag{3.6}$$

where $R(i,j)$, $G(i,j)$, and $B(i,j)$ respectively represent the values of the three RGB color components of each color pixel in the color image, while the $Gray_k(i,j)(k = 1,2,3)$ on the left represents the gray pixel value after the gray processing of the color image.

In practical application, one of the aforementioned three equations should be selected according to different scenarios.

(b) Maximum value method

The maximum value method reassigns the maximum value of the RGB color components of each color pixel to the pixel, thus obtaining the gray image. Its expression is shown as follows [45]:

$$Gray(i,j) = \max\{R(i,j), G(i,j), B(i,j)\} \tag{3.7}$$

(c) Mean value method

The mean value of the RGB color components of each color pixel is reassigned to the pixel by the mean value method, to obtain the gray image. Its expression is shown as follows [45]:

$$Gray(i,j) = \frac{R(i,j) + G(i,j) + B(i,j)}{3} \tag{3.8}$$

(d) Weighted average method

The weighted average method weighted the RGB color components of each color pixel with different weights, and then assigned the weighted average value to the pixel, to obtain the gray image. Its expression is shown as follows [45]:

$$\begin{aligned} Gray(i,j) &= \alpha^* R(i,j) + \beta^* G(i,j) + \gamma^* B(i,j) \\ s.t\ \alpha &+ \beta + \gamma = 1 \end{aligned} \tag{3.9}$$

where $\alpha + \beta + \gamma$ represent the weights of the RGB color components respectively, and they should satisfy the condition that the sum of them is equal to one.

3.3.1.2.8 Gray gradient

The gray image is regarded as a 2D function $f(x, y)$, where (x, y) is the coordinate of the pixel, and the value of the function $f(x, y)$ is the gray value of the pixel. The gray gradient is the derivative of this two-dimensional discrete function.

In the calculation of the gray gradient, the difference is usually used instead of differential. It should be noted that at a certain pixel, the gray gradient has a magnitude and direction. The equation for calculating the gray gradient is presented as follows [46]:

$$G_x(x, y) = H(x+1, y) - H(x-1, y) \tag{3.10}$$

$$G_y(x, y) = H(x, y+1) - H(x, y-1) \tag{3.11}$$

$$G(x, y) = \sqrt{G_x(x, y)^2 + G_y(x, y)^2} \tag{3.12}$$

$$\alpha(x, y) = \arctan\left(\frac{G_y(\pi x, y)}{G_x(x, y)}\right) \tag{3.13}$$

where $H(x, y)$ refers to the pixel value, $G(x, y)$ refers to the magnitude of the gradient, $G_x(x, y)$ and $G_y(x, y)$ are the components of in two directions, and $\alpha(x, y)$ refers to the direction of the gradient.

The edges in images usually have large gradients, so the calculation of gray gradients is usually used for edge detection.

3.3.1.2.9 Gray histogram

Histogram is the most basic tool for analyzing the gray image. Gray histogram reflects the relationship between the occurrence frequency of pixels of each gray level in an image, and it describes the gray distribution of the image. Its abscissa represents the gray level, and the ordinate represents the occurrence frequency.

3.3.1.3 Algorithms for feature extraction

At present, many feature extraction algorithms have been put forward, and these feature extraction algorithms vary from each other in terms of feature types, computational

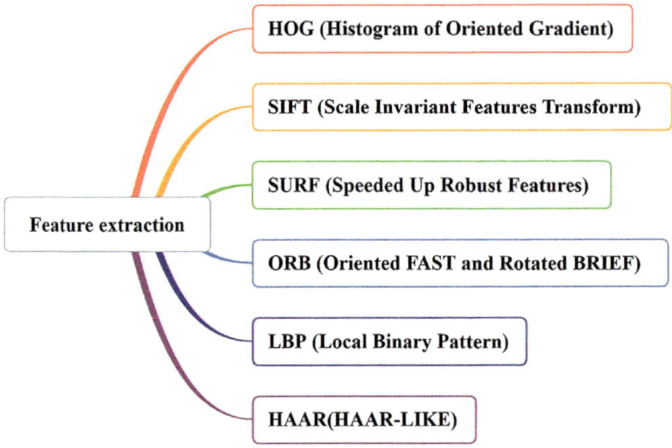

Figure 3.6
Feature extraction algorithms.

complexity, and repeatability. Some commonly used feature extraction algorithms are given for reference. They are shown in Fig. 3.6.

3.3.1.3.1 Histogram of oriented gradient

Histogram of Oriented Gradient (HOG) is a feature descriptor used for object detection in computer vision and image processing. It forms features by calculating and statistics the histogram of gradient direction in the local area of the image. The HOG was proposed by Dalal et al. [47].

At present, the combination of HOG features and Support Vector Machines (SVM) has been widely used in image recognition, especially in pedestrian detection, which has achieved great success.

Theoretically, as the pixel value (gray value) of each pixel point in the image is usually not zero, that is to say, the calculation of each central element needs to include the whole image. However, in practical application, when Gaussian blur is carried out on the image, it can be considered that the pixels beyond the distance of 3σ are useless (because most probability is concentrated within the distance of 3σ), so these pixels do not participate in convolution operation.

The flow chart of the HOG algorithm is shown in Fig. 3.7.

The steps of the HOG algorithm are as follows:

(a) Grayscale the image.

The image is regarded as a 3D image, including x, y and z coordinates (z coordinates represent grayscale).

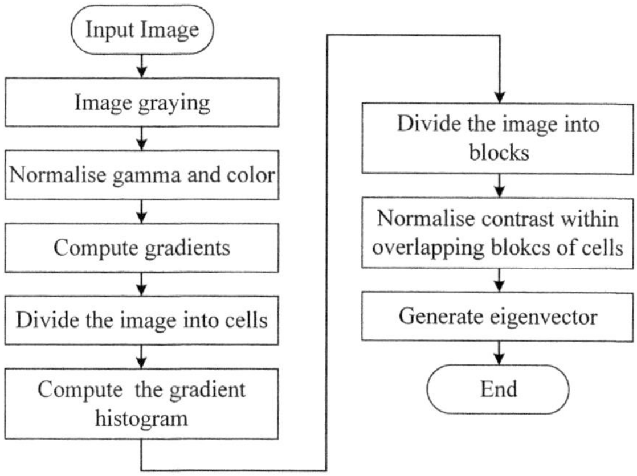

Figure 3.7
The flow chart of the HOG algorithm.

(b) Standardize (normalize) the image.

The input image was normalized by Gamma correction. The purpose is to adjust the contrast of the image, reduce the impact of local shadow and light changes in the image, and suppress the interference of noise.

(c) Calculate the gradient.

The gradient (including magnitude and direction) of each pixel of the image is calculated primarily to capture contour information while further reducing the interference of light.

(d) Divide the image into cells.

Divide the image into small cells (for example, 6*6 pixels/cell)

(e) Calculate the gradient histogram of the image.

The gradient histogram (the number of different gradients) of each cell can be counted to form the descriptor of each cell.

(f) Divide the image into blocks.

Every few cells are formed into a block (for example, 3*3 cells/block), and all the cell feature descriptors in a block are connected to obtain the HOG feature descriptor of the block.

(g) Generate eigenvectors.

The HOG descriptor of the image (the target to be detected) can be obtained by concatenating the HOG descriptor of all the blocks in the image. This is the final eigenvector for classification.

Compared with other feature description methods, the HOG has many advantages. First, since the HOG operates on the local square element of the image, it can maintain a good invariance for both the geometric and optical deformation of the image, and these two kinds of deformation will only appear in a larger space. Secondly, under the conditions of coarse airspace sampling, fine direction sampling, and strong local optical normalization, as long as the pedestrian can maintain an upright posture in general, the pedestrian can have some subtle body movements, which can be ignored without affecting the detection effect. Therefore, the HOG feature is particularly suitable for human body detection in the image.

3.3.1.3.2 Scale invariant feature transform

Scale Invariant Feature Transform (SIFT) is a machine vision algorithm used to detect and describe local features in images. The essence of the SIFT algorithm is to search for extreme points in the scale space and extract their position, scale, and rotation invariant. This algorithm was presented by *Lowe* [48].

The flow chart of the SIFT algorithm is shown in Fig. 3.8 [48].

The steps of the SIFT algorithm are as follows:

(a) Scale space and peak value selection.

The purpose of this step is to construct the Difference of Gaussians (DOG) scale space and select potential key points (or interest points) in the scale space that satisfy the scale invariance and rotation invariance. The multiscale features of the simulated image data are as follows: the large scale focuses on the general features, while the small scale focuses on the details. By constructing a Gaussian pyramid (each layer is Gaussian blurred (weighted) with a different parameter called sigma), the image is guaranteed to have corresponding

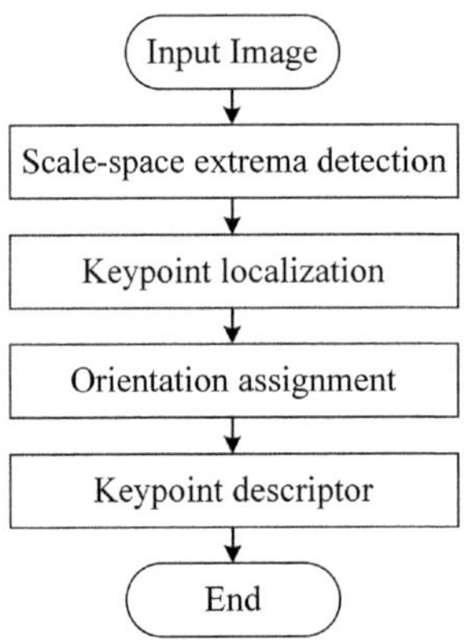

Figure 3.8
The flow chart of the SIFT algorithm.

feature points at any scale, that is, scale invariance. To determine whether the potential key points is a key point, it is necessary to compare the potential key points to all adjacent point in the image with different sigma values in the same space. If the point is max or min, it is a feature point.

(b) Positioning of key points.

The purpose of this step is to precisely locate the location of the key points and eliminate the fake key points. After all feature points are found, the points with low contrast and unstable edge effects should be removed, leaving the representative key points. The benefit of removing these points is to enhance the antinoise capability and stability of the match. Finally, these reserved discrete points were fitted to get accurate information about the location and scale of the key points.

(c) Direction distribution.

The purpose of this step is to assign directions to each key point based on its local gradient direction. To realize rotation invariance, the feature points should be assigned according to the local image structure of the detected key points. The specific approach is to use the gradient direction histogram. When calculating the histogram, each sampling point added to the histogram is weighted by the circular Gaussian function, that is, Gaussian smoothing. This is mainly because the SIFT algorithm only considers scale and rotation deformation, not affine invariance. Through Gaussian smoothing, the gradient amplitude near the key point can be weighted, which can partly compensate for the instability of feature points without considering affine deformation. A key point may have multiple key directions, which helps enhance the robustness of image matching.

(d) Description of key points.

The purpose of this step is to describe each key point with a high-dimensional vector. The key point descriptor includes not only the key point but also the pixels around the key point that contribute to it. This can make the key points have more invariable characteristics and improve the efficiency of target matching. When describing the subsampling area, bilinear interpolation after rotation should be considered to prevent white spots appearing in the rotating image. Meanwhile, to ensure rotation invariance, the angle should be rotated in the nearby field with feature points as the center, and then the gradient histogram of the sampling area should be calculated to form n-dimensional SIFT feature vector (such as 128-sift). Finally, to remove the influence of illumination variation, it is necessary to normalize the feature vector.

SIFT features are local features of images, and the description and detection of local features can help identify objects. SIFT features are independent of rotation and scaling, and have a certain degree of stability to light, microangle changes, noise, shielding, etc. SIFT has good uniqueness and abundant information, which is suitable for fast and accurate matching in the mass feature database. SIFT also has a large quantity. Even a few objects can produce a large number of SIFT feature vectors. SIFT has high real-time performance, and the optimized SIFT matching algorithm can meet the real-time requirements.

3.3.1.3.3 Local binary pattern

Local Binary Pattern (LBP) is an operator used to describe local texture features of images. It has the obvious advantages of rotation invariance and gray invariance. It was first proposed by Ojala et al. for texture feature extraction [49]. Moreover, the extracted feature is the local texture feature of the image.

The flow chart of the LBP algorithm is shown in Fig. 3.9.

The steps of the LBP algorithm are as follows:

(a) Division of area

Firstly, the detection window is divided into cells.

(b) Calculation of LBP value

For a pixel in each cell, the gray value of the adjacent eight pixels is compared with it. If the surrounding pixel value is larger than the central pixel value, the position of the pixel is marked as 1, otherwise 0. In this way, eight points in the 3*3 neighborhood can generate 8 bits binary numbers after comparison, that is, the LBP value of the pixel point in the center of the window can be obtained.

(c) Calculation of histogram

Then calculate the histogram of each cell, which is how often each number (presumably the decimal LBP value) appears. Then normalize the histogram.

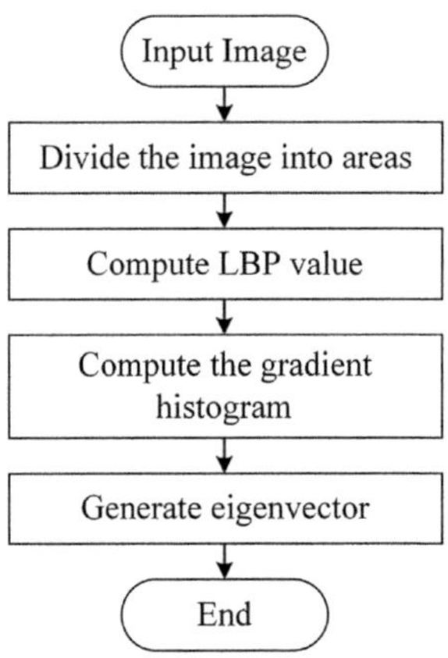

Figure 3.9
The flow chart of the LBP algorithm.

(d) Generation of eigenvectors

Finally, the statistical histogram of each cell obtained is connected to an eigenvector, which is the LBP texture eigenvector of the whole image.

3.3.2 Target detection algorithms

The task of target detection is to find out all the target objects of interest in an image, determine their position and size, and draw a boundary box around them. This usually involves two processes: target classification and target localization.

The core problems of target detection are mainly presented as follows: First, the target may appear anywhere in an image, and its size, shape, posture, angle, etc. all have great changes. Second, in many cases, the targets included in an image may also be many different kinds of targets.

In the case of collaborative robot applications, the perception of the existence and location of environmental objects is extremely important. This is related to the normal operation of its work tasks, as well as the security of its operation.

At the time of workpiece localization and grasping, the collaborative robot system needs to specify the specific position of the workpiece in space, and then drives the robotic arm and the end effector to grasp the workpiece accurately. This requires the collaborative robot to perform target detection on the work area using a camera mounted on the robot arm to confirm the position, shape, and number of the workpiece. The coordinates determined in the camera coordinate system need to be converted into coordinates in the motion reference coordinate system of the collaborative robot.

Target detection for human workers is also essential. The collaborative robot system must always be aware of the position of human workers for path planning and collision prevention.

3.3.2.1 Basic task of machine vision

Machine vision has four basic tasks: classification, localization, detection, and segmentation.

The target classification is to structure an image into a certain category, and describe the image with a predetermined category or number. It needs to find the label that best matches the target in the image in all the category labels of the training set. Target classification is the most important and most basic task.

Target localization not only needs to identify the target category but also needs to give the target's position, which is usually marked with a boundary box. Target localization generally refers to the localization of a single target or multiple targets with constant numbers in an image.

The tasks of target detection include classification and localization. It needs to mark multiple targets with multiple borders and give category labels. Compared to target localization, target detection usually refers to the classification and localization of multiple targets with varying numbers.

Target segmentation is used to assign a category (or instance) meaning to each pixel. It is used to solve the problem of which target or scene each pixel belongs to. Instead of using a bounding box to indicate the location of the target, it separates the target from the background to find the outline of the target.

3.3.2.2 Basic process of target detection

The traditional target detection process is described as follows. Region selection is performed first. Through exhaustive strategy, sliding windows of different sizes, lengths, and widths are set to traverse the image. Feature extraction is then performed using methods such as SIFT, HOG, etc. Finally, classifiers are used for classification, such as SVM, Adaboost. However, the traditional method traverses all areas of the image, resulting in a lot of redundancy and slow recognition speed.

3.3.2.3 Methods and algorithms of target detection

The current target detection algorithms are divided into three categories: traditional target detection algorithm, deep learning target detection algorithm based on region proposal and deep learning target detection algorithm based on regression method.

Before the advent of deep learning, the traditional target detection methods were roughly divided into three parts: region selection, feature extraction (SIFT, HOG, etc.) and classifier (SVM, Adaboost, etc.). The main problems of the traditional method have two aspects: on the one hand, the selection strategy of the sliding window is not targeted, its time complexity is high, and the window redundancy is large; on the other hand, manually designed features are less robust.

After the emergence of deep learning, target detection has made great breakthroughs. The two most noticeable development directions are presented as follows:

Deep learning target detection algorithm based on region proposal represented by Regions with CNN features (R–CNN). There are also SPP-Net, Fast R–CNN, Faster R–CNN, etc. These are two-stage methods. These methods need to generate the target candidate box with the algorithm first and then carry on the classification and regression of the candidate box.

Deep learning target detection algorithm based on the regression method represented by YOLO (and SSD, etc.). These are one-stage methods that use only a convolutional neural network to directly predict the categories and positions of different targets. The accuracy of the deep learning target detection algorithm based on region proposal is higher, but the

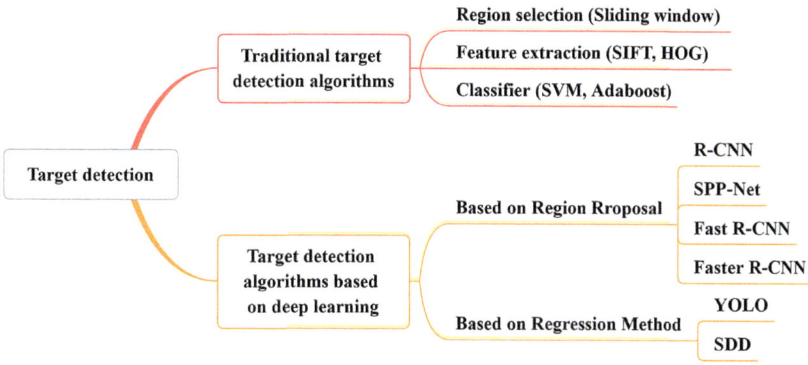

Figure 3.10
Target detection algorithms.

speed is slower. In contrast, the deep learning target detection algorithm based on regression method performs faster, but its accuracy is lower. The map of the target detection algorithm is shown in Fig. 3.10.

Since there are many kinds of target detection algorithms, it is very important to evaluate the performance of these algorithms. The evaluation metrics used to evaluate these algorithms mainly include the following: accuracy, confusion matrix, precision, recall, average precision and mean Average Precision (mAP), Intersection over Union (IoU), Receiver Operating Characteristic curve (ROC), Area Under Curve (AUC), and Nonmaximum Suppression (NMS).

3.3.3 Target tracking algorithms

The goal of target tracking is to predict the size and position of the target in subsequent frames when the target size and position in the initial frame of a video sequence are known. The data involved in this task is usually a video. To complete the task of target tracking, the target localization should be completed first. This localization can be done by the relevant content of target detection in the previous section or manually given. Then the target tracking algorithm needs to relocate the target quickly and efficiently in the subsequent frames. To avoid unnecessary repetitive calculations, it is necessary to utilize the correlation of time series, which may involve some geometric transformations (such as rotational scaling) and motion artifact processing. In simple terms, target tracking is to locate an object in successive frames of a video.

Moving target tracking is a challenging task. For moving targets, the scenes of their movements are very complex and often change, and the state of the target itself may also change constantly. The factors that will increase the difficulty of target tracking include the following: appearance deformation, light changes, fast motion and motion blur, similar background interference, scale changes, out-of-plane/in-plane rotation, obstacle occlusion, and vision departure. In many cases, the motion of the camera itself in 3D space also increases the difficulty of tracking the target. Therefore, it is a challenging task to identify moving targets in complex changing scenarios.

The collaborative space where the collaborative robot operates is dynamic. This means that operators, workpieces, obstacles, etc. are very likely to be moving. For operators, the collaborative robot always needs to keep track of its position to ensure that the minimum safe distance is not violated (if the collaborative robot works in the collaborative mode of speed and separation monitoring) and to avoid collisions. When the operator delivers the workpiece to the collaborative robot, the workpiece may be in motion, so the collaborative robot needs to track the position of the workpiece to determine its coordinate position and drive the joints of the robot arm to act in concert. The aforementioned description is equally applicable to the case where the workpiece passes sequentially on the conveyor belt. For some collaborative robots used in entertainment situations, such as the collaborative robots proposed by Al Attar et al. for air hockey games, target tracking is even more necessary [16]. Because the targets that interact with them are usually moving at high speeds, this puts higher requirements on the real-time performance of the target tracking system.

Therefore, collaborative robots have high requirements for the real-time performance of target localization, which requires the use of target tracking methods to reduce time consumption and improve real-time performance.

3.3.3.1 Differences between target tracking and target detection

If target detection is performed on every frame of a video, it is usually easy to detect the position of the target in every frame. Despite this, due to the inherent advantages of target tracking, the method of repeatedly performing target detection is not usually performed, but the target tracking algorithm is executed.

In general, the execution speed of the target tracking algorithm is faster than that of the target detection algorithm. This is because the target tracking algorithm can use the information obtained in the previous frame to predict the position of the target in the next frame. This information may include the appearance, outline, shape, and size of the target object, as well as the speed and direction of movement of the target object that is important for target tracking. When the target tracking algorithm is executed, it only needs to conduct a small-scale search near the predicted position to determine the new position of the target. For target detection, the processing of each frame needs to start from the beginning, and the amount of such calculation is very large.

When target detection fails, target tracking can provide help. If the target tracking and target detection algorithms are executed on a video at the same time, they will behave differently when the target is blocked by other objects. In general, the target detection algorithm is likely to fail because it loses the features used for detection. For an excellent target tracking algorithm, it can usually handle a certain degree of occlusion for a certain period of time. This is because the target tracking algorithm can use the position, speed, and other information provided by the previous frames to predict its new position.

Target tracking can track specific targets. Target tracking is instantiated and its target is a specific object. Using the target tracking algorithm, the motion trajectory of a specific object can be obtained. For target detection, the output of the algorithm is a series of rectangular boxes, which usually detect all objects without distinction. This means that the rectangles drawn for the same target in adjacent frames may not be the same.

3.3.3.2 Basic process of target tracking

When the target tracking algorithm is executed, the parameters of the initialized target box are first input, and then multiple candidate boxes are generated in the next frame. The features of these candidate frames are then extracted and the candidate frames are scored. Finally, the candidate box with the highest score is selected as the prediction target in these candidate boxes. It is also possible to fuse multiple predicted targets to obtain a better prediction target.

According to the aforementioned process, the research content of target tracking can be divided into the following five aspects.

(a) **Motion model:** This is part of the study of how to generate multiple candidate samples.
(b) **Feature extraction:** Study which features can represent the target.
(c) **Observation model:** Study how to score many candidate samples.
(d) **Model update:** Study how to update the model to adapt to changes in targets.
(e) **Integration approach:** Study how to integrate multiple prediction objectives to obtain better decision results.

In target tracking, the task is to find the target in the current frame that has been successfully located in all (or almost all) previous frames. Since the target has been captured in the previous frame, the information can be used to build a motion model that includes information such as the direction and speed of the target. Therefore, the model can be used to predict the possible position of the target in the next frame. In general, this predicted position is very close to the real position. Of course, this predicted position needs to be modified, generally using the location of the target's appearance model. It is easy to build an appearance model based on the information obtained in the previous frame. By using the appearance model to search near the target position predicted by the motion model, the target position can be determined more accurately.

However, the actual appearance of a target may vary greatly during the movement. Therefore, this appearance model needs to be upgraded to a classifier, and this classifier needs to be trained in an online manner. The training is performed while the algorithm is running, and only a few examples can be used for that training.

3.3.3.3 Method and algorithm of target tracking

Target tracking methods can be divided into two categories according to different observation models: generative method and discriminant method.

The generative method establishes the target model through online learning and then uses this model to search and reconstruct the image area with minimum error to complete the target localization. This kind of method does not consider the background information of the target, so the image information is not fully utilized. This approach is to model the target area in the current frame and then look for the area most similar to this model in the next frame, which is the predicted location. The concrete realization methods of this method include KF, particle filter, mean shift algorithm, etc.

The discriminant method treats target tracking as a binary classification problem. It extracts the information of the target and the background at the same time to train the classifier, then separates the target from the background of the image sequence, and finally obtains the position of the target in the current frame. The classifier in this method is trained by the machine learning method and the background information is used in the training, so the result of this method is usually better than that of the generative method. At present, the latest development of the discriminant method is the method based on correlation filtering and the method based on deep learning. The basic idea of the discriminant method based on correlation filtering is to find a filter template, and then convolve the next frame image with the filter template. The region with the largest response is the predicted target region. The tracking method based on correlation filtering and convolution neural network or the combination of both are developing rapidly.

The map of the target tracking algorithm is shown in Fig. 3.11.

3.3.4 Conclusion and outlook

This chapter first gives the basic definition of collaborative robots, and then introduces the development history, application fields, key technologies, and methods of collaborative robots. This chapter then outlines the hardware components and characteristics of collaborative robots. Finally, the basic concepts, characteristics, methods, and algorithms of feature extraction, target detection and target tracking in visual perception are introduced.

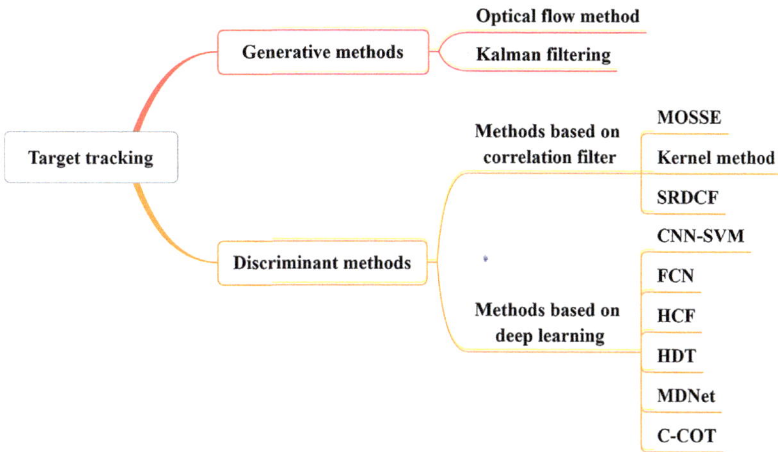

Figure 3.11
Target tracking algorithms.

(1) This chapter begins with an overview of collaborative robots. The main contents include collaborative robots, collaborative robot systems, development history, application fields, and key technologies. The key technologies and methods of collaborative robots and related knowledge mainly introduce the standards and specifications of collaborative robots, four types of collaboration, ROS systems, programming methods, collision detection algorithms, etc. In fact, as a multidegree-of-freedom robot in space, collaborative robots also involve knowledge, methods, and algorithms related to kinematics and dynamics. This mainly includes forward dynamics, backward dynamics, and trajectory planning. Since this part of the content has been introduced in other chapters, this chapter does not give a repeated introduction. At the end of Section 3.1, a system block diagram of a collaborative robot system is presented. The diagram clearly shows the general composition and state of the collaborative robot system in the collaborative space.

(2) After introducing the general situation of the collaborative robot, the hardware structure of the collaborative robot is summarized. Since the design of collaborative robots considers the inherent security, their hardware design often has different characteristics from traditional industrial robots. For example, collaborative robots are usually characterized by light weight, high flexibility, and low power. Section 3.2 roughly divides the hardware structure of the collaborative robot into three parts according to the composition of a closed-loop system, namely sensors, controllers, and actuators. The sensor part mainly introduces motion sensors, force sensors, range sensors, and vision sensors. The controller part mainly introduces the layout and characteristics of the collaborative robot controller, and gives an introduction of the HRI equipment and methods. The actuator part describes the driving devices and the end effector. Although the hardware architecture mentioned earlier also has applications on other types of robots,

sensors, and actuators for collaborative robots often have special requirements and application purposes.

(3) After the introduction of the main components is finished, the visual perception of the collaborative robot is introduced. This part mainly includes feature extraction, target detection, and target tracking. In the section of feature extraction, the most basic concepts of machine vision are also introduced, which helps to learn the subsequent methods and algorithms. Although machine vision is important for the safe and efficient operation of collaborative robots, this section only provides a basic overview of machine vision because of the wide range of knowledge involved in machine vision.

Some prospects for the future development of collaborative robots are given as follows:

(1) Collaborative robots should be just a transitional concept. In the future, all robots should have the ability to work safely with humans. They should be safe and efficient enough. Collaborative robots have the potential to replace the single-arm large robots currently used in factories and may become the dominant robotics technology of the future [2]. The collaborative robots mentioned in this chapter are generally robots with mechanical arm structures. The concepts and characteristics of collaborative robots can be widely extended to other types of robots, so that various robots may be inherently safe.
(2) The collaborative robot can be equipped with AGV to form a composite robot. The collaborative robot is mounted on an AGV for a larger working space. This combination makes the collaborative robot no longer limited to a specific work area or type of work. Under reasonable production scheduling or demand, they will be intelligently deployed between stations to perform handling tasks and specific collaborative operations.
(3) Due to the uncertainty of the work scene and task, the autonomous learning ability of a collaborative robot is very important. Through self-learning, collaborative robots can realize its rapid understanding and judgment of the environment, and it can realize some high-quality control strategies that are often difficult to implement by manual design.
(4) At present, collaborative robots have many applications in aerospace manufacturing, but their application in the field of rail transit has not been widely seen. Therefore, research on the application of collaborative robots in the rail transit industry for manufacturing, operation and maintenance, and services should be taken seriously.

In the design and development of future collaborative robots, inherent security, efficient HRI, and collaboration should continue to be guaranteed. More attention should be paid to autonomous cognition, autonomous learning and autonomous operation. This can make the collaborative robot not only a tool to reduce the labor intensity of workers and improve production efficiency, but also reduce the technical requirements of workers, and can better accomplish complex tasks under complex conditions through human—robot collaboration.

References

[1] R. Bogue, Europe continues to lead the way in the collaborative robot business, Ind. Robot Int. J. 43 (2016) 6–11.

[2] R. Bloss, Collaborative robots are rapidly providing major improvements in productivity, safety, programing ease, portability and cost while addressing many new applications, Ind. Robot Int. J. 43 (2016) 463–468.

[3] H. Ding, J. Heyn, B. Matthias, et al., Structured collaborative behavior of industrial robots in mixed human-robot environments, in: IEEE International Conference on Automation Science and Engineering (CASE), 2013, pp. 1101–1106.

[4] L. Rozo, S. Calinon, D.G. Caldwell, et al., Learning physical collaborative robot behaviors from human demonstrations, IEEE Trans. Robot. 32 (2016) 513–527.

[5] J.E. Colgate, W. Wannasuphoprasit, M.A. Peshkin, Cobots: robots for collaboration with human operators, in: International Mechanical Engineering Congress and Exposition, Atlanta, 1996, pp. 433–440.

[6] C. Connolly, Motoman markets co-operative and humanoid industrial robots, Ind. Robot: Int. J. 36 (2009) 417–420.

[7] H. Qu, K. Xing, T. Alexander, An improved genetic algorithm with co-evolutionary strategy for global path planning of multiple mobile robots, Neurocomputing 120 (2013) 509–517.

[8] A. Vysocky, P. Novak, Human-Robot collaboration in industry, MM Sci. J. 9 (2016) 903–906.

[9] I.E. Makrini, K. Merckaert, D. Lefeber, et al., Design of a collaborative architecture for human-robot assembly tasks, in: IEEE/RSJ International Conference on Intelligent Robots and Systems (IROS), 2017, pp. 1624–1629.

[10] L. Pérez, S. Rodríguez-Jiménez, N. Rodríguez, et al., Symbiotic human–robot collaborative approach for increased productivity and enhanced safety in the aerospace manufacturing industry, Int. J. Adv. Manuf. Technol. (2019) 1–13.

[11] D. Wu, A. Zhao, K. Chen, et al., A survey of collaborative robot for aircraft manufacturing application, Aeronaut. Manufact. Technol. 62 (2019) 24–34.

[12] J.R. Leiva, T. Villemot, G. Dangoumeau, et al., Automatic visual detection and verification of exterior aircraft elements, in: IEEE International Workshop of Electronics, Control, Measurement, Signals and Their Application to Mechatronics (ECMSM), 2017, pp. 1–5.

[13] A. Jubien, M. Gautier, A. Janot, Dynamic identification of the Kuka lightweight robot: comparison between actual and confidential Kuka's parameters, in: IEEE/ASME International Conference on Advanced Intelligent Mechatronics, 2014, pp. 483–488.

[14] Z. Ju, C. Yang, H. Ma, Kinematics modeling and experimental verification of baxter robot, in: Proceedings of the 33rd Chinese Control Conference, 2014, pp. 8518–8523.

[15] David Kirschner, R. Velik, S. Yahyanejad, et al., YuMi, come and play with Me! A collaborative robot for piecing together a tangram puzzle, in: International Conference on Interactive Collaborative Robotics, 2016, pp. 243–251.

[16] A. Ahmad, L. Rouillard, P. Kormushev, Autonomous air-hockey playing cobot using optimal control and vision-based bayesian tracking, in: Proc 20th International Conference towards Autonomous Robotic Systems (TAROS 2019), 2019, pp. 243–251.

[17] J. Xiao, Z. Tian, Y. Hong, et al., Collision detection algorithm for light and modular cooperative robots, J. Tianjin Univ. 50 (2017) 1140–1147.

[18] N. Sanchez-Tamayo, P. Juan, Wachs. Collaborative robots in surgical research: a low-cost adaptation, in: Companion of the 2018 ACM/IEEE International Conference on Human-Robot Interaction, 2018, pp. 231–232.

[19] International Organization for Standardization, ISO 10218-1:2011 Robots and Robotic Devices — Safety Requirements for Industrial Robots — Part 1: Robots, 2011.

[20] International Organization for Standardization, ISO 10218-2:2011 Robots and Robotic Devices — Safety Requirements for Industrial Robots — Part 2: Robot Systems and Integration, 2011.

[21] International Organization for Standardization, ISO/TS 15066:2016 Robots and Robotic Devices — Collaborative Robots, 2016.

[22] J.A. Marvel, R. Norcross, Implementing speed and separation monitoring in collaborative robot workcells, Robot. Comput. Integrated Manuf. 44 (2017) 144−155.

[23] H. Shin, K. Seo, S. Rhim, Allowable maximum safe velocity control based on human-robot distance for collaborative robot, in: 15th International Conference on Ubiquitous Robots (UR), 2018, pp. 401−405.

[24] S. Mokaram, M. Jonathan, Aitken, U. Martinez-Hernandez, et al., A ROS-integrated API for the KUKA LBR iiwa collaborative robot, IFAC-PapersOnLine 50 (2017) 15859−15864.

[25] S. Lee, K. Ahn, J. Song, Torque control based sensorless hand guiding for direct robot teaching, in: IEEE/RSJ International Conference on Intelligent Robots and Systems (IROS), 2016, pp. 745−750.

[26] B. Matthias, S. Kock, H. Jerregard, et al., Safety of collaborative industrial robots: certification possibilities for a collaborative assembly robot concept, in: IEEE International Symposium on Assembly and Manufacturing (ISAM), 2011, pp. 1−6.

[27] A. Mohammed, B. Schmidt, L. Wang, Active collision avoidance for human−robot collaboration driven by vision sensors, Int. J. Comput. Integrated Manuf. 30 (2017) 970−980.

[28] J.-B.S. Sang-Duck Lee, Sensorless collision detection based on friction model for a robot manipulator, Int. J. Precis. Eng. Manuf. 17 (2016) 11−17.

[29] Y. Tian, Z. Chen, T. Jia, et al., Sensorless collision detection and contact force estimation for collaborative robots based on torque observer, in: IEEE International Conference on Robotics and Biomimetics (ROBIO), 2016, pp. 946−951.

[30] G. Dumonteil, G. Manfredi, M. Devy, et al., Reactive planning on a collaborative robot for industrial applications, in: 12th International Conference on Informatics in Control, Automation and Robotics (ICINCO), vol. 02, 2015, pp. 450−457.

[31] S. Chen, M. Luo, O. Abdelaziz, et al., A general analytical algorithm for collaborative robot (cobot) with 6 degree of freedom (DOF), in: International Conference on Applied System Innovation (ICASI), 2017, pp. 698−701.

[32] P. Wang, Q. Zhu, X. Hu, et al., Research on interaction safety of human-robot collision based on series elastic actuator, in: 5th International Conference on Information, Cybernetics, and Computational Social Systems (ICCSS), 2018, pp. 180−185.

[33] X. Li, Y. Pan, G. Chen, et al., Adaptive human−robot interaction control for robots driven by series elastic actuators, IEEE Trans. Robot. 33 (2017) 169−182.

[34] L.C. Visser, R. Carloni, S. Stramigioli, Energy-efficient variable stiffness actuators, IEEE Trans. Robot. 27 (2011) 865−875.

[35] Le Fu, R. Wu, J. Zhao, The evolution and enlightenment of safety specification of cooperative robots: ISO/TS 15066, Robot 39 (2017) 532−540.

[36] J.A. Corrales, F.A. Candelas, F. Torres, Hybrid tracking of human operators using IMU/UWB data fusion by a Kalman filter, in: 3rd ACM/IEEE International Conference on Human-Robot Interaction (HRI), 2008, pp. 193−200.

[37] R. Li, J. Liu, L. Zhang, et al., LIDAR/MEMS IMU integrated navigation (SLAM) method for a small UAV in indoor environments, in: DGON Inertial Sensors and Systems (ISS), 2014, pp. 1−15.

[38] J. Wang, J. Steiber, B. Surampudi, Autonomous ground vehicle control system for high-speed and safe operation, in: American Control Conference, 2008, pp. 218−223.

[39] R. Meziane, P. Li, M.J. Otis, et al., Safer hybrid workspace using human-robot interaction while sharing production activities, in: IEEE International Symposium on Robotic and Sensors Environments (ROSE) Proceedings, 2014, pp. 37−42.

[40] B. Teke, M. Lanz, J. Kämäräinen, et al., Real-time and robust collaborative robot motion control with Microsoft Kinect ® v2, in: 14th IEEE/ASME International Conference on Mechatronic and Embedded Systems and Applications (MESA), 2018, pp. 1−6.

[41] A. Cherubini, R. Passama, B. Navarro, et al., A collaborative robot for the factory of the future: BAZAR, Int. J. Adv. Manuf. Technol. 105 (2019) 3643−3659.

[42] P. Barattini, C. Morand, N.M. Robertson, A proposed gesture set for the control of industrial collaborative robots, in: IEEE RO-MAN: The 21st IEEE International Symposium on Robot and Human Interactive Communication, 2012, pp. 132–137.
[43] M. Safeea, R. Bearee, N. Pedro, End-effector precise hand-guiding for collaborative robots, in: ROBOT 2017: Third Iberian Robotics Conference, 2018, pp. 595–605.
[44] F. Donadio, J. Frejaville, S. Larnier, et al., Artificial Intelligence and Collaborative Robot to Improve Airport Operations. Online Engineering & Internet of Things, Springer, 2018, pp. 973–986.
[45] S. Liu, G. Gu, J. Wang, et al., An object recognition algorithm based on machine vision of collaborative robot baxter, Electron. Optic. Contr. 26 (2019) 95–99. +105.
[46] Y. Yin, H. Zheng, T.T. Gao, et al., Algorithm of license plate recognition based on joint HOG feature, Comput. Eng. Design 36 (2015) 476–481.
[47] N. Dalal, B. Triggs, Histograms of oriented gradients for human detection, in: International Conference on Computer Vision & Pattern Recognition (CVPR '05), vol. 1, 2005, pp. 886–893.
[48] D.G. Lowe, Distinctive image features from scale-invariant keypoints, Int. J. Comput. Vis. 60 (2004) 91–110.
[49] T. Ojala, M. Pietikäinen, D. Harwood, A comparative study of texture measures with classification based on featured distributions, Pattern Recogn. 29 (1996) 51–59.

CHAPTER 4

Automatic guided vehicles (AGVs) in the rail transit intelligent manufacturing environment

4.1 Overview of automatic guided vehicles

4.1.1 Definition of automatic guided vehicles

The manufacturing industry is the foundation of the national economy and the main battleground for scientific progress. Moreover, it is also the key area of supply-side structural reform. With the development of technological innovations and information technologies, the integration and development of manufacturing and artificial intelligence are becoming increasingly intensive. The manufacturing industry is turning to intelligence, information, and automation. Intelligent manufacturing is the main direction for the future development of manufacturing. The intelligent manufacturing environment is highly integrated with machine equipment, sensor equipment, and other hardware. Besides, it integrates calculation, control, data collection, operating system, and other software.

The new scientific revolution and industrial transformation are emerging. The global industrial technology system, development model, and competition pattern are undergoing significant changes. Developed countries have introduced advanced manufacturing as the core of the reindustrialization national strategy. The United States vigorously promotes the strategic layout of intelligent manufacturing featuring the industrial Internet and the new generation of robots. Germany's Industry 4.0 plan aims to boost manufacturing competitiveness through intelligent manufacturing [1]. In the 2020 Growth Strategy, the European Union proposes to focus on developing advanced manufacturing with intelligent manufacturing technology as the core. Japan, South Korea, and other powerful manufacturing countries also put forward corresponding strategic measures to develop intelligent manufacturing [2].

Intelligent manufacturing has become a hot point for countries to develop advanced manufacturing. China stressed the importance of intelligent manufacturing in its "A Country With Strong Transportation Network" strategy. In the current scenario, characterized by medium-high speed, excellent structure, a new driving force, and multiple

Robot Systems for Rail Transit Applications. https://doi.org/10.1016/B978-0-12-822968-2.00004-8

challenges, the development of intelligent manufacturing is not only a breakthrough in the industrial transformation and upgrading of China but also a new engine to reshape the competitive advantage of the manufacturing industry. Developing intelligent manufacturing is widely believed in theory and practice to represent the future direction of manufacturing. Thus, a manufacturing revolution is brewing across China. Intelligent manufacturing is a crucial area of development in the "A Country With Strong Transportation Network" initiative, which emphasizes the use of intelligent equipment to replace human labor and realize the unmanned production process. Therefore, manufacturing, logistics, e-commerce, and other enterprises have increasingly strong demands and higher requirements for intelligent and unmanned production and transportation processes. However, there are still some traditional workshops that mainly rely on manual trucks to transport materials [2]. Efficiency is not high. To promote the development of intelligent manufacturing, the automated guided vehicle (AGV) is being adopted to replace the traditional manual truck.

The AGV, which first appeared in the 1950s, is a kind of automated unmanned intelligent transport equipment [3]. It is a mobile robot system that can follow a predetermined path and is a key piece of equipment in modern industrial automation logistics systems such as the computer integrated manufacturing system. The AGV system (AGVS) is an automatic control system with multiple AGVs. The AGVs not only greatly improve the work efficiency and reliability of the system, but also realize the intelligence and automation of the system.

The AGVs are flexible and intelligent, and can be conveniently reorganized to achieve flexible transportation of the production process. Compared with traditional manual or semimanual material transportation, AGVs reduce labor intensity, reduce risks, and improve production efficiency. Thus, AGVs can have an important role in all walks of life. With the rapid development of advanced manufacturing technologies and the increasing degree of automation, modern manufacturing has entered a new stage that integrates personalization, automation, intelligence, and flexibility [4]. As important transportation equipment in flexible manufacturing systems, AGVs have the characteristics of high adaptability, satisfactory safety, good flexibility, reliable work, and easy computer management. They are widely used in warehousing, automobile manufacturing, chemical, papermaking, electronics manufacturing, and other fields. AGVs have effectively improved the degree of automation and production efficiency of the system to meet the needs of enterprises.

This chapter first introduces the development progress of the AGVs in the intelligent manufacturing environment; the main components of the AGVs are introduced. After that, the key technologies of the AGVs are introduced. This part includes the navigation technologies, path planning technologies, human–robot interaction technologies, and the

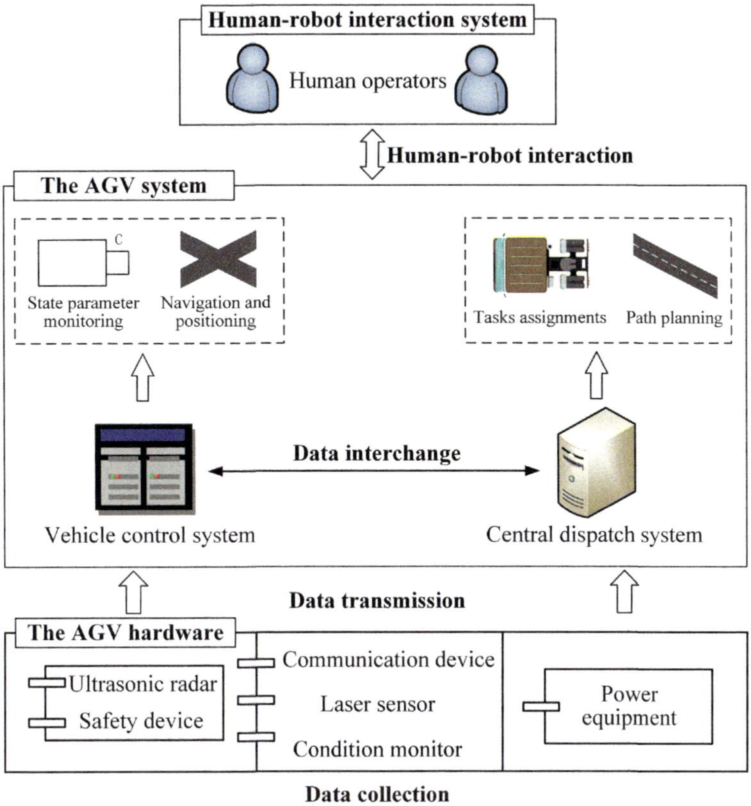

Figure 4.1
Main content diagram of automatic guided vehicles (AGVs).

task assignment of the AGVs in the rail transit manufacturing environment. Then, the hybrid model-based path planning of AGVs in the rail transit manufacturing environment is introduced. The main content diagram of the AGVs is shown in Fig. 4.1.

In this diagram, sensors in the hardware structure of AGVs are used to collect external information. After that, navigation, positioning, and state parameters of the AGVs are monitored through the vehicle control system of the AGVs. Then, path planning and task assignment are carried out for AGVs through the central dispatching system. Finally, a series of human–robot interaction operations including gesture recognition control, motion recognition control, and speech recognition control can be accomplished through the human–robot interaction system.

4.1.1.1 Development progress of automatic guided vehicles

In the 1950s, the United States successfully developed the world's first AGV. It was a trailed trolley system used to transport materials in the storage system [5]. The AGV was guided along a certain path through a wire to accomplish simple tasks. This AGV could be automatically connected with other logistics systems, which greatly improved the automation level of loading, unloading, and handling of the entire storage system. In 1954, Britain produced the first electromagnetic induction-guided AGV, which was widely promoted and applied [4]. Although these early AGVs were bulky and energy-intensive, they had extraordinary implications for the future of the AGVs. From the 1960s to the early 1970s, AGVs rapidly grew in Europe. In 1973, flexible assembly lines with multiple AGVs were developed for cars [6]. The application range of the AGVs has been widened and the assembly time and failure rate have been greatly reduced. The economic benefit has apparently been elevated. As a result, many European enterprises adopted this production method, which greatly promoted the development of AGVSs. By the end of the 1970s, Europe had about 520 AGVSs, with a total of 4800 [7]. By 1985, there were about 10,000 AGVs. Since 1976, the development of AGV industries has attracted great attention in Japan [7].

Based on the extensive application and research of AGV technologies in Europe, American and Japanese enterprises have developed AGVS to a more advanced level. In the late 1980s, AGVs in other countries were developed to a relatively mature stage with more comprehensive function [8]. AGVs were widely used in automobile manufacturing, mechatronics, warehousing, transportation, and other industries.

Throughout the development process of AGVs, AGV technologies are mainly divided into two types. The is fully automatic AGV technology. This technology is aimed at automating AGVSs and minimizing human intervention while operating. This technology can execute complex and changeable path planning and task scheduling tasks. Moreover, AGVs with fully automatic technologies have strong adaptability and can be used in a variety of transport environments. This type of AGV has advanced technologies and complete function. A modular design is adopted to reduce production costs of this type of AGV. The second kind of technology is simple AGV technology. This type of AGV is designed for simple production applications with a single driving path and a fixed execution flow. Because of the low technical threshold and production cost, this type of AGV has been widely used.

4.1.1.2 Types of automatic guided vehicle

The AGVs can be divided into several categories in industrial applications. These categories include the loaded AGVs, the traction AGVs, the forklift AGVs, the pallet AGVs and the special AGVs [7].

4.1.1.3 Loaded automatic guided vehicles

Similar to electric cars, loaded AGVs can carry cargo. Loaded AGVs place pallets, material racks, material boxes, and other goods on the AGV body for handling. After that,

AGVs complete material delivery through a lifting action [7]. Loaded AGVs can be used in many industries such as mechanical processing, automobile manufacturing, port freight, paper making, power plant, and the ultraclean workshop of the electronic industry.

The structure of a loaded AGV contains a chassis and power, control, and safety devices [3]. The loaded AGV has four wheels and a steel chassis. The AGV implements forward, backward, turning, parking, and other actions through the control device. The AGV can adopt rear-wheel driving and front-wheel steering modes, a differential drive mode, or a four-wheel full drive mode.

There are many dangerous factors in operating a loaded AGV. Therefore, its safety device is important; it guarantees not only its own safety during operation but also guarantees that of personnel and equipment in the factory. In the software, when a loaded AGV encounters a dangerous situation, the AGV immediately senses it through anticollision sensors and then stops or moves backward. A loaded AGV is usually equipped with safety lights and sound devices to warn workers who are close.

4.1.1.4 Traction automatic guided vehicles

The traction AGV itself has no carrying capacity; it is mainly used for the traction operation of vehicles, which is similar to the locomotive of a train. The world's first AGV was successfully developed in 1953 and was modified from a tractor [3]. It can transport goods in a general store along overhead navigation lines. The production modes of traction AGVs can be divided into two types: newly manufactured AGVs and those that were modified from ordinary tractors.

The main structure of traction AGVs includes a frame, door frame, and working device. Self-made traction AGVs have a poor running speed and higher requirements for the software control system owing to the short development time of AGV manufacturers and the poor match of high-precision machines. European AGV executive components are basically self-made and can be matched according to their technology, weight, customer materials, and working conditions.

4.1.1.5 Forklift automatic guided vehicles

In China, AGVs are widely used in the auto parts industry and to stack containers in ports [3]. To realize the function of loading or stacking in the warehouse, it is necessary to combine the automatic navigation function of AGVs with the lifting, stacking, forward unloading, and other functions of a traditional forklift. Many countries' forklift enterprises are in the process of exploring AGV technologies, and their technologies are not mature. It is still difficult to master this technology in a short time. Therefore, some countries' forklift enterprises choose to establish strategic alliance relationships with international leading enterprises through technical cooperation [9].

Forklift AGVs combine the technologies of mechanization, intelligence, and information [10]. These are also an important part of an intelligent factory and warehouse. Although storage forklifts have a high growth rate, the demand for intelligent storage forklifts is also becoming increasingly high. Intelligent logistics warehousing needs AGVs to build and improve logistics efficiency.

4.1.1.6 Pallet automatic guided vehicles

Pallet AGVs are compact and flexible for handling [11]. When a pallet AGV carries goods, the goods are placed on a pallet and the AGV is guided by a magnetic stripe; then, landmarks are read and the goods are transported to the designated position. When several AGVs are used together, a multi-AGV assembly line can result in a flexible production system.

4.1.1.7 Special automatic guided vehicles

Special AGVs can be divided into the following categories [12]: for assembly, for extraheavy goods, for extralong goods, for surface mount technology, for handling radioactive goods, for ultraclean rooms, and so on [13]. Each type of AGV is designed for a particular purpose.

4.2 Main components of automatic guided vehicles

Typically, AGVs are battery-powered and equipped with automatic navigation devices such as electromagnetic or optical devices. AGVs are capable of autonomous addressing independently and are controlled by computer devices to complete unmanned driving and operations [14]. Therefore, AGVs have the structural characteristics of electric vehicles. The main components of AGVs used in intelligent manufacturing environment are a chassis and driving, power, control, and safety devices.

4.2.1 Chassis

The chassis, which is the basic part of an AGV, is composed of a frame, shell, control room, driving devices, and so on [15]. The chassis has the structural characteristics of electric vehicles and the special requirements for an unmanned automatic operation. The frame commonly uses welded steel components. The lower the gravity center is, the better the antitipping performance can be. Load transfer devices, electronic control devices, buttons, and display screens are often placed on the chassis. The frame of the chassis is usually made of steel and requires good strength and stiffness.

The driving devices consist of wheels, reducers, brakes, motors, governors, and so on [15]. The driving devices are often controlled by speed and are servo-driven. The driving devices can drive the AGV and have speed control and braking capabilities. Meanwhile, according to the different AGV operation modes, common AGV steering mechanisms have the form of hinged shaft, differential, and full wheel steering. With different steering mechanisms, the AGV can move forward, backward, longitudinally, transversely, and obliquely and achieve rotational motion in all directions.

4.2.2 Power devices

The battery energy has been widely used in various industries. Battery power energy is also used in the field of AGVs [16]. Battery power not only simplifies the AGV energy devices, it has better control flexibility in power output and energy recovery. It also reduces exhaust emission and noise pollution. There are two kinds of power batteries: lead-acid and lithium. Lead-acid batteries have a large energy density, bulkiness, and relatively short service life. Their technologies are relatively mature and procurement costs are low. Lithium batteries have a high energy density and long life and are lightweight. However, the technology of lithium batteries is complex and costly. The performance of lithium batteries in different materials has various energy densities, charge and discharge ratios, cycle lives, working temperatures, and other aspects.

Energy drive devices are used to drive AGVs and their accessories, such as control, communication, and safety equipment [17]. AGV peripheral facilities use general industrial power according to different requirements such as the recharging room frequency generator, automatic door, computer room, and communication device, and the power of the device required by the working environment. The most critical factors are the length of recharging time and the ease of maintenance. Quick recharging is suitable for large current recharging. Professional recharging equipment is generally adopted for AGVs. They must have a recharging limiting device and safety protection equipment.

4.2.3 Control devices

AGV control devices consist of onboard controllers and ground controllers (outside the vehicle). The devices all use microcomputers and are connected by the communication device. Generally, the controllers on the ground issue control instructions, which are input to the communication device; the output is transferred to the controller on the vehicle to control AGV operation. Modern AGVs have microprocessor-centered controllers [18]. Microprocessors are used to position the AGVs, control their movement, and handle goods along the specified path. Microprocessors on the AGVs also communicate with the controllers or devices outside the vehicles and receive current status and scheduling and working instructions of the AGVs.

Controllers on the AGVs can control the AGVs' movement, start the AGVs' peripheral devices, monitor batteries' states, limit steering, release the brake, control drive and steering motors, and monitor recharging contactor [18]. The main hardware of the controller includes a variety of sensors, such as navigation and distance sensors, a charge-coupled device camera, and an encoder. Through multisensor fusion, the information is complementary, redundant, real-time, and low in cost. Therefore, it can reflect the environmental characteristics completely and accurately to make correct judgments and decisions. The controllers ensure the rapidity, accuracy, and stability of the AGVs. Meanwhile, this improves the reliability and autonomy of the AGV decision methods. The control units are the core of the AGVs' control devices [19]. The control units are mainly used for the AGVs to process information collected by sensors, communicate with external dispatching devices, control the attitude of the AGVs, and control the action of the actuators. The operation, monitoring, and realization of various intelligent controls of the AGVs all need to be achieved through the control units.

The ground controllers can dispatch the AGVs, send the control instructions, and receive AGV operation status information. The centralized control and management of AGVs generally use workstations or high-speed microcomputers equipped with 32-bit operating devices. Because of the rapid development of personal computer hardware and software, their performance has approached or exceeded the previous level of minicomputer workstations and the price is much lower than that of workstations. Therefore, there is a trend toward replacing workstations with personal computers.

Communication technologies are the bridge between AGVs and ground control devices to exchange information or command [19]. Because wireless communication is free from obstacles, wireless communication is generally adopted between ground control devices and AGVs. Wireless communication technologies in AGVs include a wireless local area network (LAN) and wireless data transmission module. However, because the wireless data transmission module does not support a full-duplex mode and it has a long wait time for input and output switching, communication between AGVs and the upper computer usually employs a wireless LAN module.

4.2.4 Safety devices

Safety devices of AGVs are vitally important. It is necessary to ensure the safety of the AGV itself during operation, as well as that of field personnel and all kinds of equipment. Therefore, safety and anticollision are important issues to ensure the normal operation of AGVs. Safety devices can be divided into contact and noncontact types [20].

Noncontact safety devices include infrared sensing, ultrasonic, microwave, laser scanning, and laser radar. In general, AGVs adopt multilevel hardware and software for security monitoring. For example, the front end of the is equipped with a contactless anticollision sensor and contactless anticollision sensor, and the top of the AGV is equipped with an eye-catching signal lamp and sound alarm device to remind surrounding operators. For AGVs that require forward and backward operation or lateral movement, anticollision sensors should be installed on all sides of the AGVs [21]. Once a fault occurs, AGVs will automatically issue a sound and light alarm and notify the AGV monitoring device using the wireless communication method.

To ensure the safety of people and objects in the operating environment, obstacle contact safety devices, such as buffers or force sensors, can be set on the bodies of AGVs [22]. An obstacle contact buffer is a kind of compulsory parking safety installation. The premise of its action is to contact other objects and cause them to deform, triggering the corresponding limit device. An obstacle contact buffer is a terminal security protection barrier of AGVs. An anticollision tension sensor is usually combined with an elastic-plastic plate in front of the vehicle body to form an anticollision device, which can detect any slight collision from different directions and brake urgently without causing collision damage to people or objects.

Contactless safety devices can detect obstacles around AGVs. It is a safety device that takes effect before the obstacle contact buffer. Noncontact obstacle approach detection devices consist of laser, ultrasonic, infrared, and other types.

4.3 Key technologies in automatic guided vehicles

Common AGV technologies will be introduced in this section, including navigation, path planning, human–robot, and task assignment.

Navigation technologies can let AGVs move according to the desired direction and path. Common navigation methods include electromagnetic, optical, laser, and path planning. Electromagnetic navigation is similar to magnetic navigation. It is also designed to drive a route on the ground in advance, with wires buried under the corresponding route. When high-frequency current flows through the wire, corresponding electromagnetic fields are generated around it, and then AGVs are guided. Optical navigation is the continuous laying of a strip of luminescent material on the ground, or the application of luminescent material on a specified running path. The method of laser navigation is to install a rotating laser scanner on AGVs while ensuring that the walls or pillars along the path are equipped with specular mirrors, and then send back signals based on signs from all directions. Visual navigation is to set up the image data of the path around the AGVs [23]. In the process of driving, image information about the surroundings is obtained through the onboard cameras and sensors. The obtained images are compared with the database to determine the corresponding position and make the next decision.

Path planning technologies refer to finding a collision-free path from the initial state to the target state in an environment with obstacles [24]. There are many mature algorithms for path planning in the known environment, which can reach the target location without collision. However, in the unknown environment, how to make path planning based on local environmental information detected by sensors of AGVs in real time is challenging. The intelligent manufacturing industry is developing rapidly. Path planning as a key technology of autonomous navigation is a research hot spot for intelligent manufacturing.

Path planning technologies based on a template matching path can compare the current state and environment of AGVs with the past. Then, a similar state can be found and modified. After that, a new path will be made. The idea of path planning based on an artificial potential field is to treat the movement of AGVs in the environment as a kind of movement in a potential field [24]. The obstacle has a repulsive force on AGVs whereas the object has a gravitational force on AGVs. In this way, the control force on AGVs will be generated under the control of gravitational force and repulsive force. In this way, under the control of gravity and repulsive force, AGVs will be controlled to avoid obstacles and reach the target position. Path planning technology based on map theory searches for obstacles and other information received by the AGVs' sensors [24]. Then, the environment around the AGVs is divided into different grid spaces. The computer determines the optimal path according to the existence of obstacles in the grid space. Finally, the optimal path is determined according to certain planning and decisions. Artificial intelligence path planning technology applies modern artificial intelligence technology to the path planning process of AGVs. Common algorithms include artificial neural networks, evolutionary computation, fuzzy logic, and information fusion.

Human–robot interaction methods have received increasing attention [25]. They belong to an emerging research field. These methods mainly research interaction between humans and AGVs. The methods break from the traditional operation mode of AGVs and allow humans to work with AGVs to accomplish complex tasks. Many fields have used human–robot interaction strategies to solve complex problems. With the development of the world rail transit equipment manufacturing industry, the demand for rail transit equipment manufacturing has becoming great and the manufacturing environment has become increasingly complex. Therefore, it is of great importance and significance to research the application of human–robot interaction methods in the rail transit equipment intelligent manufacturing environment. In designing human–robot interaction software for the AGVs, different types of interactive software are developed according to different requirements in practical applications. Human–robot interaction is mainly performed on platforms such as gesture recognition control, motion recognition control, and voice recognition control.

Task assignment methods are the basic function of AGVSs [26]. Task assignment of AGVS refers to accepting tasks requested by other software systems and assigning tasks to specific AGV according to a certain plan. The purpose is to maximize the efficiency of the whole system. Task assignment methods can be divided into two categories. The first is that there is a parallel relationship between tasks, each of which is independent of the other. The second one is a collaboration task, which is related and requires multiple AGVs to work together. Excellent task assignment methods can make AGVS work efficiently and smoothly. The main goal of the multi-AGV task assignment is to improve the overall performance of the AGVS. Methods include optimizing time or the path, minimizing the number of AGVS, maximizing the operating efficiency, or minimizing operating costs. In practical applications, a large number of AGVSs have fixed nodes and paths, and path optimization within the task assignment limits improving efficiency. Thus, optimizing task assignment methods of AGVs and shortening the empty driving distance between tasks can effectively improve system efficiency.

4.3.1 Navigation

AGVs detect information about the external environment according to the environmental detector. After that, AGVs adjust their motion postures in real time and finally reach the target position.

As is shown in Fig. 4.2, AGV navigation methods include electromagnetic navigation, optical navigation, laser navigation, visual navigation, and simultaneous localization and mapping (SLAM).

4.3.1.1 Electromagnetic navigation

Electromagnetic navigation is a relatively traditional navigation method. By embedding metal wires in the AGVs' driving path and loading low-voltage currents, magnetic fields can be generated around the wires. The induction coils achieve AGV navigation by identifying and tracking the strength of the navigation magnetic field.

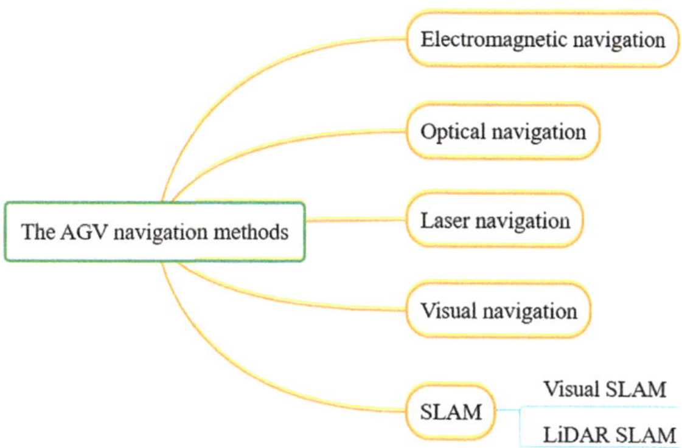

Figure 4.2
Navigation methods. *AGV*, automatic guided vehicle; *LiDAR*, light detection and ranging; *SLAM*, simultaneous localization and mapping.

The electromagnetic navigation system for AGVs is mainly composed of the computer controller, sensor, navigation signal generator, guidewire, and magnetic sign [27]. The detailed process of electromagnetic navigation can be simply explained: (1) a fixed frequency current is transmitted to the guideline by the navigation signal generator; (2) the AGVs can obtain road condition information through the electromagnetic sensor and transmit it to the AGV computer controller; (3) the AGV controller analyzes the road information, controls the course of the vehicle, and completes line tracking; and (4) AGVs can obtain the magnetic signal installed at the start−stop point through the magnetic sensor and execute start−stop action. This is a traditional navigation method. The advantage of electromagnetic navigation is that the guidance line is not easily polluted and damaged. The guiding principle is simple and reliable, and is an easily controlled communication. However, it is difficult to change or expand the path and to lay the guidance line.

4.3.1.2 Optical navigation

The detailed process of optical navigation is given in Urban et al. [28] and can be simply explained: (1) the AGV trajectory is painted or pasted with ribbon; (2) the color band image signal collected by the camera is simply processed; and (3) the AGVs complete automatic navigation. This navigation method includes ribbon tracking navigation, quick response code recognition, and other functions. Because the navigation method is more flexible and the ground route setting is simple, optical navigation technology is widely used in the world. However, this method also has some disadvantages. It is sensitive to pollution and mechanical wear of the ribbon. Moreover, it has high requirements for the environment.

4.3.1.3 Laser navigation

The process of laser navigation is given in Jung et al. [29] and can be simply explained: (1) the laser reflectors (three or more) are installed around the AGVs' trajectory; (2) the AGVs emit a laser beam through laser scanners; (3) the AGVs collect laser beams reflected by the reflectors to determine the current position and course; and (4) AGV navigation is completed by continuous trigonometric operations.

Laser navigation has the advantages of high positioning accuracy. There is no need for other positioning facilities on the ground and the driving path is flexible. Thus, it is the advanced navigation method preferred by many AGV manufacturers in the world. However, its disadvantages are obvious. The manufacturing cost of laser navigation is high. Requirements for the environment are rigorous, so it is unsuitable for the outdoor environment.

4.3.1.4 Visual navigation

Through the installed camera, AGVs can collect image information about the surrounding environment and conduct positioning navigation by matching it with the existing environment image database [30].

A common process of visual navigation is: (1) the visual sensor acquires the surface feature image of the object; (2) the surface feature image is converted to a digital signal by a high-speed visual image acquisition system; (3) a high-speed visual image processing system completes high-speed bottom processing of the visual digital image and extracts image coordinates of feature information; and (4) the computer completes fast calculation of space geometry parameters, the measured object's position, and attitude parameters.

For the AGV visual navigation system, visual sensors provide original and direct visual information, commonly called visual images. The premise and foundation of the application of the AGV visual navigation system are to extract visual images for processing and feature information.

4.3.1.4.1 Visual image pretreatment

Visual imagery is often polluted by strong noise that needs to be smoothed to reduce or eliminate it. Commonly used image smoothing filters include a mean filter, median filter, and Gaussian transform and wavelet transform. In 1995, Donoho et al. proposed a nonlinear wavelet transform threshold denoising method [31]. This method retains the wavelet coefficients of low-frequency components and sets a threshold to process the coefficients in high-frequency components. In 2002, Do et al. proposed a contour transformation method [32]. This method has a multidirectional feature and effective edge capture feature that can satisfy the anisotropic scale relation of the curve. This method can complete the rapid decomposition of curved waves. In 2004, Candès et al. proposed a curvelet firming frame method [33], which directly makes a multiscale analysis of the frequency domain. This method can be used to find the optimal sparse representation problem of the sectionalized C^2 edge.

4.3.1.4.2 Visual image feature extraction

Visual image feature extraction is an important method for image recognition and classification. Moreover, it is the basis of image information understanding, processing, and decision. In visual pictures, the image edge is the main feature of information. In 1975, Fram et al. evaluated the effectiveness of edge detection operators [34]. In 2006, Chen et al. built a mathematical model of the distribution of gray levels by analyzing the image features of the grid target point [35]. On this basis, a pixel-level extraction method of the grid target point was proposed.

4.3.1.4.3 Visual positioning method

There are two main methods of visual navigation: visual odometry and the visual navigation method based on ground punctuation matching.

Visual odometry is a common navigation method. First, it uses a monocular or binocular camera to get an image's sequence. Then, it estimates carrier motion information by feature extraction, matching, and tracking. In 1987, Matthies et al. proposed the concept of visual odometry and designed a theoretical framework [36]. This method can estimate carrier motion information from image feature extraction, matching, and tracking. It completed AGV movement navigation. It also provided the realization procedure of the monocular vision odometer and stereovision odometer. Moreover more, this method laid the foundation for study of the nonomnidirectional visual range meter.

In the visual navigation method based on ground punctuation matching, landmarks are marked out of special objects in the environment. Through the coordinates and shapes of these ground markers, AGVs can detect these ground markers to determine their location. In 1987, Kabuka et al. proposed a visual navigation method based on artificial marks matching [37]. In this study, this method completed the calculation of position using selected ground marks. AGVs updated the ground marks gradually as they went forward. This method is affected only by the focal length of the camera and is simple and flexible. In 2002, Olson et al. proposed a navigation and positioning method by choosing the best ground marks [38]. This method is based on the maximum likelihood positioning algorithm to determine the position of the AGV. This method not only increases positioning accuracy but enhances the stability of the system as well.

4.3.1.5 Simultaneous localization and mapping

In the process of studying AGVs' navigation methods, it is hoped that AGVs will know not only the positions but also the distribution of environmental objects in an unfamiliar environment without prior knowledge. The system equipped with sensors that can realize simultaneous localization and mapping is called the SLAM [39]. Sensors are an essential part of the SLAM. Sensors are a kind of devices that can perceive the external environment and obtain environmental information in a certain form. Different sensors observe environmental information in various ways. There are two categories of SLAM: that based on vision and that based on light detection and ranging (LiDAR).

4.3.1.5.1 Visual simultaneous localization and mapping

Visual SLAM completes simultaneous localization and map construction by vision sensors. The detailed process of visual SLAM is given in Yousif et al. [40]. It can be simply explained: (1) visual SLAM gets information from the visual sensor; (2) the visual odometry estimates camera motion between adjacent images; (3) loopback detection determines whether the camera has been in the previous position; (4) visual odometry and loop detection are fed into the back end for optimization; and (5) scene reconstruction can be based on the estimated camera track and attitude.

The visual SLAM is mainly used in drones, driverless cars, augmented reality, and virtual reality [40]. In addition, it is also widely used in AGVs. Classical visual SLAM methods include feature-based, optical flow, and direct approaches. With the development of information technologies, visual SLAM based on deep learning has been paid more attention. To improve the navigation accuracy of visual SLAM, the combination of visual SLAM and intelligent algorithms has also become an important research content of visual SLAM.

The methods of camera pose regression are different according to the image features of frame estimation. The classical methods of interframe estimation mainly include feature point matching, optical flow, and direct methods. Feature point methods extract feature points from adjacent frame images by the method of multiview geometry and return the camera's pose from the geometric relationship between feature points [41]. Optical flow methods are used to track optical flow and assist the extraction of feature points. Direct methods return the camera pose directly according to the brightness information of pixels in the picture.

In addition to classical algorithms, more and more deep learning algorithms are used in interframe estimation. Deep learning is implemented in an end-to-end manner and is largely automated. Interframe estimation based on deep learning is more practical. It does not depend on the selection of key points and the description of features. Moreover, it can avoid the influence caused by the change in direction, shape, and illumination in the image. Therefore, deep learning-based visual SLAM has attracted great attention from researchers [42]. Interframe estimation based on deep learning is a feature learning process from low-level to high-level features. Compared with the classical interframe estimation method, interframe estimation based on deep learning requires a large number of image databases for training and a large training time. At the end of the training, the neural network is obtained, which is implemented in an end-to-end structure. Owing to a large number of network structure parameters, deep neural networks are easy to overfit.

Loop closure detection is used to determine whether it has reached a location where it has visited before. Correct loop detection is helpful to obtain a consistent map and correct pose estimation. However, the wrong loopback detection results may destroy the existing map and cause the map creation to fail. This problem has attracted wide attention from SLAM researchers.

There are two kinds of methods for camera-based positioning [43]. The first is to obtain the position based on the image and generate a three-dimensional (3D) structure according to the current scene. After that, it can generate different views of the current scene and the corresponding camera in the database. The second is to estimate the camera pose incrementally without prior scene information, while map construction and positioning are carried out simultaneously. The second is the realization of visual SLAM [40], but in reality, noise will bring errors to the pose estimation. It will make the composed map uncertain. To reduce errors generated in this process, it is necessary to design a framework. This framework can optimize the process of SLAM to ensure the accuracy of state estimation over a long time. Error generation has uncertainty. Thus, the core problem of SLAM optimization is to solve the uncertainty. Optimization of SLAM can be divided into two types: that based on a filtering scheme and that based on a nonfiltering scheme.

Mapping is the description of the environment by a moving process; the forms of constructed maps can be divided into measurement and topology maps. Measurement maps are divided into sparse and dense maps. Sparse maps represent the environment in a certain abstract way and cannot express information about the surrounding environment [44]. They are usually used for fast positioning or navigation and have relatively fast calculation ability. Dense maps are divided into 2D and 3D maps. 2D dense maps are composed of small grids with a certain resolution. 3D dense maps consist of squares or 3D point clouds with a certain resolution, which are mainly used for 3D reconstruction. Topology maps are composed of two elements: nodes and edges. The main purpose of topology maps is connectivity between elements.

4.3.1.5.2 Light detection and ranging simultaneous localization and mapping

LiDAR SLAM is derived from the early location method based on distance measurement [45]. It has an advantage in that it can be measured at a distance without contact and the measurement is accurate. The reliability of LiDAR SLAM is high. LiDAR can collect a series of points with precise angle and distance information. This information is called the point cloud. Generally, the relative motion information of LiDAR can be obtained by comparing information about two-point clouds at different times in LiDAR SLAM. The point cloud map contains geometric information. Because of this, path planning and navigation are easily understood. It can be well-applied to the AGV navigation in a structured environment. However, this method has some disadvantages: LiDAR is often affected by the range and is more expensive.

The LiDAR SLAM framework is generally divided into four key parts [46]: front-end scan matching, back-end optimization, closed-loop detection, and map construction. Front-end scan matching is the core step of LiDAR SLAM. In this part, the position of the current frame is estimated by the known position of the previous frame and the relationship between adjacent frames. In 2009, Olson et al. proposed the correlation scan match (CSM) algorithm [47]. The height of the likelihood field model matched by this algorithm is nonconvex and there are many local extremums. Because the algorithm subdivides the environmental resolution, precision is limited by the resolution. Besides, this algorithm can reduce the computational burden by the accelerating strategy, such as the branch and bound method. Front-end scan matching can generate the pose and map in a short time. However, owing to the inevitable accumulation of errors, back-end optimization can optimize the speedometer and map information after long-time incremental scan matching.

In LiDAR SLAM, there are two ways to solve residual local error accumulation [48]. Common methods are filter-based back-end and graph optimized back-end. Particle filtering is often used as a framework for mathematical optimization in filter-based LiDAR SLAM. In 2014, Tipaldi et al. proposed a raster map that did not require each particle [49]. This method can be used to express smaller dimensional environments. Graph theory is used to represent the process of LiDAR SLAM, in which the pose is represented by a node and the spatial constraint relationship between nodes is represented by an edge. In this way, resource demand volume can be reduced. AGVs will accumulate errors as they build a diagram. In other words, the optimal method considers all interframe constraints at the same time and iteratively linearizes the solution. Closed-loop detection is responsible for reducing global map drift by detecting the closed-loop.

Then, the global consistency map can be created. The closed-loop problem is an important kind of data association problem used to obtain a topologically consistent trajectory map. 2D laser data are used for automatic estimation transformation between closed-loop constraints. Based on a heuristic search algorithm and distance measure of special modeling outliers, robust transformation estimation can be provided even when approximate initial conditions are unknown.

Map construction is responsible for generating and maintaining global maps. The map of LiDAR SLAM is a model of surroundings. It is the premise of AGV navigation and path planning. In the LiDAR SLAM domain, there are three main categories of constructed maps: scale, topology, and semantic [50]. The map construction module usually uses the raster building algorithm with low computational costs, which is based on the Bayesian inference method. The purpose of the algorithms is to emphasize the real-time capability of mapping based on LiDAR SLAM systems.

According to the mathematic optimal frame, LiDAR SLAM can be summarized into two categories: the filter-based and graph-based. In 1988, Smith et al. proposed the extended Kalman filter SLAM (EKF-SLAM) method [51]. This method used the maximum likelihood algorithm for data correlation. Disadvantages of this method include the complexity of calculation and poor robustness. The constructed map is feature maps rather than raster maps and cannot be applied to navigation obstacle avoidance. In 2003, aiming at the deficiency of the EKF-SLAM method, Montemerlo et al. proposed a fast-SLAM method. In this method, the SLAM problem was decomposed into the localization problem and the composition problem based on the known pose. It is the first LiDAR SLAM method that can output a raster map in real time. This method uses a particle filter to estimate pose. Each particle is propagated using kinematic models.

After that, the weights of propagated particles are calculated by the observation model and the map is constructed according to the estimated position and pose. There are two problems with the method. The first is that it consumes a lot of internal storage. Each particle contains the trajectory and corresponding environment map. For large-scale environments, if the odometer error is large, the difference between the predicted distribution and the real distribution is large. Thus, more particles are needed to represent the posterior probability distribution of pose. The second is that it heavily influenced the construction of maps. Because of the randomness of resampling, the diversity of particles dissipates with the increase in resampling times. In 2010, Blanco et al. proposed a more optimized Rao-Blackwellized particle filter method [52]. In this method, when it propagates in the odometer model, one particle gets N particles at a time. The optimal particle will be chosen from N particles. For this true propagation, it is equivalent to giving each particle N chances. In this way, degradation of optimal particles to noisy particles will be greatly reduced.

In the domain of LiDAR, Lu et al. first proposed optimizing the SLAM problem using a graph-based optimization mathematical framework [53]. The nonlinear least square method is used to optimize errors. However, this method has some problems. Sparsity of the system is not recognized and this method deals with the SLAM problem only offline. In 2010, Konolige et al. proposed the Karto SLAM [54]. This scheme is the first open-source scheme based on a graph optimization framework. The scheme recognizes system sparsity. To some extent, the scheme replaces the filter-based LiDAR SLAM scheme. The disadvantage of this scheme is that it takes a long time to construct subgraphs before using local subgraph matching.

Moreover, if the global matching method is adopted, the speed will slow when the search area is large. The Cartographer open-source scheme of Google is an optimal solution for the Karto SLAM. The core content is the local subgraph creation of fusion multisensor data and the scan matching strategy for closed-loop detection. In this scheme, the front-end scan matching algorithm is realized by combining the CSM with gradient optimization. Thus, the speed of closed-loop detection can be guaranteed. The drawback of this method is that the closed-loop test results are not verified. Thus, in a geometrically symmetric environment, it is easy for false closed loops to occur.

4.3.2 Path planning

4.3.2.1 Introduction

The definition of the AGV path planning can be found in Sariff and Buniyamin [55], and is simply explained as follows. The AGVs' position in the work environment is known. According to single or multiple programming requirements, AGVs can find an optimal or better collision-free path from the starting point to the target point. Common requirements include the shortest path, optimal movement time, and minimum energy consumption. AGV path planning mainly solves the following problems: (1) AGVs can go from the starting point to the target point; (2) AGVs will not encounter obstacles while moving; and (3) according to the need, it is decided whether to optimize the AGV path.

Path planning is a key technology of AGVs [56]. The essence of path planning technologies for AGVs is to achieve the optimal or feasible solution under several constraints. Merits and demerits of path planning results will directly affect the real-time performance of AGV tasks. According to the information characteristics of the studied environment, path planning can be divided into the discrete and continuous domains. According to the discovery order of the path planning algorithm, it can be divided into the traditional and modern intelligent algorithms. Moreover, according to the unknown degree of environment, it can be divided into global path and local path planning. In AGV path planning, a combination of global and local path planning is needed. The former aims to find a globally optimized path, whereas the latter aims to avoid obstacles in real time. The most important part of AGV path planning is to select an algorithm. A good algorithm has a vital role in path planning.

4.3.2.1.1 Time-optimal path planning

AGV time-optimal path planning refers to the track that takes the shortest time for AGVs to complete the same path under certain constraints [57]. The purposes of time-optimal path planning are to improve the working efficiency of AGVs and to complete the same path as much as possible at the same time. Research on time-optimal path planning is the earliest and most popular type. More and more practical applications indicate that path planning should not only focus on operational effectiveness and task requirements but also consider efficiency, energy consumption, and stability.

4.3.2.1.2 Minimum energy consumption

Although time-optimal path planning is widely used owing to its high efficiency in industrial production, minimum energy consumption path planning is more applicable in some specific cases, especially in many energy-intensive AGVs [58]. Because energy storage devices carried by AGVs, in this case, are usually limited, it is necessary to achieve minimum energy consumption. The minimum energy consumption path planning of AGVs refers to the path with the minimum energy consumed by AGVs. On the one hand, minimum energy consumption path planning tries to find the most suitable path to reduce energy consumption of AGVs; on the other hand, it also optimizes energy distribution by optimizing the whole path.

4.3.2.1.3 Shortest path

The shortest path planning problem is a classical problem in graph theory and algorithm design [59]. The shortest path refers to finding the shortest path between two nodes in the graph. The method has been widely used in many fields. The shortest path method has been paid close attention by researchers in other fields. Shortest path planning methods include accurate shortest path methods and approximate shortest path methods. Accurate shortest path methods can calculate the exact path value between nodes in the graph and calculate the optimal solution. Classical algorithms such as Dijkstra and Floyd are fundamental to the study of shortest path problems [60]. They are also the experimental reference object for comparing performance detection. However, with a more complex network structure and increasing data scale, current accurate shortest path methods still cannot meet real needs. In practical applications, it is not necessary to solve the accurate shortest path in most cases. An approximate result is enough for path planning. Approximate shortest path methods refer to finding better search results within an acceptable error range [61]. Although search results may not be optimal, they need less time and less space to build indexes than the accurate shortest path method.

4.3.2.2 Global path planning algorithms

Common global path planning algorithms are shown in Fig. 4.3. Global path planning belongs to static planning. It is generally applied to the situation in which obstacle information has been fully grasped in the operating environment. Common global path planning algorithms include graph search, random sampling, and intelligent algorithms.

4.3.2.2.1 Graph search algorithm

As can be seen in Fig. 4.3, graph search algorithms mainly include two algorithms, which have their advantages and disadvantages.

The A* algorithm is the most effective method to solve the shortest circuit in a static road network. Compared with the Dijkstra algorithm, the A* algorithm increases heuristic estimation, reduces search volume, improves efficiency, and can ensure path optimality. However, when the environment is complex and the scale is large, the efficiency of the algorithm is still low.

In 2014, aiming at the problem of a long computation time, Duchoň et al. proposed a path planning method for mobile robots based on an improved A* algorithm [62]. Several different improved A* algorithms were compared in the study, including the Basic Theta*

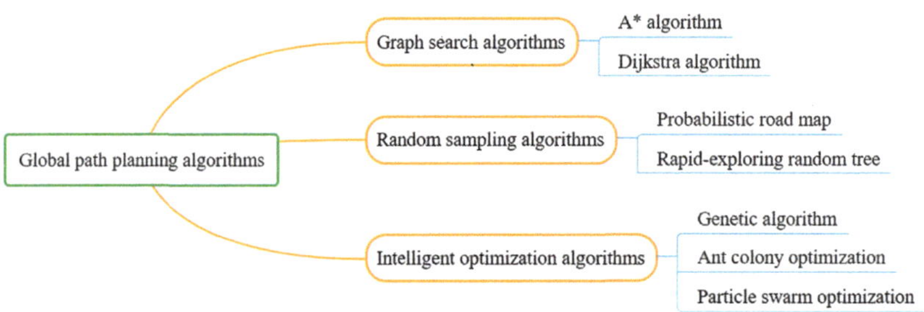

Figure 4.3
Global path planning algorithms.

algorithm, Phi * algorithm, and jump point search (JPS) A* algorithm. This report analyzes which improved algorithm is applicable to different scenarios and environments with different complexity. The JPS A* algorithm is the best in the fast-path search by comparative trial. However, the disadvantage is that when the path is too long, the searched path is not the shortest. Thus, it is suitable for the application when the path needs to be found quickly. The Basic Theta* algorithm is more applicable if the calculation time is not high and the shortest path is required. This ability of the Basic Theta* is especially prominent in the low symmetric environment. In 2015, Peng et al. proposed an improved method for storing arrays of A* [63]. The method can solve the problem that the A* algorithm takes a long time to traverse open and closed tables. It also retains the advantages of the algorithm effectively and can improve the operation effect.

4.3.2.2.2 Random sampling algorithm

Random sampling class algorithms mainly include the probabilistic road map (PRM) and rapid-exploring random tree (RRT). These algorithms are widely used in dynamic obstacles, high-dimensional state space, and complex environments. However, these algorithms have some problems, such as high cost and poor real-time performance, and the planned path may not be the optimal one.

The reliability of heuristic planning cannot be fully guaranteed. The PRM is more reliable than earlier path planning methods. The PRM is a graph-based search method. Even in a single query planning problem that is easily solved by stochastic potential field calculation, the PRM is favored by researchers because of its reliability. In 1996, Kavraki et al. proposed classical PRM [64]. This method has high efficiency in the high-dimensional space environment. After that, many scholars studied the PRM, which made the forms of this method diverse. Because of the poor completeness of the PRM algorithm in dealing with narrow channels, many scholars have improved this algorithm. Many algorithms have been widely used in dynamic obstacles and AGV path planning. However, the algorithms fail to solve the problem of differential constraints in the motion of AGVs. Therefore, the rationality of the result is poor.

The basic idea of the PRM method is: (1) the collision-free area in the space is randomly sampled with a certain probability. These points are connected to form a probability landmark according to a certain rule. There will be a probability signpost. (2) The graph search problems in the road sign algorithm will be transformed into the probabilistic map search problems. Compared with the complete planner, the PRM involves many parameters in high-dimensional space. Calculation of the PRM is not only expensive but slow.

The RRT is a data structure and algorithm. Its unique advantage can be directly applied to the nonintegral programming and kinematic mechanics programming. Therefore, it is fast and efficient in AGV path planning. Many scholars have proposed path planning methods based on RRT and achieved remarkable results. RRT has made considerable progress in the AGV path planning field. The method still has some shortcomings, such as a low node use rate and an unstable path. This method can solve the high computational complexity of the RRT and can guide AGVs to avoid obstacles online. The method reduces the complexity of the traditional RRT and can be used for real-time planning. On this basis, selective parameter deviation was introduced to improve uniform probability in the traditional algorithm, accelerate the convergence speed of the RRT, and make the algorithm more targeted in path planning.

In 2009, Kang et al. proposed integrating the improved RRT into the improved path planning algorithm based on the scroll window [65]. Aiming at the problem of local minimization, the regression analysis method was used to screen new nodes, which retained the feature of an intelligent approach to the end point of the algorithm and improved the trend of exploring unknown space. This algorithm not only maintains the random search characteristics of the RRT algorithm, it reduces the computation time of the algorithm by using the convergence standard function of heuristic evaluation to guide search tree growth.

In 2015, aiming to reduce the complexity of path planning calculation, Gîrbacia et al. proposed a path planning method for the AGV in a virtual environment with an improved fast random search tree algorithm [66]. By setting the maximum and minimum distance between obstacles of the feasible path in the virtual environment, this method can reduce the complexity of simulation and make the optimal path easier to find. The computation steps of the proposed algorithm can be explained as: (1) the artificial potential field method was used to predict a path for the initial environment in path planning; and (2) if a new obstacle appears and blocks the path, the RRT is used to plan the current local path. This algorithm effectively solved the problem of local minimum value easily caused by the RRT and greatly improved planning efficiency.

4.3.2.2.3 Intelligent optimization algorithm

Intelligent optimization algorithms mainly include the genetic algorithm (GA), ant colony optimization (ACO), and particle swarm optimization (PSO). The GA and ACO are suitable for solving and optimizing complex problems and have potential parallelism. However, these methods have a slow operation speed. Compared with these two methods, the PSO has the advantage of fast convergence.

The GA is a global optimization search algorithm with excellent self-organization and self-learning habits; it has an excellent searching ability for the optimal path in path planning. The algorithm mainly simulates crossover, variation, and heredity in natural selection. After that, the natural law of survival of the fittest is integrated. According to the results, the candidate solutions of each generation are obtained. Finally, the optimal solution is obtained from the obtained candidate solution. The biggest advantage of this algorithm is that it can be used well with other algorithms while giving full play to its iterative advantages. Meanwhile, this method guarantees good global optimization. Disadvantages of the GA are the low search efficiency and that it is easy to get into a local optimal solution. When the algorithm is running, some unneeded population will increase the difficulty for subsequent calculation, which can lead to low operation efficiency and a slow convergence rate.

The GA is not good at online path planning. Use of the GA for AGV path planning has been widely studied by scholars. In 2016, to solve the shortest collision-free problem of AGVs, Wang et al. proposed a double global optimum GA-PSO algorithm [67]. Combined with other algorithms, the traditional algorithm is optimized and the running efficiency of the algorithm is improved.

The ACO was proposed by Dorigo et al. [68]. It was inspired by exploration of the foraging behavior of ants. The ACO simulates the foraging behavior of ants through iteration to realize the optimal path selection. Each ant leaves a certain concentration of pheromones along the way as it forages. The pheromone concentration is high in the shortest path at the same time because the ant traverses many times. Later ants choose their paths based on pheromone concentrations. The higher the pheromone concentrations, the greater the probability that the path can be chosen. The algorithm has the advantages of good global optimization ability, strong robustness, and easy implementation. It has good application value in fields such as AGV path planning. However, there are many problems such as large calculations and local optimal solutions.

The PSO is a new parallel metaheuristic algorithm first proposed by Eberhart and Kennedy in 1995 [69]. Experts and scholars have done a lot of in-depth research on the traditional PSO and successfully apply the algorithm to the actual AGV path planning environment.

The basic PSO can effectively deal with the low-dimensional optimization problem, but it is easy to fall into local optimization in the process of dealing with the complex optimization problem. Besides, parameter selection of the PSO has a great influence on the performance of the algorithm. Therefore, improvement of the PSO has become a hot topic in the field of evolutionary algorithms. Among these evolutionary algorithms, relevant improvement studies mainly focus on population topology, adaptive PSO, a particle velocity update strategy, and the hybrid algorithm. The topological structure in the PSO often directly affects the update speed of particles. Thus, the appropriate topological structure can maintain the diversity of particles in the population and improve the search performance of the algorithm. Meanwhile, because different intelligent optimization algorithms have their search characteristics, the hybrid algorithm combined with two or more different algorithms can give full play to their advantages and compensate each other for their shortcomings. The hybrid algorithm based on the PSO and differential evolution algorithm has been widely studied and applied.

4.3.2.3 Local path planning algorithms

Common local path planning algorithms are shown in Fig. 4.4. The local path planning algorithm belongs to dynamic planning. Local dynamic path planning requires only the real-time acquisition of environmental information by sensors and determines environment map information. After that, location of the current map and distribution of local obstacles are obtained. Finally, the optimal path from the current node to a subgoal node can be obtained.

4.3.2.3.1 Artificial potential field method

The artificial potential field method is based on the virtual field scheme. The basic idea is to make the working environment of the AGVs become virtualized into a potential field, which can be divided into gravity and repulsion. The force generated by the target point is the gravitation of the AGVs, which increases as the distance between the target point and the AGVs decreases. The force generated by the obstacle is the repulsive force to the AGVs. The whole potential field is superposed by the vector of gravitation and repulsion. Motion of the AGVs is controlled by the resultant force of force, to avoid obstacles, reach the target point, and find the path. This method is simple in structure and convenient for real-time control and can save a lot of computation work. However, this method also has some defects. When passing through the narrow passage or near the obstacle, AGVs will oscillate. When there is a target point near the obstacle, the target is unreachable and there are some local optimal solutions. This will make AGVs stagnate on the local optimal solution before reaching the target point, and the path obtained may not be the optimal one. To solve these problems, many scholars have done a lot of research.

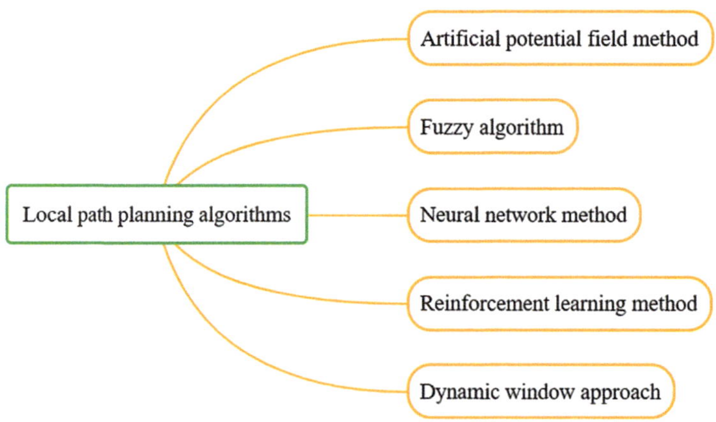

Figure 4.4
Local path planning algorithms.

In 2017, Matoui et al. proposed a decentralized architecture [70]. In a dynamic path planning environment, the AGVS is regarded as a network with a decentralized architecture. Each AGV plans its path according to its actual position, the positions of other AGVs, and the positions of obstacles. Changes in the workspace are reflected in real time by updating the system equations associated with each AGV.

4.3.2.3.2 Fuzzy algorithm

The concept of fuzzy sets was first proposed by Zadeh in 1965 [71]. The basic idea is to make absolute membership in the ordinary set become flexible. The value of the element can be expanded from {0,1} to any value in the interval [0,1]. Studies have found that with increasing complexity of the environment, adjusting the membership functions and fuzzy rules that describe the behavior of the system becomes tedious. It is difficult to establish an accurate mathematical model for path planning and obstacle avoidance of AGVs. Therefore, scholars focus on the fuzzy algorithm. Fuzzy algorithms combine the robustness of fuzzy control with sensor action based on physiology and bring a new direction to developing path planning of AGVs in the complex intelligent manufacturing environment. This method avoids the defects of the traditional algorithm and has great advantages and strong real-time performance in the path planning of AGVs in the complex intelligent manufacturing environment.

Many improvement schemes have emerged to enhance the performance of fuzzy algorithms. In 2015, to solve the time-optimal solution, Monfared et al. proposed a generalized intelligent water droplet algorithm based on a fuzzy local search [72]. In this method, the map was divided into several equal parts and the paths in each part were compared with those in other parts by the fuzzy inference algorithm. The fuzzy number was assigned to each part and the fuzzy factor determined the value of the solution in an interval. The path with the lower value was improved, and the new heuristic factor was used in the part that needed improvement. In 2016, Fakoor et al. proposed a new path planning method [73]. This method improved the fuzzy inference system by mixing the Markov decision process. In the absence of accurate estimation of the distance and shape of the obstacle, the reward function was calculated and the Bellman equation was solved by a numerical iterative method in real time. The method can avoid collision effectively. In this method, because of the combination of fuzzy control and physiology, the dependency on environment information was effectively reduced. Advantages of this method are good robustness, fast convergence, and better real-time feedback.

4.3.2.3.3 Neural network method

Artificial neural networks (ANNs) are derived from brain science because the structure and working mechanism of the neural network are based on the organizational structure of the human brain and activity. This method reflects certain features of the human brain, but it is not a true representation of it.

The ANN is an extensive parallel network composed of a large number of adaptive processing elements (i.e., neurons). ANNs are an artificial method of biological neural network simulation. They have good parallel distributed processing ability. The inherent nonlinear characteristics of the networks make them suitable for AGV path planning, especially in the case of unknown information about the environment. The relationship between the input and output of the neural network is complex. The ANNs' parallel distributed processing can improve computing speed. Scientists have been working on applying machine learning technologies to the AGVs. Relying on intelligent AGV technologies, the development of machine learning is promoted. ANNs have made brilliant achievements in navigating AGVs and AGV path planning. The basic principle is to take environmental barriers as the input layer information of neural networks. After parallel processing by the neural network, the neural network output layer outputs the desired turning angle and speed. This information will guide AGVs to avoid obstacles until they reaches their destination. The advantages of the ANN method are the high efficiency of parallel processing and learning function.

In 2011, Singh et al. designed an algorithm based on the neural network method [74]. This algorithm enabled the AGV to navigate safely in an unknown environment. The input of the neural network is the distance between the AGV and the forward, left, and right obstacles, and the angle between the AGV and the target point. The output of the neural network is the steering angle of the AGV. The obstacle's information can be acquired by sensors installed in front of the AGV.

4.3.2.3.4 Reinforcement learning method

The reinforcement learning (RL) method, also known as the reinforced learning method or the evaluation learning method, was first proposed by Minsky et al. [75]. RL uses a trial-and-error approach similar to the human mind to find optimal behavior. RL is based on previous results to learn. After that, the method can achieve the maximum return by modifying the policy.

The basic elements of the method include policy, value function, reward function, and environment model [76]. According to different solving approaches, the methods can be divided into value function and direct policy search.

In value function methods, the value function is used to define advantages and disadvantages of the AGVs' selection policy in the current state. The goal of the value function method is to obtain the optimal policy by maximizing the value function of each state. According to whether the condition transition probability is known, the value function method can be divided into two types: model-based RL models and model-free RL models. To obtain the learning policy, model-based RL models use the acquired learning experience to construct the environment model learning value function. The model-free Q-learning algorithm can guarantee convergence without knowing the model, which is one of the most effective RLs applied in the AGV path planning. In 2013, Bianchi et al. proposed a heuristic accelerated Q-learning algorithm that used heuristic factors to influence action selection [77]. In this method, the heuristic factors enhanced a link of the algorithm and finally achieved a better effect. Moreover, heuristic factors can alleviate the delayed reward problem of RL.

Direct policy search methods parameterize policy and search for optimal parameters through the idea of gradient ascending. Direct policy search methods include imitation learning (IL) models, policy gradient (PG) models, and RL models based on evolutionary algorithms [78]. The first two kinds of algorithms are mainly used in AGV path planning.

IL models are inspired by the teaching process of human society [79]. These methods solve the problem by introducing mentors. IL algorithms can be divided into inverse RL models and direct policy learning models. Inverse RL models learn the reward function from the demonstration data. With the reward function, inverse RL models learn the optimal policy through the traditional RL algorithm. Direct policy learning models learn the reflection function from state to action directly from the demonstration data.

PG models optimize the policy with parameters by gradient ascending and maximize the accumulation of expected rewards [80]. PG models are based on gradient ascending. This kind of method optimizes the leaderboard policy by maximizing the accumulation of expected rewards. The action space of AGVs in path planning is generally discrete. The superiority of the PG search cannot be reflected. Therefore, the PG method is seldom used in the AGV path planning.

4.3.2.3.5 Dynamic window approach

The dynamic window approach (DWA) is a commonly used collision avoidance algorithm proposed by Fox et al. [81]. It is an autonomous barrier algorithm that searches directly in the required area for the best control instruction to maximize the objective function. The DWA is divided into two steps: (1) the search space is limited to the possible command area; and (2) using the objective function to test each command in the possible command area, the command with the largest objective function value is selected as the optimal command. In short, the DWA simultaneously considers different objective functions to obtain the best control instructions, including the moving direction of AGVs, the distance to the nearest obstacle, and the moving speed. In many cases, AGVs can use the DWA to perform local obstacle avoidance successfully for the target position. However, the objective function of the traditional method considers only obstacles on the planned path, which makes it possible for AGVs to have a trajectory near the obstacle. If the obstacle measurement position is inaccurate, the probability of AGVs colliding with the obstacle is higher with traditional algorithms.

In 2013, Saranrittichai et al. proposed the field DWA [82]. This method improved the disadvantage of the DWA. This method considers not only obstacles along the trajectory but also those near the trajectory. With this algorithm, AGVs can avoid obstacles that might be close but not on the trajectory. The method has good robustness, and AGVs can travel more safely when involving near-trajectory obstacles. In 2014, Inigo-Blasco et al. proposed a shared control DWA [83]. Because this method accepts user commands through the control interface, it provides the most appropriate, dynamically feasible trajectory.

4.3.2.4 Human–robot interaction methods

Human–robot interaction has received increasing in recent years [84]. It is free from traditional robots in isolation space, allowing humans to work with robots to complete complex tasks. It mainly designs and studies interaction between humans and robots. They can communicate with each other and interact with physical space. Human–robot interaction is extensive and diverse and has certain challenges because of the complexity of humans and the diversity of interactions.

As urbanization continues to deepen, the urban population continues to increase and the pace of urban rail transit construction accelerates [85]. Research into human-robot interaction can improve the working safety and efficiency of robots.

Human–robot interaction is mainly realized on such platforms as gesture recognition, motion recognition, and speech recognition control. Human–robot interaction can be represented by the system block diagram shown in Fig. 4.5.

4.3.2.5 Gesture recognition control

The gesture is one of the most important nonverbal types of communication between humans and one of the most important ways of human–robot interaction. Gesture recognition is a key part of robot gesture interaction [86]. Gesture recognition requires accuracy and speed, which affects the efficiency of human–robot interaction. It conforms to the general trend of human–robot natural interaction and has a wide range of applications.

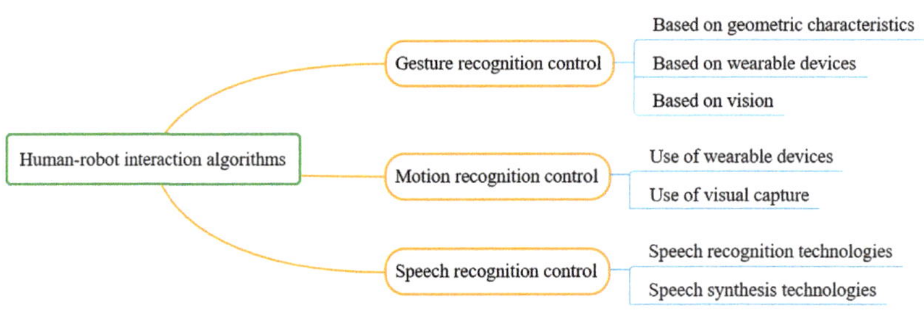

Figure 4.5
Human–robot interaction algorithms.

Because gestures have a strong information representation function, the information transmitted is richer and human manipulation is the main way for people to interact with the world. Gesture is divided into communication and operation [87]. Communication gestures are mostly used for information representation and are divided into symbols and actions. The symbols are divided into reference and pattern, and actions are divided into actions and indicative. Operational gestures are natural behaviors of humans. With the development of human—robot interaction, they are closer to the interaction of natural behaviors. Vision-based gesture interaction mainly solves the problem of recognizing the opponent's shape and tracking the gesture [87]. Traditional methods can be divided into representation-based and model-based methods. The recognition of the shape of the opponent is divided into two levels according to the representation-based method: extraction of image features and of semantic features. The model-based approach searches for matching models.

In the human—robot interaction process, especially under the condition of machine vision of the human—robot interaction, gesture recognition is based on machine vision image acquisition and image feature extraction. Human—robot interaction gesture feature recognition algorithms are designed by using a combination of 3D human body model reconstruction technology and visual feature technology [88]. Typical human—robot interaction gesture feature recognition algorithms include local linear interpolation gesture feature recognition, autocorrelation feature extraction, edge contour feature extraction, and wavelet analysis. Through the image acquisition technologies of machine vision, human—robot interaction gesture image is acquired visually. The collected human—robot interaction gesture image is fused with edge information and reconstructed with features. It improves the feature analysis ability of the human—robot interaction gesture feature and motion feature.

Current studies of gesture recognition can be divided into three directions: gesture recognition methods based on the geometric characteristics, gesture recognition methods based on wearable devices, and gesture recognition methods based on vision. Gesture recognition methods based on geometric features define a gesture category feature set based on a large number of features extracted from gesture action images. Geometric feature information extracted from the collected human hand images can be divided into two types: overall contour feature information and local feature information. The overall contour and local feature information of the hand images can be used to track and recognize gestures. In gesture recognition methods based on wearable devices, the wearable device collects gesture data through sensors and transmits data to the computer using a wireless transmission method. Gesture recognition methods based on vision mostly use monocular or multicamera cameras to collect gesture action pictures. Recognition matching results are obtained by matching the collected gesture action pictures with the preestablished gesture action template library.

There are many types of studies on human–robot interaction based on visual technologies. Porta et al. provided a detailed summary of the human–robot interaction of color cameras and some representative applications [89]. Multiple video and image processing areas are involved, such as target detection, tracking, various gestures, and face recognition. The use of the robot for visual capture and multiinformation interaction is generally achieved by integrating user motion, hand skin color, and different gesture information.

4.3.2.6 Motion recognition control

Human motion is a natural and intuitive means of human–robot interaction [90]. It is a convenient way for people of different ages and levels of knowledge to learn. It is an important and advanced direction in image processing, machine learning, and human–robot interaction. Human motion recognition is mostly based on skeletal data, which provides 3D coordinates of 20 nodes in the body. The 3D coordinates of these nodes change as people move; they are not directly used as models to describe motion; they first turn into features. The method of motion recognition is based on the bone node. The two eigenvector waveforms of the same motions are similar, but the waveforms of different motions are different. The recognition function is to solve the sequence similarity problem of spatial eigenvectors on the time axis.

Human motion recognition methods are divided into two types [91]. The first is the use of wearable devices to obtain human motion information, such as clothing embedded with sensors. The accuracy is relatively high, but it increases the burden on the human body, motion is restricted, and the operation is complex. The second type is current mainstream methods, which use visual capture technology. This type of method overcomes the disadvantages of high cost and the small range of wearable devices, using cameras to collect real-time human motion information. Cameras for motion recognition contain red-green-blue (RGB) color cameras and 3D depth sensors. RGB color cameras are commonly used to collect information and analysis to identify motion. 3D depth sensors overcome the vulnerability of traditional RGB cameras to light and provide 3D depth information.

Human motion recognition algorithms for robots are based on semantic analysis, template matching, and probability statistics [92]. Based on the semantic recognition algorithm, each movement of the human body is represented by a series of symbols. Each symbol is atomic and cannot be decomposed again. By decomposing a motion into a symbol set and then identifying according to the symbol, the final action recognition can be achieved. Based on the template recognition algorithm, features extracted from the motion are compared with templates in the standard template library according to the similarity. The template with the smallest difference and less than the similarity threshold is selected as the recognition result [93]. Based on the recognition algorithm of probability statistics, the body's limb movement is modeled dynamically. Movement in the model consists of a series of continuous attitude states, and a certain probability transition is carried out between states. Commonly used probability statistical models include the hidden Markov model, dynamic Bayesian network, support vector machine, and backpropagation.

4.3.2.7 Speech recognition control

The study of speech recognition is mature and has been applied in many fields. Language is the most commonly used and effective way of communication in human daily life. Directly inputting language for human–robot interaction is highly efficient [94].

Speech human–robot interaction technologies mainly include speech recognition and speech synthesis technologies [95]. Speech recognition technologies convert natural speech signals into computer-readable text files or executable commands. The realization of speech recognition needs three parts: speech feature extraction, pattern matching, and a reference pattern library. The method of speech interaction is not restricted by spatial factors and does not need high attention from people in real time. The human–robot interaction system based on speech control includes speech recognition and speech synthesis. In speech recognition, (1) the speech signal is preprocessed to improve the signal–noise ratio, (2) speech features are extracted through the speech frame analysis, (3) the features are matched with speech pattern information in the reference pattern library, and (4) the optimal matching result is selected according to the similarity.

Speech synthesis refers to outputting the standard, smooth speech. Commonly used speech synthesis methods can be divided rule-based speech and data-based speech synthesis methods [96]. Rule-based speech synthesis methods can simulate the physical process of human pronunciation according to a series of rules. Data-based speech synthesis methods can use statistical methods to synthesize speech according to the features.

4.3.3 Task assignment

In practical applications, a large number of AGVSs have fixed nodes and paths, and path optimization within tasks has limited efficiency improvement. Task assignment is important for AGVSs to improve transportation efficiency [97]. Optimizing the AGVS scheduling scheme and shortening the no-load distance between tasks can effectively improve system efficiency. Under constraints, AGVSs need to perform tasks and match tasks with corresponding AGVs so that the entire system runs smoothly with maximum efficiency and completes the specified transportation tasks. The task assignment problem in AGVs refers to the rational assignment of a plurality of known tasks to AGVs in the system. The function of the task assignment includes accepting tasks requested by other software systems and assigning tasks to specific robots according to a certain plan.

Task assignment methods are divided into two main categories according to the control schemes: offline task assignment and online task assignment [98]. Offline task assignment means that all transportation needs are known until the task is assigned down. Each AGV travel path is optimized before the task is performed. The objective function is usually to minimize the total travel time or maximize AGV use. The online task assignment is mainly used when the environment has uncertainty. This kind of task assignment method can reduce the possibility of failure of entire AGVSs owing to some unexpected situations, such as vehicle failure and task delay.

Task assignment methods can be categorized by the number of tasks completed by a single AGV, including single-task and multitask methods. Single-task methods stipulate that once the AGV starts to perform a certain task, it is locked. The AGV cannot perform other tasks at the same time until the task is completed. Single-task methods are simple but they cause a lot of waste and increase the transportation cost [99]. For many AGVs, path coordination can be difficult with single-task methods. In the multitask model, after the task is generated, tasks are grouped according to certain principles. The generated task group is dispatched to an AGV for execution [100]. With multitask methods, efficiency is improved by performing multiple tasks in one transport process, and the corresponding transportation costs are reduced. In single-task methods, to ensure the timeliness of execution, a larger number of AGVs than in the multitask methods are required. It is more likely to cause road congestion and a longer time to complete the same number of tasks with single-task methods. However, task grouping and redistribution may affect the optimal task grouping with multitask methods. The optimal solution at the first moment may be degraded with the task created the next time.

Task assignment methods are divided into two kinds according to the relationship between tasks [101]. The first kind of command is parallel. Each task for an AGV is independent. The second kind of command is collaborative. Commands for the AGVs are related, and multiple AGVs collaborate for a common task.

Many scholars have studied it and proposed a lot of task scheduling algorithms. Current study in the AGV task assignment algorithm is mainly of the Sufferage algorithm, simulated annealing algorithm, Min-Min algorithm, Max-Min algorithm, ant colony algorithm, GA, and so on.

(a) Sufferage algorithm
The Sufferage algorithm is a kind of comprehensive performance good scheduling strategy. The difference between the second time and the minimum time is an important basis for task computing node matching [102]. If a task cannot be assigned to the processor on which it obtains the earliest completion time, it is assigned to another processor. It indicates the extent to which the final completion time of the task is affected. It has the advantage of better comprehensive performance. However, it has disadvantages. Only the execution time is considered when calculating the task completion time.

(b) Simulated annealing algorithm
Simulated annealing is a single solution heuristic technique based on the metallurgical annealing process [103]. It is also a stochastic optimization algorithm based on probability. The simulated annealing algorithm is mainly used to simulate the process of heating to isothermal and then cooling to zero. The advantage of this algorithm is that it avoids falling into the local optimal solution [104]. It is less constrained by initial conditions and it is easier to search for the global optimal solution. The disadvantages are a slow convergence speed, low efficiency, and long running time when the problem size is large; moreover, it easily affected by parameters.

(c) Ant colony optimization
ACO is a bionic algorithm based on biological ant colonies [68]. The algorithm makes full use of information transfer between ant colonies. The distributed positive feedback mechanism is used to find the shortest path from nest to food. As a global optimization algorithm, it is suitable for various combinatorial optimization problems. The algorithm has the advantages of great searchability, good parallelism, effective shortening of search time, and strong global searchability; it is also easily combined with other algorithms. Although the ACO algorithm has low computation efficiency, it has a high degree of robustness in solving nonlinear problems and can search for solutions in global multipoints simultaneously. In 2018, Riahi et al. combined ACO with a simulated annealing algorithm [105]. It improved the updating method of the pheromone path. Some different areas of the search space were examined and the best solution was selected. The new method of local search in ACO is effective for solving the problem of the no-waiting flow shop.

(d) Min-min algorithm
The basic process of the min-min algorithm is [106]: (1) the expected completion time of each scheduling task is calculated; (2) the earliest and least time-consuming task among all the tasks is located; (3) the appropriate resource is assigned, and (4) after the assignment is completed, the already assigned task can be deleted from the task set. By repeating these steps, all tasks in the task set and the assignment of corresponding execution resources can be completed. The basic idea of the min-min algorithm is to execute the fast resource to deal with the small task and complete the heuristic selection by obtaining the earliest task time and the fastest task completion. Its advantages are simple and easy to execute, and its disadvantages are a long waiting time for large tasks, easy execution of server resources, and poor processing capacity. However, because of the short waiting time of small tasks and the better corresponding resources, the system load is uneven.

(e) Max-min algorithm
The basic idea of the max-min algorithm is similar to that of the min-min algorithm. It also calculates the shortest predicted completion time for all tasks in the unassigned

tasks [107]. The difference is that it prioritizes scheduling large tasks and assigns them to the resource that minimizes task completion time. After the task assignment is completed, the expected ready time for the physical resource is updated and the completed tasks are deleted from the task set. This scheduling method is easy to realize the simultaneous execution of large and small tasks. Unfortunately, because it prefers to deal with large tasks first, it will delay the scheduling of small tasks to some extent.

(f) Genetic algorithm
The GA is a random find algorithm based on biological evolution and genetics; it is also a parallel search algorithm based on the laws of biology and natural genetic mechanism [108]. This algorithm mainly exchanges search methods among groups and individual information in groups. It is often used to solve nonlinear problems that are difficult to solve by traditional search methods and production line optimization problems. In the GA, chromosomes are randomly generated numeric codes that make up the initial population. The parameters used to evaluate each individual are fitness functions. The lower the fitness function, the more difficult it is to survive. Individuals with low fitness values are eliminated, whereas individuals with high fitness values participated in genetic manipulation. The task of the GA is to evaluate all chromosomes generated in the algorithm process and then screen the chromosomes according to the fitness value. Those with higher fitness values have a better chance of reproducing. This algorithm combines the advantages of directional search and random search. It is easily combined with other algorithms. However, its local search capacity is not good and its feasible solution speed is slow. It is complex to operate and has many parameters that are not easy to choose.

4.4 Automatic guided vehicle path planning application in the rail transit intelligent manufacturing environment

The AGV hybrid path planning model refers to the combination of global static path planning and local obstacle avoidance path planning [109]. Global static path planning refers to path optimization in the case of a known environment, in which the minimum length of the path is required. Local obstacle avoidance path planning is when there is an obstacle that is not marked in the global map during operation of the AGV. It is necessary to replan the path. Usually, only a small section of the road needs to be changed, and the length of the calculation path is required. The structure of the path planning hybrid model is shown in Fig. 4.6.

4.4.1 Global static automatic guided vehicle path planning

The global static path planning model of AGVs is established in an intelligent environment that has obtained environmental information. After an AGV gets the start and

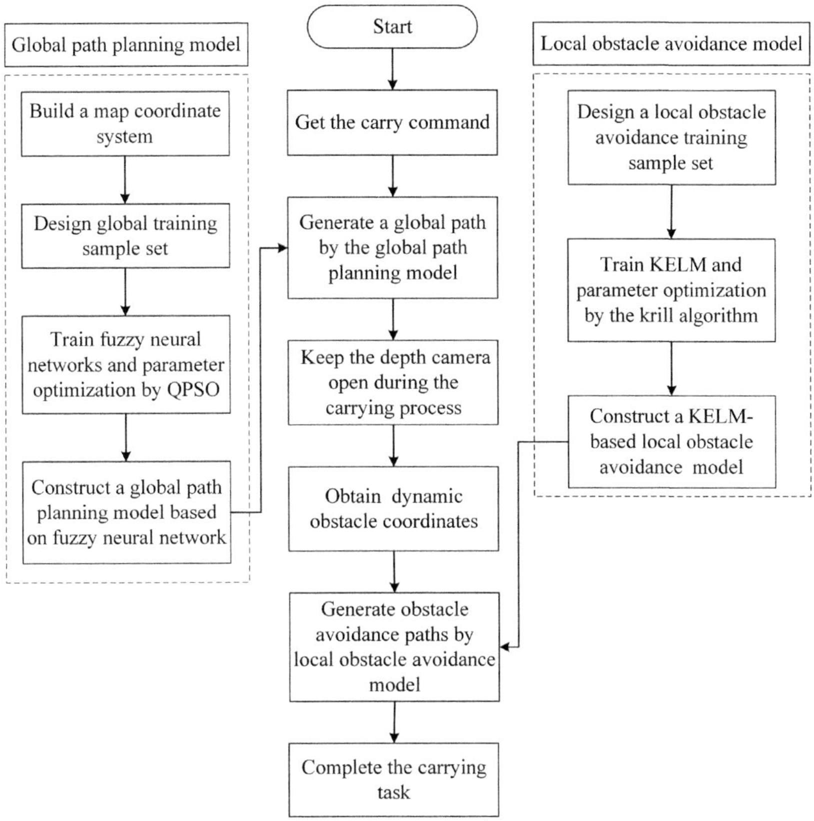

Figure 4.6
Flowchart of the hybrid path planning model. *KELM*, kernel-based extreme learning machine; *QPSO*, quantum particle swarm optimization.

end points of the transportation task, an optimal path should be found between the start and end points. This optimal path requires no collisions and the shortest length. The main contents of global AGV path planning can be explained simply [110]: (1) the path planning map environment is established, (2) objects are mapped in the environment, (3) the environment is transformed into a model that can be processed by algorithms, (4) the path planning algorithm is used in the model for optimization, and (5) the results of path planning are used to guide the actual movement of the AGV. Specific steps are:

Step 1: Construct a global map 3D coordinate system for the AGV's carrying area and obtain the feasible area coordinates in the global map 3D coordinate system.

Taking the ground center point of the carrying area as the origin, the east direction is the X-axis, the north direction is the Y-axis, and the vertical ground direction is the Z-axis. The AGV carrying area is all of the floor connecting areas in the working space and the feasible area refers to the area where the obstacles are deleted. In the global map 3D coordinate system, the 2D plane coordinates of the floor connecting area of each floor are the same, and the Z coordinate corresponds to the height.

Step 2: Obtain a training sample set.

In the global map 3D system, 200 groups of point-to-point optimal design global paths in the feasible region are designed. Each optimal design global path is used as a training sample.

Step 3: Construct a global static path planning model for the AGV.

Using the starting and end points' coordinates of each global sample as the input data, the corresponding optimal design global path is used as the output data. The fuzzy neural network is trained to obtain a global static path planning model [111]. The Gaussian function is selected as the membership function. In the training process, the weights, membership functions mean, and variance in the fuzzy neural network are obtained by quantum particle swarm optimization (QPSO). The process is:

Step 3.1: Initialize the parameters.

In the QPSO, position vector parameters of the quantum particle population are initialized to a random number within $[-1\ 1]$, and iteration time is initialized as $t = 0$. In the fuzzy neural network, the weights of the network, the mean values, and variances of the membership functions are initialized with the position vector of each quantum particle in the quantum particle group.

Step 3.2: Calculate the fitness function.

Parameters in the position vector of the quantum particle are assigned to the global static path planning model based on the fuzzy neural network. The global path of all global training samples is obtained by the fuzzy neural network. The average value of the paths' lengths is defined as the fitness function.

Step 3.3: Calculate the variance of the population fitness of each quantum particle group and perform premature convergence judgment.

If the variance of the quantum particle group population fitness is smaller than the premature convergence judgment threshold γ, the $\delta\%$ of the worst-fit particles and the group extremum particles in the quantum particle group are mutated, and the current best-fit particles are used as the global optimal.

Step 3.4: Form the elite population.

When the number of iterations is greater than that of the elite population, the elites of the various groups are extracted by information sharing between populations, and the process proceeds to Step 3.8. Otherwise, the process proceeds to Step 3.5.

Step 3.5: Update various group particle parameters.

Step 3.6: Recalculate and compare the fitness value of each particle.

If the fitness value is better than the current individual extremum, the individual extremum is updated. The global extremum particles are compared. If there is a particle fitness value that is better than the current group extremum, the global extremum particles are updated, $t = t + 1$, and the iteration goes to Step 3.3.

Step 3.7: Output the fuzzy neural network.

The elite population continues to evolve, whether the maximum number of iterations is satisfied is judged. If yes, the iteration is stopped; otherwise, $t = t + 1$, and the iteration continues to Step 3.3 until the global optimal value is found.

Step 4: Complete path planning.

The coordinates of the starting point and the end point in the transportation task are input into the global static path planning model based on the fuzzy neural network. The neural network generates the optimal planning path of the corresponding AGV. The AGV walks according to the optimal global path and completes the transportation task.

4.4.2 Local dynamic automatic guided vehicle obstacle avoidance

During actual operation of the AGV, owing to environmental changes and staff walking, many unobstructed obstacles in the global map will be encountered and the AGV cannot travel normally. Therefore, the local path planning obstacle avoidance model based on stereovision is designed.

Using the starting and ending coordinates of each sample of the local obstacle avoidance training sample set as the input data, the corresponding optimal design obstacle avoidance path is used as the output. The multikernel support vector machine is trained to obtain the local obstacle avoidance model based on the multikernel support vector machine [112]. The local obstacle avoidance training sample set is obtained in the global map 3D coordinate system. Two hundred groups of obstacles are designed in the feasible areas, and the optimal obstacle avoidance path is planned.

The multikernel support vector machine selects a linear kernel function and a polynomial kernel function to form a mixed kernel function. Among them, the value range of the kernel function weight d is [0, 1], and the value of d is optimized by the krill herd algorithm. During the training process, the penalty coefficient, kernel function, and kernel function weight d of the multikernel support vector machine are determined by the krill herd algorithm [113].

Step 1: Initialize the krill herd parameters.

Taking the individual position of krill as the penalty coefficient, kernel function, and kernel function weight d of the multikernel support vector machine, the krill population is randomly initialized. Iteration time is initialized as $t = 0$.

Step 2: Build fitness functions.

A fitness function is set to determine an initial optimal krill individual position. The values of the krill individual position vector are assigned to the local obstacle avoidance model based on the multikernel support vector machine. The local obstacle avoidance model determined by the krill algorithm is used to output the obstacle avoidance training samples. The results of the trained obstacle avoidance training samples are multiplied by the length of the designed path, and the inverse is taken as the fitness function.

Step 3: Induced movement, foraging activity, and random diffusion.

The induced movement, foraging activity, and random diffusion are carried out. The position of the krill individual is updated. The current optimal krill position is determined according to the fitness function.

Step 4: Output the multikernel support vector machine.

Whether the maximum number of iterations is satisfied is determined. If yes, the weights of the optimal krill individual are assigned to the obstacle avoidance model. If not, $t = t + 1$, and return the iteration to Step 3 until the maximum number of iterations is satisfied. The optimal krill population is output to optimize the penalty coefficient, kernel parameter. and kernel function weight in the optimized multikernel support vector machine.

4.4.3 Hybrid path planning application

After receiving the task, the AGVs start path planning. When the AGVs perform tasks across floors, the global path plan is decomposed into separate global path plans for two floors and an elevator path. Global path planning uses a global static path planning model based on fuzzy neural networks. The starting point of the first part of the global path planning is the starting point of the transportation task, and the ending point is the way point of the elevator on the first floor. The starting point of the second part of the global path planning is the way point of the elevator on the second floor, and the ending point is the target point of the transportation task. In processing, AGVs in the intelligent environment needs to interact with elevators or electrically controlled doors. If the task received by AGVs does not cross the floor, the starting point and target point of the task are directly used as the input of the global static path planning model based on the fuzzy neural network.

During execution of the mission, AGVs always monitor the surrounding environment in real time with the onboard stereovision sensor. The AGVs observe real-time dynamic obstacles that are not indicated in the global map on the route of travel, such as people temporarily occupying aisles, chairs, and other carrying AGVs. Once the obstacle is found, AGVs scan the contour position of the obstacle and transmit it to the onboard computer. The onboard computer adds obstacle contour locations to infeasible areas and calculates the best obstacle avoidance path. After AGVs go through the obstacle according to the best obstacle avoidance path, the global path is continued to reach the target point.

4.5 Conclusion and outlook

AGVs are the product of mechatronics, intelligence, and information. They are an important part of an intelligent factory and warehouse. The demand for the AGVs has maintained rapid growth. The rail transit intelligent manufacturing environment needs intelligent AGVs to improve logistics efficiency and improve the accuracy of logistics. It is believed that with improvements in the intelligent level of industrial products in the world, AGVs will break through their original functional limitations and be more widely used in rail transit intelligent manufacturing environments.

This chapter analyzes the overview, key technologies, and main components of AGVs. First, this chapter introduced the progress in development and the types of AGVs in the rail transit intelligent manufacturing environment, and then introduced the main components of AGVs. The AGVs' main components are divided into four parts: the chassis, power equipment, control devices, and safety devices. The key technologies components of AGVs were also introduced. This part includes the navigation system, path planning, human–robot interaction, and task assignment of AGVs. Finally, AGV path planning applications in rail transit intelligent manufacturing environment were introduced.

Research on the AGV technologies has always been a focus across the world. Navigation and positioning, path planning, and task assignment of AGVs in the intelligent manufacturing environment are key research contents for relevant scholars. However, there are still some shortcomings in important AGV technologies.

The first is navigation technologies. With the continuous improvement of computer processing speed and ongoing development of signal processing technology, AGV navigation technologies have become increasingly mature. However, because of the complexity of the rail transit manufacturing environment, there are still some major problems in navigation technology to be solved.

The second is the path planning algorithms of AGVs. Although research on intelligent path planning for AGVs has achieved important results, there are still limitations. For example, the GA and ACO make it easy to fall into local optimization, and the neural network algorithm needs a large number of training samples. Current improved algorithms are mainly based on a combination of various algorithms and hierarchical optimization. Although they make up for the shortcomings, there are many bottlenecks in development, such as the high algorithm complexity and slow convergence speed.

The third is the task assignment methods of AGVs. As a core technology of the AGV automatic transportation system, task assignments should further strengthen research on AGV scheduling algorithms and scheduling strategies to provide theoretical support for improving system efficiency and reducing operating costs.

Because of the deficiencies of AGV technologies, there are some prospects for future development:

(a) Adding optical flow field distribution as an auxiliary feature can improve the accuracy of positioning and attitude estimation. Integrating the input of the auxiliary features to enrich the neural network is an effective way to conduct interframe estimation based on deep learning. If one or several modules in SLAM are replaced by deep learning methods, the accuracy and robustness of estimating the position and attitude of the camera based on deep learning can be improved.

(b) In the process of the formation, transmission, reception, and processing of visual images, there will be noise interference. Therefore, research into image denoising technology has an important meaning. Basic problems of wavelet denoising technology are how to find the optimal wavelet coefficient model for different models and how to model non-Gaussian noise. Another development trend of image denoising technology is that different filtering methods can be combined according to specific images to achieve better results than a single denoising method.

(c) The hybrid algorithm of path planning is an effective combination of each algorithm. Any single algorithm is not enough to solve all path planning problems in practical problems, especially for some new problems in interdisciplinary fields. It is difficult to create a new algorithm, and the complementary advantages of the path planning algorithm can effectively provide a new idea to solve the problem. Some functional algorithms such as swarm intelligence algorithms, reinforcement learning algorithms, fuzzy control algorithms, neural network algorithms are gradually being introduced into the path planning problem. This kind of complementary hybrid algorithm promotes the fusion and development of various methods. Combinations of artificial intelligence methods, new mathematical methods, and bionic methods have fast development prospects.

(d) The task assignment algorithms should be optimized. With the optimization and improvement of intelligent algorithms such as ACO and GA, the scheduling system can more reasonably allocate a large number of tasks generated in a short time to

AGVs. To improve the AGV task assignment, it is also possible to adopt the method of optimizing system communication. For a large manufacturing space and a large number of AGVs, smooth communication between AGVs and the scheduling system is also an important factor that determines the stability of the system. It can increase information interaction between AGVs, reduce the workload of the scheduling system, and thus improve system stability and operational efficiency.

(e) Existing excellent path planning algorithms should be improved. Any algorithm will face many difficulties in the process of practical application, especially problems caused by its limitations. Targeted improvement in specific application directions can effectively improve the performance of the algorithm and solve practical problems.

References

[1] J. Mehami, M. Nawi, R.Y. Zhong, Smart automated guided vehicles for manufacturing in the context of Industry 4.0, Procedia Manuf. 26 (2018) 1077–1086.

[2] A. Gunasekaran, Agile manufacturing: a framework for research and development, Int. J. Prod. Econ. 62 (1999) 87–105.

[3] G. Ullrich, The history of automated guided vehicle systems, in: Automated Guided Vehicle Systems, Springer, 2015, pp. 1–14.

[4] F. Kimura, H. Lipson, M. Shpitalni, Engineering environments in the information age: research challenges and opportunities, CIRP Ann. 47 (1998) 87–90.

[5] G. Hammond, Evolutionary AGVS–from concept to present reality, in: Proceeding of 6th International Conference on Automated Guided System, 1987, pp. 25–26.

[6] E. Bordelon Hoff, B.R. Sarker, An overview of path design and dispatching methods for automated guided vehicles, Integrated Manuf. Syst. 9 (1998) 296–307.

[7] I. Fa Vis, Survey of research in the design and control of automated guided vehicle systems, Eur. J. Oper. Res. 170 (2006) 677–709.

[8] G. Giralt, R. Chatila, M. Vaisset, An Integrated Navigation and Motion Control System for Autonomous Multisensory Mobile Robots. Autonomous Robot Vehicles, Springer, 1990, pp. 420–443.

[9] R. Bostelman, Towards improved forklift safety: white paper, in: Proceedings of the 9th Workshop on Performance Metrics for Intelligent Systems, 2009, pp. 297–302.

[10] K. Ct Vivaldini, J.P. Galdames, T.B. Pasqual, et al., Intelligent Warehouses: focus on the automatic routing and path planning of robotic forklifts able to work autonomously, Mechatron. Series Intell. Transport. Veh. 1 (2011) 115–145.

[11] M. Seelinger, J.-D. Yoder, Automatic pallet engagment by a vision guided forklift, in: Proceedings of the 2005 IEEE International Conference on Robotics and Automation, 2005, pp. 4068–4073.

[12] S. Li, X. Hou, Research on the AGV based robot system used in substation inspection, in: International Conference on Power System Technology, 2006, pp. 1–4, 2006.

[13] J.R. Minifie, Survey of robotics in the health care industry, Eng. Costs Prod. Econ. 17 (1989) 223–230.

[14] Z. Leng, M.-Z. Dong, G.-Q. Dong, et al., A rapid path planner for autonomous ground vehicle using section collision detection, J. Shanghai Jiaot. Univ. (Sci.) 14 (2009) 306–309.

[15] S. Kaliappan, J. Lokesh, P. Mahaneesh, et al., Mechanical design and analysis of AGV for cost reduction of material handling in automobile industries, Int. Res. J. Automot. Technol. 1 (2018) 1–7.

[16] X. Liu, W. Li, A. Zhou, PNGV equivalent circuit model and SOC estimation algorithm for lithium battery pack adopted in AGV vehicle, IEEE Access 6 (2018) 23639–23647.

[17] J. Ploeg, A.C. Van Der Knaap, D.J. Verburg, ATS/AGV-design, implementation and evaluation of a high performance AGV, in: Intelligent Vehicle Symposium, vol. 1, 2002, pp. 127–134.

[18] K.-Y. Lou, C.-H. Goh, Design and real-time implementation of a control system for port AGV application, in: Proceedings 2003 IEEE International Symposium on Computational Intelligence in Robotics and Automation, vol. 2, 2003, pp. 527–532.

[19] Y. Pang, F. Lu, F. Wang, et al., The structural design and control research of plant AGV, Mod. Machinery 5 (2014) 1–4.

[20] R.V. Bostelman, T. Hong Hong, R. Madhavan, Towards AGV safety and navigation advancement obstacle detection using a TOF range camera, in: Proceedings, 12th International Conference on Advanced Robotics, 2005, pp. 460–467.

[21] W.-C. Chiang, D. Chandra Ramamurthy, T.N. Mundhenk, et al., Range detection for AGV using a rotating sonar sensor, in: Intelligent Robots and Computer Vision XVII: Algorithms, Techniques, and Active Vision, vol. 3522, 1998, pp. 435–443.

[22] M. Watanabe, M. Furukawa, Y. Kakazu, Intelligent AGV driving toward an autonomous decentralized manufacturing system, Robot. Comput. Integrated Manuf. 17 (2001) 57–64.

[23] F. Fraundorfer, D. Scaramuzza, Visual odometry: part i: the first 30 years and fundamentals, IEEE Robot. Autom. Mag. 18 (2011) 80–92.

[24] M. Davoodi, F. Panahi, M. Ali, et al., Multi-objective path planning in discrete space, Appl. Soft Comput. 13 (2013) 709–720.

[25] E. Clarkson, R.C. Arkin, Applying heuristic evaluation to human-robot interaction systems, in: Flairs Conference, 2007, pp. 44–49.

[26] MHFBMd Fauadi, H. Lin, T. Murata, Dynamic task assignment of autonomous AGV system based on multi agent architecture, in: IEEE International Conference on Progress in Informatics and Computing, vol. 2, 2010, pp. 1151–1156, 2010.

[27] Q. Li, Z. Hu, G. Liang, AGV design using electromagnetic navigation, Mod. Electron. Tech. 12 (2012) 79–81.

[28] W. Urban, U. Andersson, K. Hyyppä, AGV navigation by angle measurements, in: International Conference on Automated Guided Vehicle Systems, 1988, pp. 199–212.

[29] K. Jung, J. Kim, J. Kim, et al., Positioning accuracy improvement of laser navigation using UKF and FIS, Robot. Autonom. Syst. 62 (2014) 1241–1247.

[30] H.-B. Zhang, K. Yuan, Q.-R. Zhou, Visual navigation of an automated guided vehicle based on path recognition, in: Proceedings of 2004 International Conference on Machine Learning and Cybernetics, vol. 6, 2004, pp. 3877–3881.

[31] D.L. Donoho, De-noising by soft-thresholding, IEEE Trans. Inf. Theor. 41 (1995) 613–627.

[32] M.N. Do, M. Vetterli, Contourlets: a directional multiresolution image representation, in: Proceedings International Conference on Image Processing, vol. 1, 2002, pp. 357–360.

[33] E.J. Candès, D.L. Donoho, New tight frames of curvelets and optimal representations of objects with piecewise C^2 singularities, Commun. Pure Appl. Math. 57 (2004) 219–266.

[34] J.R. Fram, E.S. Deutsch, On the quantitative evaluation of edge detection schemes and their comparison with human performance, IEEE Trans. Comput. 100 (1975) 616–628.

[35] D. Chen, Y. Wang, G. Zhang, A new sub-pixel detector for grid target points in camera calibration, in: ICO20: Optical Information Processing, vol. 6027, 2006, pp. 675–680.

[36] L. Matthies, S. Shafer, Error modeling in stereo navigation, IEEE J. Robot. Autom. 3 (1987) 239–248.

[37] M. Kabuka, A. Arenas, Position verification of a mobile robot using standard pattern, IEEE J. Robot. Autom. 3 (1987) 505–516.

[38] C.F. Olson, Selecting landmarks for localization in natural terrain, Aut. Robots 12 (2002) 201–210.

[39] H.-M. Shim, E.-H. Lee, J.-H. Shim, et al., Implementation of an intelligent walking assistant robot for the elderly in outdoor environment, in: 9th International Conference on Rehabilitation Robotics, 2005, pp. 452–455.

[40] K. Yousif, A. Bab-Hadiashar, R. Hoseinnezhad, An overview to visual odometry and visual SLAM: applications to mobile robotics, Intell. Ind. Syst. 1 (2015) 289–311.
[41] D.G. Lowe, Distinctive image features from scale-invariant keypoints, Int. J. Comput. Vis. 60 (2004) 91–110.
[42] G. Costante, M. Mancini, P. Valigi, et al., Exploring representation learning with cnns for frame-to-frame ego-motion estimation, IEEE Robot. Autom. Lett. 1 (2015) 18–25.
[43] H.S. Dulimart, A.K. Jain, Mobile robot localization in indoor environment, Pattern Recogn. 30 (1997) 99–111.
[44] M. Bosse, P. Newman, J. Leonard, et al., Simultaneous localization and map building in large-scale cyclic environments using the Atlas framework, Int. J. Robot Res. 23 (2004) 1113–1139.
[45] G. Pascoe, W. Maddern, P. Newman, Direct visual localisation and calibration for road vehicles in changing city environments, in: Proceedings of the IEEE International Conference on Computer Vision Workshops, 2015, pp. 9–16.
[46] H. Alismail, L.D. Baker, B. Browning, Continuous trajectory estimation for 3D SLAM from actuated lidar, in: IEEE International Conference on Robotics and Automation, 2014, pp. 6096–6101.
[47] E.B. Olson, Real-time correlative scan matching, in: IEEE International Conference on Robotics and Automation, 2009, pp. 4387–4393.
[48] G. Grisettiyz, C. Stachniss, W. Burgard, Improving grid-based slam with rao-blackwellized particle filters by adaptive proposals and selective resampling, in: IEEE International Conference on Robotics and Automation, 2005, pp. 2432–2437.
[49] G.D. Tipaldi, M. Braun, K. O Arras, Interest regions for 2D range data with applications to robot navigation, in: Experimental Robotics, 2014, pp. 695–710.
[50] Z. Lu, Z. Hu, K. Uchimura, SLAM estimation in dynamic outdoor environments: a review, in: International Conference on Intelligent Robotics and Applications, 2009, pp. 255–267.
[51] R. Smith, M. Self, P. Cheeseman, Estimating uncertain spatial relationships in robotics, Mach. Intell. Pattern Recogn. (1988) 435–461.
[52] J.-L. Blanco, J. González, J.-A. Fernández-Madrigal, Optimal filtering for non-parametric observation models: applications to localization and SLAM, Int. J. Robot Res. 29 (2010) 1726–1742.
[53] F. Lu, E. Milios, Globally consistent range scan alignment for environment mapping, Aut. Robots 4 (1997) 333–349.
[54] K. Kurt, G. Grisetti, R. Kümmerle, et al., Efficient sparse pose adjustment for 2D mapping, in: IEEE/RSJ International Conference on Intelligent Robots and Systems, 2010, pp. 22–29.
[55] N. Sariff, N. Buniyamin, An overview of autonomous mobile robot path planning algorithms, in: The 4th Student Conference on Research and Development, 2006, pp. 183–188.
[56] C. Wang, L. Wang, J. Qin, et al., Path planning of automated guided vehicles based on improved a-star algorithm, in: IEEE International Conference on Information and Automation, 2015, pp. 2071–2076.
[57] J.J. E Slotine, H.S. Yang, Improving the efficiency of time-optimal path-following algorithms, IEEE Trans. Robot. Autom. 5 (1989) 118–124.
[58] T. Chettibi, L. He, M. Haddad, et al., Minimum cost trajectory planning for industrial robots, Eur. J. Mech. A Solid. 23 (2004) 703–715.
[59] T. Akiba, Y. Iwata, Y. Yoshida, Fast exact shortest-path distance queries on large networks by pruned landmark labeling, in: Proceedings of the 2013 ACM SIGMOD International Conference on Management of Data, 2013, pp. 349–360.
[60] E.W. Dijkstra, A note on two problems in connexion with graphs, Numer. Math. 1 (1959) 269–271.
[61] S. E Dreyfus, An appraisal of some shortest-path algorithms, Oper. Res. 17 (1969) 395–412.
[62] F. Duchoň, A. Babinec, K. Martin, et al., Path planning with modified a star algorithm for a mobile robot, Procedia Eng. 96 (2014) 59–69.

[63] J. Peng, Y. Huang, G. Luo, Robot path planning based on improved A* algorithm, Cybern. Inf. Technol. 15 (2015) 171−180.
[64] L.E. Kavraki, P. Svestka, J.-C. Latombe, et al., Probabilistic roadmaps for path planning in high-dimensional configuration spaces, IEEE Trans. Robot. Autom. 12 (1996) 566−580.
[65] L. Kang, C.-X. Zhao, J.-H. Guo, Improved path planning based on rapidly-exploring random tree for mobile robot in unknown environment, Pattern Recognit. Artif. Intell. 3 (2009) 337−343.
[66] T. Gîrbacia, G. Mogan, An improvement of the rapidly-exploring random tree method for path planning of a car-like robot in virtual environment, Appl. Mech. Mater. 772 (2015) 471−476.
[67] X. Wang, Y. Shi, D. Ding, et al., Double global optimum genetic algorithm−particle swarm optimization-based welding robot path planning, Eng. Optim. 48 (2016) 299−316.
[68] M. Dorigo, M. Birattari, Ant Colony Optimization, Springer, 2010.
[69] E. Russell, J. Kennedy, Particle swarm optimization, in: Proceedings of the IEEE International Conference on Neural Networks, vol. 4, 1995, pp. 1942−1948.
[70] F. Matoui, B. Boussaid, M. Brahim, et al., Path planning of a group of robots with potential field approach: decentralized architecture, in: The International Federation of Automatic Control 50, 2017, pp. 11473−11478.
[71] L.A. Zadeh, Fuzzy sets, Inf. Contr. 8 (1965) 338−353.
[72] H. Monfared, S. Salmanpour, Generalized intelligent water drops algorithm by fuzzy local search and intersection operators on partitioning graph for path planning problem, J. Intell. Fuzzy Syst. 29 (2015) 975−986.
[73] M. Fakoor, A. Kosari, M. Jafarzadeh, Humanoid robot path planning with fuzzy Markov decision processes, J. Appl. Res. Technol. 14 (2016) 300−310.
[74] M.K. Singh, D.R. Parhi, Path optimisation of a mobile robot using an artificial neural network controller, Int. J. Syst. Sci. 42 (2011) 107−120.
[75] M. Lee Minsky, Theory of Neural-Analog Reinforcement Systems and its Application to the Brain Model Problem, Princeton University, 1954.
[76] R. Martinez-Cantin, N. De Freitas, E. Brochu, et al., A Bayesian exploration-exploitation approach for optimal online sensing and planning with a visually guided mobile robot, Aut. Robots 27 (2009) 93−103.
[77] R. Ac Bianchi, M.F. Martins, C.H. Ribeiro, et al., Heuristically-accelerated multiagent reinforcement learning, IEEE Trans. Cybern. 44 (2013) 252−265.
[78] D.E. Moriarty, A.C. Schultz, J.J. Grefenstette, Evolutionary algorithms for reinforcement learning, J. Artif. Intell. Res. 11 (1999) 241−276.
[79] B.D. Argall, S. Chernova, M. Veloso, et al., A survey of robot learning from demonstration, Robot. Autonom. Syst. 57 (2009) 469−483.
[80] J. Peters, S. Schaal, Policy gradient methods for robotics, in: IEEE/RSJ International Conference on Intelligent Robots and Systems, 2006, pp. 2219−2225.
[81] D. Fox, W. Burgard, S. Thrun, The dynamic window approach to collision avoidance, IEEE Robot. Autom. Mag. 4 (1997) 23−33.
[82] P. Saranrittichai, N. Niparnan, S. Attawith, Robust local obstacle avoidance for mobile robot based on dynamic window approach, in: The 10th International Conference on Electrical Engineering/Electronics, Computer, Telecommunications and Information Technology, 2013, pp. 1−4.
[83] P. Inigo-Blasco, F. Diaz-Del-Rio, S. Vicente Diaz, et al., The shared control dynamic window approach for non-holonomic semi-autonomous robots, in: The 41st International Symposium on Robotics, 2014, pp. 1−6.
[84] M.A. Goodrich, A.C. Schultz, Human−robot interaction: a Survey, Hum. Comput. Interact. 1 (2007) 203−275.
[85] Z. Shi, Planning for the construction of urban rail transit, Urban Rapin Rail Transit 2 (2004) 12−15.

[86] Y. Wu, T.S. Huang, Vision-based gesture recognition: a review, in: International Gesture Workshop, 1999, pp. 103–115.
[87] J. Suarez, R.R. Murphy, Hand gesture recognition with depth images: a review, in: The 21st IEEE International Symposium on Robot and Human Interactive Communication, 2012, pp. 411–417.
[88] H.D. Yang, A. Yeon Park, S.-W. Lee, Gesture spotting and recognition for human–robot interaction, IEEE Trans. Robot. 23 (2007) 256–270.
[89] M. Porta, Vision-based user interfaces: methods and applications, Int. J. Hum. Comput. Stud. 57 (2002) 27–73.
[90] C. Ott, D. Lee, Y. Nakamura, Motion capture based human motion recognition and imitation by direct marker control, in: 8th IEEE-RAS International Conference on Humanoid Robots, 2008, pp. 399–405, 2008.
[91] J. Wang, Z. Liu, Y. Wu, et al., Mining actionlet ensemble for action recognition with depth cameras, in: IEEE Conference on Computer Vision and Pattern Recognition, 2012, pp. 1290–1297, 2012.
[92] L. Wang, W. Hu, T. Tan, Recent developments in human motion analysis, Pattern Recogn. 36 (2003) 585–601.
[93] H. Jhuang, T. Serre, L. Wolf, et al., A biologically inspired system for action recognition, in: IEEE 11th International Conference on Computer Vision, 2007, pp. 1–8.
[94] A. Graves, A. Mohamed, G. Hinton, Speech recognition with deep recurrent neural networks, in: IEEE International Conference on Acoustics, Speech and Signal Processing, 2013, pp. 6645–6649, 2013.
[95] M. Omar, C. Bartneck, L. Feijs, et al., Improving speech recognition with the robot interaction language, Disruptive Sci. Technol. 1 (2012) 79–88.
[96] Y. Tabet, B. Mohamed, Speech synthesis techniques. A survey, in: International Workshop on Systems, Signal Processing and Their Applications, 2011, pp. 67–70.
[97] C.G. Co, J.M.A. Tanchoco, A review of research on AGVS vehicle management, Eng. Costs Prod. Econ. 21 (1991) 35–42.
[98] L. Sabattini, V. Digani, C. Secchi, et al., Optimized simultaneous conflict-free task assignment and path planning for multi-AGV systems, in: IEEE/RSJ International Conference on Intelligent Robots and Systems, 2017, pp. 1083–1088.
[99] D.B. Kim, H. Hwang, A dispatching algorithm for multiple-load AGVs using a fuzzy decision-making method in a job shop environment, Eng. Optim. 33 (2001) 523–547.
[100] C.-H. Hu, P.J. Egbelu, A framework for the selection of idle vehicle home locations in an automated guided vehicle system, Int. J. Prod. Res. 38 (2000) 543–562.
[101] B. Reddy, C. Rao, A hybrid multi-objective GA for simultaneous scheduling of machines and AGVs in FMS, Int. J. Adv. Manuf. Technol. 31 (2006) 602–613.
[102] N.M. Reda, A. Tawfik, M.A. Marzok, et al., Sort-Mid tasks scheduling algorithm in grid computing, J. Adv. Res. 6 (2015) 987–993.
[103] H. Martinez-Alfaro, D.R. Flugrad, Collision-free path planning for mobile robots and/or AGVs using simulated annealing, in: Proceedings of IEEE International Conference on Systems, Man and Cybernetics, vol. 1, 1994, pp. 270–275.
[104] S.Z. Selim, K.1 Alsultan, A simulated annealing algorithm for the clustering problem, Pattern Recogn. 24 (1991) 1003–1008.
[105] V. Riahi, M. Kazemi, A new hybrid ant colony algorithm for scheduling of no-wait flowshop, Oper. Res. 18 (2018) 55–74.
[106] M.-Y. Wu, W. Shu, H. Zhang, Segmented min-min: a static mapping algorithm for meta-tasks on heterogeneous computing systems, in: Proceedings of the 9th Heterogeneous Computing Workshop, 2000, pp. 375–385.
[107] K. Etminani, M. Naghibzadeh, A min-min max-min selective algorithm for grid task scheduling, in: The 3rd IEEE/IFIP International Conference in Central Asia on Internet, 2007, pp. 1–7.

[108] T. Shima, S.J. Rasmussen, A.G. Sparks, et al., Multiple task assignments for cooperating uninhabited aerial vehicles using genetic algorithms, Comput. Oper. Res. 33 (2006) 3252–3269.
[109] G. Zhang, X. Hu, J. Chai, et al., Summary of path planning algorithm and its application, Mod Mach 5 (2011) 85–90.
[110] Yanfei, L.L. Hui, J. Huang, K. Jin, Multi-neural network control planning method for robot path in intelligent environment, National Intellectual Property Administration, China (2017).
[111] Y. Yao, G. Chen, J. Jia, Study on the robot path planning based on fuzzy neural network algorithm, J. Sichuan Univ. Sci. Eng. 27 (2014) 30–33.
[112] Z. Bao, D. Pi, Y. Sun, Nonlinear model predictive control based on support vector machine with multi-kernel, Chin. J. Chem. Eng. 15 (2007) 691–697.
[113] A.H. Gandomi, A.H. Alavi, Krill herd: a new bio-inspired optimization algorithm, Commun. Nonlinear Sci. Numer. Simulat. 17 (2012) 4831–4845.

CHAPTER 5

Autonomous Rail Rapid Transit (ART) systems

5.1 Overview of ART

5.1.1 Development progress of ART

In 1879, trams first appeared at the Berlin Expo. The first tram line was constructed and operated in the suburbs of Berlin in 1881 [1]. At the beginning of the 20th century, trams were very popular in Europe, America, Oceania, and some Asian cities. Nevertheless, with the popularity of private cars, buses, and other road traffic in the 1950s, many tram systems were dismantled in the middle of the 20th century. As time goes by, problems began to emerge including the energy crisis, environmental pollution, land shortage, and road traffic congestion. To solve these problems of urban transportation, people began to consider upgrading the previous trams. The new generation of trams can do a better job in terms of comfort, safety, energy conservation, environmental protection, and green efficiency. This can not only minimize the adverse impact on the environment of the trams, but also complete the task of high-capacity passenger transport. At present, a new generation of passenger trams and freight trams, as well as metro trams and other means of transportation are widely used around the world.

Although the new generation of trams has been well balanced in terms of energy-saving and high-capacity passenger transport, the functions of trams have begun to show some shortcomings since the second half of the twentieth century. With the sharp increase in the number of private cars, the problem of urban traffic congestion is reappearing and becoming more difficult to solve. People started to find new solutions to improve transportation efficiency. Later, the emergence of Autonomous rail Rapid Transit (ART) seemed to alleviate the above problems.

ART is used to solve transportation problems in large urban suburbs and small urban areas. The superiority of ART has begun to appear. ART is a new type of all-electric vehicles. It uses new rubber wheels instead of steel wheels. The operation of ART does not depend on existing tracks, and it can run on roads paved with landmarks. Based on image recognition, vehicle-to-ground communication, and steering-by-wire technologies, ART can automatically drive in the direction of the ground sign. Based on the above

Robot Systems for Rail Transit Applications. https://doi.org/10.1016/B978-0-12-822968-2.00005-X

advantages, the popularity of ART can greatly reduce the waste of human and material resources. At the same time, automatic driving can make it easier to schedule trains uniformly to reduce traffic congestion.

5.1.2 Advantages and characteristics of ART

Under the condition of limited urban road resources, but the number of urban vehicles is increasing. The existing urban rail transit has the following disadvantages: long construction period, high cost and inflexible operation. In this context, a new type of medium volume rail transit is urgently needed. The emergence of ART undoubtedly conforms to this demand. The characteristics of ART can be summarized as follows:

(1) ART has low construction costs

The train bodyweight and uniaxial load are very low. It runs on the existing road surface without structural reinforcement. So the construction of ART can be made base on the existing road traffic management system and the existing signal system. At present, the cost of common subway lines in China is around 400–700 million yuan/km, and the cost of modern tram lines is around 150–200 million yuan/km [2]. ART can be put into use with a simple road modification. Surprisingly, the entire road can be paved at a fifth of the cost of a modern tram. In other words, building a 10 km line of ART would save at least 1000 million yuan compared with modern trams. The popularization of ART will greatly reduce the government's investment in transportation infrastructure. ART has a huge cost advantage.

(2) The infrastructure construction period of ART is short

ART replaces the rail laying by drawing the ground marking line, so the construction period is short. ART can operate on the existing road of the city and shorten the period of road planning, demolition, and construction. It is estimated that it will take only 12 months from project approval to full load operation. Compared to modern trams, ART is designed for a top speed of 70 km/h [2]. Fortunately, ART does not rely on rail, so the construction takes only 1 year and can be put into use quickly.

(c) ART is flexible and adaptable

There is no need to lay rails for ART, and shared road rights can be used. So the operation lines are flexible. In the case of a traffic jam or accident, ART can take a detour and adopt flexible scheduling. ART has a small turning radius and runs in both directions. It can run on relatively narrow roads in the old city. With a length of more than 30 m, ART is large but flexible and highly compatible with complex roads [3]. The secret lies in the intelligent following control for the virtual track by the design method of the multiaxis steering system. The design allows ART to make the same turn using only the turning radius of a city bus. ART occupies a smaller aisle width than the common bus. It solves the overlength problem. At the same time, ART adopts a double-headed design similar to that of high-speed trains, which solves the trouble of turning around.

(d) The capacity of ART is high

According to statistics, China will add about 80 prefecture-level cities with a population of more than one million by 2020 as urbanization progresses [2]. However, in the overall urbanization progresses, 80% of small and medium-sized city governments use traditional public transportation because they are unable to withstand the high construction costs and long construction period of existing rail transit equipment. This has led to more severe traffic congestion in cities [3]. ART is tailored products to solve the "urban disease." It can not only be used as a supplement for rail transit in first-tier cities, but also as the main passenger transport in second and third-tier cities. It can also undertake the line transport of the new area to the new area and the center to the tourist area. And it can be fully combined with the existing public transportation system to create a three-dimensional (3D) underground, ground, and air transportation network. The emergence of ART provides a new solution to the problem of modern urban transportation.

In general, ART system has the advantages of the short construction time, low cost, suitable for most cities and large passenger throughput. It is a medium-traffic rail transit system solution that combines capacity and investment.

ART runs on the virtual track. It uses the "virtual track following control" technology independently developed by the innovative team of China CRRC Zhuzhou Institute Co., Ltd. It identifies the road virtual track line by various sensors on the vehicle, then transmits the operational information to the train "brain" (central control unit). According to the operating instructions of the central control unit, the train runs along the virtual track line accurately. This driving feature allows the driver to make emergency braking and other safety measures according to the surrounding environment. ART also combines some more advanced technologies, including autonomous driving and big data, etc. The technological innovation of ART subverts the concept of traditional rail transit. By integrating various functions, it has the advantages of energy-saving, environmental protection, high capacity, and punctuality. It also has the advantage of flexible routes and low cost of city buses. In general, based on safe, efficient and green travel, ART can bring people more high-tech application enjoyment [2]. The advantages of ART trains are shown in Fig. 5.1.

ART uses the low-floor design that is currently popular in rail transit. It is a new rubber-wheeled train where the multi-group rubber tire has two-way driving. ART has the all-electric drive and trajectory following capability. ART is guided by a virtual track, in which only the road surface is painted with a track mark and no physical track is required.

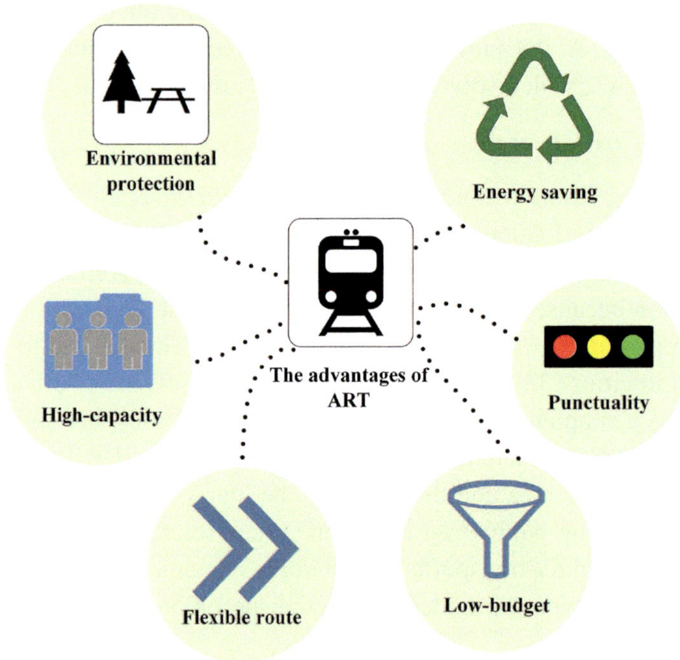

Figure 5.1
The advantages of Autonomous rail Rapid Transit (ART).

ART can adapt to a variety of road rights. Under semiproprietary road rights, ART can be manned assisted driving based on image recognition, vehicle-to-ground communication, and steering-by-wire. The driver can also manipulate the steering wheel to track the ground trajectory. Under the exclusive right of the road, the vertical and horizontal control of the vehicle can be realized based on the visual, radar, laser and other sensors. Then the Automatic Train Operation (ATO) with driver can be realized. As a new urban transportation solution, ART has different characteristics from other rail transportation modes such as urban rail transit and modern trams.

ART line is laid along the ground and crosses with public transportation on the ground. Therefore, the signal system of the traditional rail transit is not applicable. The driving safety should be adopted. ART signal system mainly includes the scheduling management system, the vehicle radio communication system, the intersection priority control system and the vehicle signal system.

By analyzing the needs of ART signal system, it can be concluded that ART signal system is more streamlined than the modern tram signal system. In the case of semiproprietary road rights, the core of ART signal system lies in the way of train positioning and intersection priority. A comparison of the main technical solutions between ART signal system and the modern tram signal system is shown in Table 5.1.

Table 5.1: Comparison of main technical solutions of ART and modern tram signal system.

Serial number	Technical solutions	ART	Tram
1	Interval positioning	There is no physical track or switch along the whole route of the intellectual track. Its full range of positioning: based on BD/GPS satellite positioning + electronic map	(1) Satellite positioning + beacon based on BD/GPS (2) vehicle coding range + fixed transponder (3) axle counting equipment (4) track circuit
2	The positioning mode of stations, ramps, and intersections		(1) axle counting equipment (2) track circuit (3) induction loop (4) ground beacon
3	Driving mode	Manual driving or automatic driving	Manual driving or automatic driving
4	An adapted form of road rights	All	All
5	The way the main train enters the road	Digital map	Under the condition of normal automatic handling fault, the driver can remote control switch handling road or manual on-site handling road
6	Vehicle wireless communication scheme	(1) LTE (2) WLAN (3) Rent a local communication service network	(1) LTE (2) WLAN (3) Rent a local communication service network
7	Switch control mode	No	(1) Centralized control (2) Decentralized control
8	Intersection priority and implementation method	Traffic signal control system can be connected by hard wire or network interface	Traffic signal control system can be connected by hard wire or network interface
9	Vehicle depot system	No	Computer interlocking + signal centralized monitoring

With the acceleration of urbanization, the problem of urban traffic congestion is increasingly prominent. Although traditional urban rail transit and other transportation systems can alleviate the pressure of urban transportation, these methods have higher construction costs and longer infrastructure construction cycles. Therefore, ART has become a new hotspot in the transportation field due to its advantages such as low investment, short construction period and flexible operation organization.

The structure of Chapter 5 is shown in Fig. 5.2. This chapter mainly describes the hardware, technology systems, and pedestrian detection algorithms of ART.

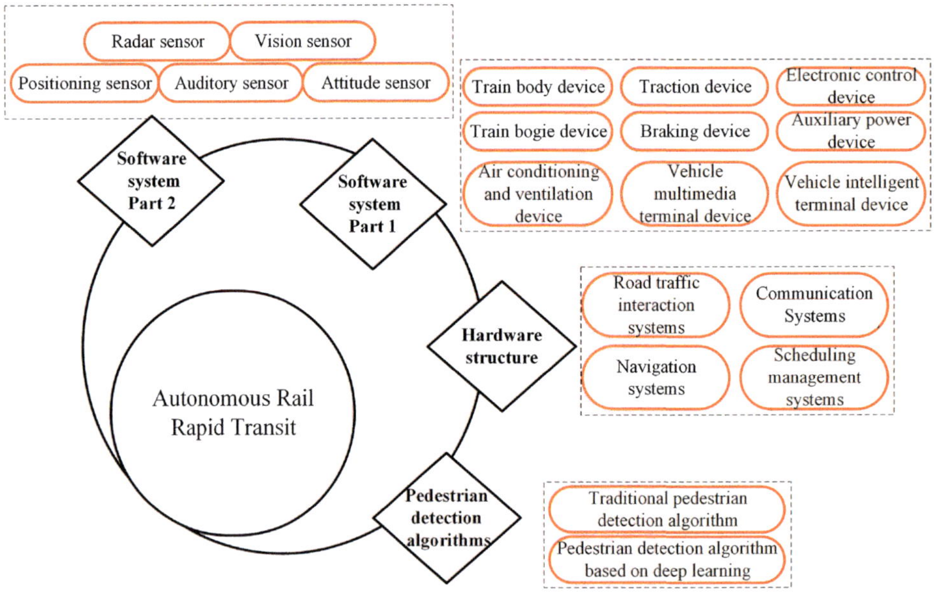

Figure 5.2
The overview diagram of Autonomous rail Rapid Transit (ART).

5.2 Main components of trams and ART

5.2.1 Structures of trams

ART is an advanced mode of transportation based on the traditional tram. It has the characteristics of environmental protection, resource-saving, smaller curve radius, and larger slope operation. The one-way operation of modern rail trains can meet the demand for passenger flow from 5,000 to 15,000 person-times per hour, and the design speed can reach 70–80 km/h [3]. In general, the noise in operation is lower than the urban background traffic. ART can provide customized services for urban public transportation systems with ground lanes as the main part.

ART adopts new low-floor multimodule articulation, flexible wheels, electric traction, and various braking methods such as resistance and hydraulics. ART consists of a vehicle, line, traction, and signal control system. ART is an updated version of the modern tram. The trams and ART have many similarities in the hardware structure. For example, they all have train bodies, wheels, traction devices, terminal connection equipment, electronic control systems, auxiliary systems, air conditioning, ventilation systems, multimedia equipment and so on. However, there is a big difference between the trams and ART in terms of steering hardware. ART adopts a nonbogie structure with independent wheel pairs, while conventional railway train bogies are used on trams. Since the manufacturing company does not disclose its hardware structure, the basic hardware structure of modern trams is introduced to help understand ART. The modern tram is composed of the following nine major hardware structure components.

5.2.1.1 Train body device

5.2.1.1.1 Train driver's cab

The modern tram body consists of two motorized modules with cabs. The intermediate modules are powered bogie modules and unpowered bogie modules. The cab is a separate module structure. The equipment of the modern tram's cab is shown as follows:

(1) Computer station and information display.
(2) Train remote control devices for track switch machines.
(3) The main controller is integrated on the console.
(4) Smart touch screen dashboard.
(5) Ergonomically comfortable seat with air conditioning, cup holder and locker.

5.2.1.1.2 Train passenger room

The passenger room of the modern tram is set as follows:

(1) Install appropriate windows on both sides of the passenger room. The window is fixed. An openable and closeable eyebrow window can be arranged on the upper part of the window.
(2) Arrange the appropriate number of passenger seats in the passenger room. The seat design meets ergonomic requirements.
(3) The number of passenger rooms is sufficient. The columns and handrails are firm and beautiful, and an appropriate number of rings can be added as needed.
(4) The passenger room should have sufficient lighting.
(5) A through-passage shall be provided at the joint of two adjacent modules. The through-passage should be sealed, fireproof, waterproof, insulated, and soundproof. The through-passage board should be wear-resistant, smooth, nonslip, and antipinch. The sealing material for the through-passage should have sufficient tensile strength, which is safe, reliable, and not easy to age.
(6) At least one wheelchair-specific location should be provided for each passenger room, and there should be a grab bar or fixture for wheelchair users.
(7) Set the passenger-oriented display in the passenger room.
(8) The passenger room is provided with a through-passage, and the width of the through-passage is not less than 1300 mm [4]. A pass gate is provided between the driver's cab and the passenger room.
(9) The vehicle has good fireproof performance.

(10) Both the internal noise level and the external noise level of the vehicle are in compliance with the standard.
(11) Set the vending machine in the passenger room.

5.2.1.1.3 Train interior decoration

The internal structures of the modern tram body are shown as follows [5]:

(1) The inner and outer wall panels of the train body structure should be laid between the bottom frame and the floor with low hygroscopicity and low expansion ratio. The performance of cold, heat, and sound insulation materials is stable.
(2) The inner wall panel should be made of flame-retardant material that is easy to clean and decorative. The floor should be abrasion resistant, nonslip, waterproof, and antistatic. The seat, decoration, etc. of the passenger room should be made of high flame-retardant materials.
(3) The interior of the vehicle is designed as a space-filling type, which can fully meet the customer's requirements. The flexible new concept interior is designed to readjust the space for future traffic changes. The interior of the vehicle can be selected according to different needs, such as making passengers feel comfortable and injecting new elements into the city.
(4) The roof of the train is equipped with an air conditioning outlet, fluorescent lamp belt, train horn, and other facilities. Passenger room seats are arranged horizontally and vertically. Seat structure and size meet the design requirements of human engineering. The fabrics are packaged with high-grade decorative fabrics. The ground inside the train is equipped with high-grade wear-resistant floor cloth, which is easy to clean and not easy to wear.

5.2.1.1.4 Train roof equipment layout

The equipment on the top of the train body mainly includes the following parts: traction inverter, brake resistor, cooling unit, auxiliary inverters, 24 V DC battery, pantographs, high voltage box, passenger room air conditioning unit, and driver air conditioning unit [4].

5.2.1.1.5 Train door

The modern tram is equipped with no less than four pairs of doors per side of each train, and the door height is not less than 1800 mm. The double-page net opening is 1300 mm, and the single-page net opening is 800 mm [6].

The opening and closing of the side door of the passenger room are generally controlled by electric power and powered by electric power or compressed air. And its transmission and control should be safe and reliable. The opening and closing of the side door are controlled by the driver. The side door of the passenger room should have zero-speed protection function. The electric interlocking device of the side door should be fully locked to ensure the locking of the door is correct during driving.

A single side door should have a system isolation function that can be removed from the gating system in the event of a fault. There should also be the function of manually unlocking the doors in the passenger room. At least one door on each side of the vehicle can be opened and closed using the key from the outside. When the side door is closed, it should have a buffering action and antipinch function.

The door should provide enough stroller or wheelchair space for passengers to get on and off. Each door has a separate door control and monitoring. The door has functions such as interlocking, switch, parking request and start. The door is equipped with the following diagnostic functions [6].

(1) Opening and closing methods: controlled by driver and flight attendant.
(2) Safety measures: it has an antipinch reopening and mechanical locking device.
(3) 5 km speed protection device: when the vehicle speed is above 5 km, the driver and attendant cannot open the door.

5.2.1.1.6 Train coupler

Automatic couplers or semiautomatic couplers can be provided at both ends of the vehicle. The foldable coupler is usually placed in the front cover of the train. The specific requirements for the train coupler settings are given as follows [7]:

(1) Foldable couplers are placed under each driver's cab.
(2) Under normal circumstances, the hook is hidden in the front cover to ensure the safety of pedestrians and the appearance of the vehicle.
(3) Open the front cover before opening the coupler, and the front cover should be easily opened.

5.2.1.1.7 Train articulated device

The articulated devices of the train are set as follows [8]:

(1) Articulated devices shall be installed between vehicle modules.
(2) The fixed articulated device composed of upper and lower single hinges should only allow two adjacent train bodies to move horizontally. The longitudinal shock absorber is suitable for the movable articulated device composed of the lower hinge and the horizontal control bar which restricts the lateral movement of two adjacent vehicles. The equipment is used to mitigate the impact and vibration of the train body.
(3) The articulated device shall have the function of relative motion. Its horizontal angle and longitudinal angle shall meet the operation requirements of the minimum curve radius and vertical curve radius.

(4) There are three forms of articulation between modules of the train body: elastic articulation, free articulation, and lower articulation. The upper part of elastic articulation is connected between two modules so that there is only a horizontal rotation movement between two adjacent modules. The upper part of free articulation is connected between two modules so that adjacent modules can be nodded, rotated horizontally, and have other movements. The lower articulation, which belongs to fixed articulation, is the main transmission mechanism of longitudinal force, in which both ends are cast steel mounting blocks and the middle is spherical bearing structure.

Fixed articulation and elastic articulation are combined to restrict the floating, sinking motion and side rolling motion between adjacent vehicle bodies. This setting allows only the relative head shaking of adjacent vehicles. Free articulation does not restrict the vertical and longitudinal translation between adjacent vehicles. Free articulation does not bear the vertical and longitudinal forces, but only bears the transverse forces. Free articulation is mainly used to limit the relative rolling between vehicles.

5.2.2 Train bogie device

(1) The frame is the skeleton of the bogie. Various devices and components of the bogie are combined into one piece through the frame. Since the frame is subjected to and transmits various loads and forces, the frame must have sufficient strength and rigidity. The power bogie is mainly composed of side beams, beams, and end beams.
The power bogie adopts the central bearing method. In addition, it also includes a two-stage elastic suspension device, motor drive, gear transmission, welded structural frame, shaft-mounted disc brake, side beam-mounted magnetic actuator, and other modules.
(2) The nonpowered bogie is mainly composed of a frame, a shaft bridge wheel set, a second-line elastic suspension device, a basic brake device, and a lateral hydraulic shock absorber. In addition to the traction motor and drive unit, the nonpowered bogie must be designed in a concave shape between the two wheels. To achieve a 100% low-floor structure for vehicles, the floor of the train needs to sink a part [3]. In addition, the axle structure of the nonpowered bogie adopts an intermediate concave axle structure, the wheel rotates relative to the axle and the axle does not rotate. This structure is called an independent wheel.
(3) Spring suspension device. Due to road irregularities, ballasts, rail surface defects, etc., the vehicle must have impact and vibration when driving on the track. To improve the stability of the vehicle operation and ensure the ride comfort, a spring suspension device must be provided in the running part of the vehicle.
(4) Wheel set. The wheel pair of the power bogie is composed of an axle, a wheel core, a tire wheel, and a driven gear. The axle part of the nonpowered bogie is an axle bridge

wheel set. The axle bridge is composed of an end shaft, a connecting shaft, a vertical plate, a cover plate, and the like. The welded parts are provided with a series of spring seats on both sides of the left and right connecting shafts.

(5) Axle box. The axle box and the bearing assembly are linked to the frame (or side frame) and the movable joint of the wheel pair so that the rolling of the wheel pair is converted into the translation of the body along the rail.

5.2.3 Traction device

The modern tram traction system includes the power receiving device, input circuit, inverter, braking resistor, and motor.

The power receiving devices of the traction device can work normally in the voltage range of DC 500 V to DC 900 V [6]. Its input circuit mainly includes the main circuit breaker, lightning arrester, line filter, preelectric circuit, line current, voltage sensor, etc. The main circuit of the traction device uses a traction VVVF main inverter. One inverter unit controls one or two traction motors. The form of traction motor is three-phase four-wire squirrel cage asynchronous motor, each power bogie is equipped with two traction motors. The design and test of the braking resistor should meet the requirements.

With the rapid development of the modern tram, higher and stricter requirements have been put forward for the safety of the modern tram. The traction motor is one of the main components of electric drive locomotive and vehicle. As the main motor driving the locomotive wheel shaft, its running performance directly affects the traction performance of the locomotive and its quality directly affects the normal operation of the locomotive.

The driving device consists of a reduction gearbox and a gear coupling. The details can be explained as follows:

(1) The reduction gearbox is composed of the box body, drive gear, driven gear, and two pairs of rolling bearings.

(2) The gear coupling is composed of gear sleeve, half coupling, locating baffle, spring, and fastening nut.

5.2.3.1 Braking device

The running route of the modern tram is not dedicated. There are level intersections and mixed-road surfaces. The vehicle is required to have a high braking speed reduction (2.5 m/s^2 or above) [6]. Braking and easing control is the most frequent in the course of rail train running, and it is often necessary to continuously brake and ease control every few seconds. To stop at any time, the short braking distance requires the braking system to have very high sensitivity and very short idle time. At the same time, the modern tram uses a composite braking method, so it must have good air-electric combined braking performance. Another characteristic is that there must be perfect backup braking measures. The modern tram is equipped with a braking control system. Once a fault occurs, it means that all the braking systems fail. Therefore, there must be perfect backup braking measures to ensure parking.

The braking system of the modern tram adopts microcomputer control direct electric air braking mode, which is composed of air braking and regenerative braking. It mainly includes braking command generation and transmission system, braking control system, regenerative braking device (belonging to traction system), basic braking device, antiskid system, wind source system, and pneumatic system accessories.

Regenerative braking is controlled by a traction inverter. Taking the advantage of the reversible characteristics of the motor, the motor converts the kinetic energy of the vehicle into electrical energy, which will be fed back to the power grid for use by other vehicles on the line or absorbed by the braking resistance of the vehicle. Therefore, the electric braking force can be generated only on the moving shaft. Electromagnetic braking is nonadhesive braking, which does not rely on the adhesion between the wheel and rail. It is a connector that transmits the torque of active side to the passive side. The connector can be freely combined, cut off and braked as required. It uses electromagnetic force to the brake, also known as electromagnetic clutch or brake. Electromagnetic force has the advantages of high response speed, simple structure, etc. It can also avoid wheel abrasions caused by emergency braking. The principle is that a magnetic field is formed when the excitation coil is energized. The armature on the brake shaft rotates to cut the magnetic force line and generates an eddy current. The eddy current in the armature interacts with the magnetic field to form the braking torque.

5.2.3.2 Electronic control device

The electric control system of a modern tram is composed of the control circuit, the main circuit, and the auxiliary circuit. The circuits are connected to realize automatic or indirect control and coordination of work by electromagnetic, electric, electromechanical transmission, etc. This keeps the driver safe and makes it easier to operate the vehicle.

The control circuit is composed of the low-voltage electrical appliance of the driver controller in the main circuit, the interlocking for the electromagnetic coils of the electrical appliances and the electrical appliances in the auxiliary circuit. The actions of the electrical appliances in the main circuit and the auxiliary circuit are controlled by the control circuit.

The main circuit is mainly composed of high-power electrical components, auxiliary measurement components, and protection components. It is used to complete the conversion of electrical energy and mechanical energy to generate traction and braking force.

The main circuit can be roughly divided into a traction circuit, an electric brake circuit, a protection circuit, and a measurement circuit.

The auxiliary circuit consists of an auxiliary power supply and various auxiliary electrical equipment. The auxiliary circuit can guarantee the main circuit can exert its power and achieve its performance.

5.2.3.3 Auxiliary power device

The modern electric tram's auxiliary power supply unit is composed of a battery pack, a low-voltage power supply device, and a static inverter. The advantages of auxiliary power systems are centralized power supply, small size, intelligent operation, and high reliability.

The low-voltage electrical appliances adopt various voltages such as DC 24 V, AC 380 V, AC 220 V, etc. The equipment power ranges from 5 to 16 kW [6]. To this end, the auxiliary power supply adopts new technologies such as the chopper voltage regulation circuit, full-bridge power conversion circuit, forced air cooling, microcomputer control, and protection.

5.2.3.4 Air conditioning and ventilation device

The air conditioning and ventilation systems of the modern tram are necessary for different climates throughout the year. The air conditioning and ventilation system is also an important part of the modern tram. The modern tram is separated by cabs and passenger rooms while meeting different needs of drivers and passengers [9].

The vehicle terminal equipment is responsible for the basic operation control tasks of the entire vehicle, and its control performance directly affects the running performance of the vehicle.

The vehicle terminal device is mainly composed of vehicle two-way wireless communication device, vehicle antenna, host computer, Global Positioning System (GPS) terminal, display unit, vehicle multimedia terminal and terminal device of signal priority system on the vehicle. In addition, public transportation card point of sale (POS) machines and coin boxes for tolls are included. The in-vehicle device receives the operation interval plan of the control center in real-time. The in-vehicle device also displays information such as the current tram position, the distance between the front and rear vehicle, the vehicle speed, the positioning status of way indicator, and the ballast in real-time. When the current rear distance and speed do not meet the set driving safety requirements, an alarm will be given.

5.2.3.5 Vehicle multimedia terminal device

The train multimedia terminal realizes the release of video information on the train. A mobile digital set-top box is installed on each vehicle to enable reception and playback based on mobile digital signals. The coordinated control of the mobile digital set-top box and the media player can be described as follows: When switching to mobile digital TV program playback content, the media player is closed; when switching to playing media player playback content, the mobile digital TV program is closed. The mobile digital TV is controlled by the GPS vehicle control terminal. The playback control of the mobile digital TV is controlled by the GPS vehicle terminal. A liquid crystal screen is installed on each train to realize the playback of synchronized video information [10].

5.2.3.6 Vehicle intelligent terminal device

The vehicle intelligent terminal subsystem device includes a GPS terminal and an antenna. The GPS terminal is composed of a GPS module, a speed direction sensor, a wireless communication module, a power module, and an interface device. A set of vehicle intelligent terminal subsystem device is provided in each cab at both ends of the train.

The vehicle intelligent terminal subsystem can accurately determine the position of the vehicle. Thereby the real-time position monitoring of the vehicle is realized. In addition, it can provide auxiliary services for operation scheduling and emergency handling (emergency repair and emergency command scheduling). At the same time, it can ensure that the trams conform to the traffic plan by controlling the travel interval of the trams. It also provides real-time vehicle timetable information for passengers waiting at the station. The vehicle intelligent terminal includes a vehicle host, a human-robot interface, and a supporting cable.

The functions of the vehicle intelligent terminal are given as follows: using GPS as the main positioning method, the auxiliary positioning module can be automatically switched when entering the blind zone; using General Packet Radio Service (GPRS) as the communication platform, the data (vehicle positioning, vehicle operation, vehicle violation, security, etc.) collected in real-time is transmitted to the control center server. In addition, the in-vehicle intelligent terminal simultaneously receives various messages and instructions transmitted by the control center server to implement a two-way communication function. The in-vehicle intelligent terminal supports voice and text mode intelligent reporting station. It also has functions such as service terms, violation prompts, voice prompts and electronic stop signs. The vehicle intelligent terminal receives and displays single item scheduling and cluster scheduling instruction information from the system platform. Real-time information communication coordination can be achieved when dangerous situations occur, such as overspeeding, stagnation, abnormal opening and closing, abnormal conditions, road congestion, accidents, faults, disputes, etc. [11].

5.2.3.7 Sensors structures of ART

The development of modern rail trains in hardware has reached a very high level. But rail trains also have their limitations. ART came into being to find a cross-era, but new, means of transportation. It can maintain the advantage of high capacity like rail trains without building proprietary rails, and it can also be a flexible urban vehicle such as a bus. It combines the advantages of modern rail trains and buses. ART subverts people's traditional cognition of urban traffic, and it provides a new choice for residents in large and medium-sized cities to travel. The biggest advantage of ART is that it can be driven along a paved virtual track. However, some equipment is required to realize traveling along a virtual track. The equipment mainly includes a large number of various sensors such as positioning sensors, laser radar, and other types of radar.

ART can adapt to a variety of road rights. Under semiproprietary road rights, ART can be manned assisted driving based on various sensors, image recognition, vehicle-to-ground communication, and steering-by-wire. ART can also be driven by the driver to track the ground virtual track. Under proprietary road rights, vertical and horizontal control of vehicles can be realized based on visual sensors, radar, laser sensors, etc., so as to achieve ATO with a driver.

(1) Semiproprietary road-right for ART

Based on the semiproprietary right-of-way, the technical route of ART is mainly realized by various sensors identifying the laid virtual track. The virtual track following control technology is realized by identifying various road virtual track lines using various sensors on the vehicle. The sensor transmits operational information to the train "brain" (central control unit).

The "brain" instruction ensures the train to achieve normal operations such as traction, braking, steering, etc. ART can accurately control the train to travel on the established "virtual track" to achieve intelligent operation. ART is guided by a virtual track, without requiring a physical track.

The basic principle of multisensor information fusion technology is just like the process of human brain comprehensive processing information. It combines various sensors with multilevel and multispace information complementation and optimization. Finally, it produces a consistent interpretation of the observation environment. Multisource data are fully utilized during this process. The ultimate goal of information fusion is realized based on the separated observation information obtained by each sensor. More useful information is obtained through the multilevel and multifaceted combination of information. It not only takes advantage of the synergy for multiple sensors but also comprehensively processes the data of other information sources to improve the intelligence for the entire sensor system.

A schematic diagram of sensor fusion is shown in Fig. 5.3. The hardware equipment of ART under semiproprietary road rights mainly includes positioning sensors, radar sensors, auditory sensors, visual sensors, and attitude sensors.

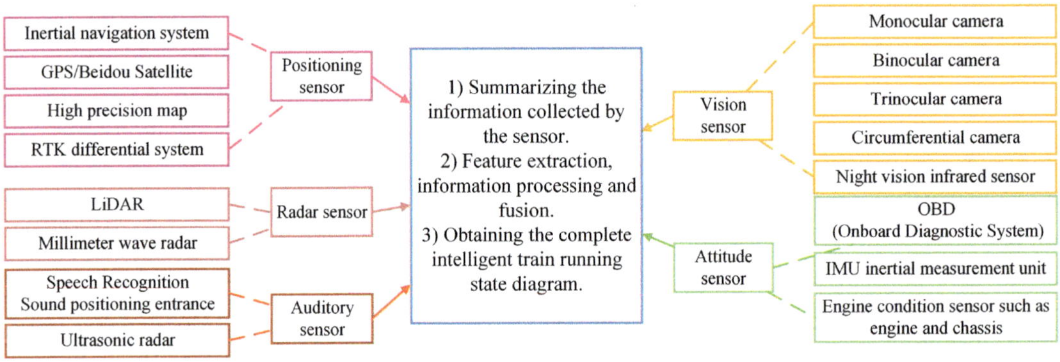

Figure 5.3
Schematic diagram of ART sensor fusion.

5.2.3.8 Proprietary road-right for ART

Based on proprietary road rights, ART has a development method that is very close to the unmanned driving of the train. The method realizes the fully automatic driving of ART mainly through multiple types of cameras, light detection and ranging (LiDAR), the Global Navigation Satellite System, ultrasonic sensors and infrared sensors. The positioning accuracy of the Global Navigation Satellite System [12] cannot reach the desired precision in centimeters, so other auxiliary means are needed. Among them, sensors play a very important role in improving train running accuracy through visual reconstruction. The details are shown as follows:

5.2.3.8.1 Camera

There are mainly four kinds of cameras according to the lens and the arrangement: monocular camera, binocular camera, trinocular camera, and circumferential camera.

(1) Monocular camera

The monocular camera module contains only one camera and one lens. The image of the camera is a perspective. The farther away the object is, the smaller the image. The same object at a distance may be described in just a few pixels. This property causes pixels in distant places to represent long distances. Therefore, for a monocular camera, the farther the object is, the lower the accuracy [13].

(2) Binocular camera

Due to the defect of the monocular camera in ranging, the binocular camera came into being. When two similar cameras shoot one object, the pixel offset of the same object in the imaging plane of the camera is obtained. The camera focal length, the actual distance of the two cameras and the distance of the object can be obtained by mathematical conversion with the information of the pixel offset. Although the binocular camera can obtain high precision ranging results and provide the ability of image segmentation, the field of vision of the lens is completely dependent on the lens, which is just like a monocular camera. The binocular ranging principle requires more installation position and distance of two lenses, which will bring troubles to camera calibration [14].

(3) Trinocular camera

Because both monocular and binocular cameras have some defects, the trinocular camera is the widely used camera scheme in unmanned systems. A trinocular camera is a combination of three monocular cameras with different focal lengths [15]. Depending on the focal length, each camera has a different range of perception. For the camera, the range of perception either loses sight or distance. The trinocular camera can make up for the problem of perception range. Therefore, it is widely used in the industry. Because the field of vision for the trinocular camera is different, the field of vision that each type for cameras can capture is also different. It can be divided into a wide field of vision camera, the main field of vision camera and narrow field of vision camera. The combination of the three fields of vision obtained can obtain better shooting results. Under such conditions, each camera can exert its greatest advantages.

Trinocular cameras also have some drawbacks. Firstly, calibrating three cameras at the same time can lead to a surge in workload. Secondly, the algorithm requirements are also very high because the software part needs to correlate the data of the three cameras.

(4) Circumferential camera

The lens of the circumferential camera is a fisheye lens and the mounting position is toward the ground. Some premium models have a "360-degree panoramic display" feature, which uses a circumferential camera. The circumferential camera is mounted to capture images on the front of the vehicle, under the rearview mirrors of the vehicle and at the rear of the vehicle. To obtain a sufficiently large field of view, the fisheye camera is costly with severe distortion of the image. Through the calibration value, the image projection transformation can be carried out to restore the image to the top view. After that, the images in the four directions were assembled. A top view of the train was placed in the middle of the four images to achieve the effect of looking down from the top of the train. The perception range of the circumferential camera is not large, and it is mainly used for obstacle detection within 5–10 m of the body and identification of the parking position line.

5.2.3.8.2 Light detection and ranging

LiDAR can complete the detection and measurement of light [16]. It is a system that combines laser, GPS and Inertial Measurement Unit (IMU). The LiDAR is used to obtain data and generate an accurate Digital Elevation Model (DEM). The combination of these three technologies can highly accurately locate the spot of the laser beam on the object, and the ranging accuracy can reach the centimeter level. The biggest advantages of laser radar are "precision" and "fast and efficient operation." It is a sensor for accurately obtaining 3D position information. Its function in the machine is equivalent to the human eye. It can determine the position, size, external appearance, and even material of the object.

The LiDAR system measures the distance and direction of each pixel in the 3D space to the transmitter, and create a real-world complete 3D model through the sensor. The basic way to operate a LiDAR system is to shoot a laser and measure the signal that the light bounces off the surface. The time required for the LiDAR module to receive the reflected signal provides a means of directly measuring the distance between the LiDAR system and the object. Additional information about the object, such as its rate or material composition, can also be determined by measuring certain characteristics of the reflected signal, such as induced Doppler shift. Finally, by manipulating the emitted light, many different points in the environment can be measured to create a complete 3D model.

5.2.3.8.3 Millimeter-wave radar

Millimeter-wave radar refers to a radar operating in the millimeter-wave band. Generally, the millimeter wave refers to an electromagnetic wave in the frequency domain of 30–300 GHz (wavelength of 1–10 mm) [17]. The wavelength of the millimeter wave is between the centimeter wave and the light wave, so the millimeter-wave radar has the advantages of microwave navigation and photoelectric navigation. Millimeter waves have a wide range of applications in 5G communications, satellite remote sensing, missile navigation, and electronic countermeasures. In recent years, related technologies such as circuit design and antenna technology have been increasingly developed and matured with the continuous improvement of the level of components. The application of millimeter-wave radar in railway trains has also been greatly developed.

Millimeter-wave radars can be classified into Long-Range Radar (LRR) and Short-Range Radar (SRR). Since the millimeter wave is weakly attenuated in the atmosphere, it is possible to detect a farther distance. Among them, LRR can achieve more than 200 m of sensing and detection [17]. The advantages of millimeter-wave radar make it occupy a large part in the vehicle anticollision sensor. According to IHS data, the millimeter-wave/microwave radar + camera accounts for 70% of the vehicle anticollision sensor. ART also uses the same millimeter-wave radar method.

Compared with ultrasonic radar, millimeter-wave radar has the characteristics of small size, lightweight, and high spatial resolution. Millimeter-wave radar has a strong ability to penetrate fog, smoke, and dust when compared with optical sensors such as infrared sensors, laser sensors, and cameras. Millimeter-wave radar has all-weather all-day characteristics. In addition, the antiinterference ability of millimeter-wave radar is better than other on-board sensors.

5.2.3.8.4 Ultrasonic radar

Ultrasonic radar is an extremely common sensor. It has a more popular name: reverse radar. It is a safety aid when the train is parking or reversing. The device can tell drivers about obstacles around them by sound or visual display. The reverse radar removes the trouble caused by the front, rear and left side visitation of the driver when parking, reversing and starting the vehicle. The reverse radar helps the driver clear the blind angle of vision and blurred vision.

Ultrasonic radar works by sending out ultrasonic waves through the ultrasonic transmitter and measuring the distance by the time difference between receiving and sending ultrasonic waves through the receiver. At present, the working frequency of commonly used probes for ultrasonic radar is 4 Hz, 48 MHz and 58 kHz [18]. Generally speaking, the higher the frequency is, the higher the sensitivity, but the smaller the horizontal and vertical detection angle. So the probe of 40 kHz is generally used. Ultrasonic radar is waterproof and dustproof, even if there is a small amount of sand shielding does not affect. The detection range is between 0.1 and 3 m, and the accuracy is high. So ultrasonic radar is very suitable for parking.

5.2.4 Key technologies in ART

The key technologies of ART can be summarized into eight parts [19–21].

The first part is the technique of trajectory following control. ART detects vehicle attitude, coordinates, and other information by installing inertial sensors or angle sensors on the vehicle. The use of such sensors can increase the coincidence rate between the rear wheels and the front wheels in the forward direction of the vehicle. In addition, it can also help reduce the steering "inner wheel difference" and reduce the influence of "blind angle of sight." It can also ensure the overall passage and the steering performance of ART. Sensor technology accurately controls the train's intelligent operation on a given "virtual trajectory." Therefore, ART replaced the traditional steel wheel rail with a rubber wheel instead of a special physical track.

The second part is the vehicle system integration technology. The modularization and design functions of each subsystem for the vehicle are planned one by one. In addition, the technology constructs the logic control function of the intelligent track express system in the road running process. To adapt to the needs of urban rail transit, modular integration of trains can be carried out to realize a combination of all trains.

The third part is the intelligent driving. ART adopts artificial intelligence technology, including high-precision positioning. ART enables driver assistance through fast communication, while also achieving a safer driving experience.

The fourth part is the technology of active security. Other vehicles likely invade the running route in the course of train operation, so ART must have corresponding security measures. To ensure the safety of the train body itself, the technology uses image recognition, image dynamic mosaic, and sensor fusion technology. It realizes the constraint of ART on the vehicle itself without physical track. It is also a kind of protection for the surrounding intrusions. If ART deviates from the virtual track without authorization or an external object invades the vehicle limit, the driver can use the blocking power, emergency braking, and other technical means to avoid the accident.

The fifth part is traction brake cooperative control technology. The power of the train is based on the permanent magnet drive, and coordinated control is realized by distributed dynamic cooperative control technology. This technology enables ART to have a stronger climbing ability, which is much higher than that of traditional trams.

The sixth part is the technology of power supply without a network. The train is powered by batteries. It supports multiple modes of power supply. Shorter charging times allow longer driving range.

The seventh part is the multitasking train control and service network (TCSN) control technology. It is the world's most advanced network technology for vehicles. The use of broadband technology enables ART to carry both functions such as control and monitoring of vehicle equipment.

The eighth part is the vehicle-ground-person signal-coupling technology. In the case of limited road resources, technology can allow trains to enjoy the right-of-way at intersections and achieve the purpose of fast transportation.

In this chapter, the software systems of ART are introduced in detail. The software systems of ART consist of road traffic interaction systems [22], navigation systems [23], communication systems [24], and scheduling management systems [25].

Among them, road traffic interaction systems involve the allocation of two road rights when the streetlight intersection is involved. It requires the use of pedestrian detection algorithms for reasonable obstacle avoidance and special crossing signal timing schemes to adapt ART to the normal operating rules of road traffic. Navigation systems involve the automatic monitoring of ART. The positioning systems can ensure safe, punctual, reliable and efficient operation of ART.

At the same time, it completes the train scheduling and emergency handling of the train. Communication systems involve the communication between the vehicles of ART and the communication between the vehicle and the control center. Scheduling management systems involve the following points. First, manage the effectiveness of the vehicle schedule. Second, detect the location information of the other vehicles. Third, monitor the running status and line conditions of the train throughout the jurisdiction. It updates the real-time station passenger information system to finally complete the communication with the vehicle. Scheduling management systems are equivalent to the integration of interactive systems, navigation and positioning systems, and communication systems. Scheduling management systems belong to the cross-cutting field. The scheduling management system is the necessary guarantee to ensure the orderly and efficient operation of urban traffic.

5.2.4.1 Road traffic interaction

The main contents of the road traffic interactive systems are given as follows: (1) formulate the principle and plan for the timing of road traffic signals at intersections; (2) according to the driving characteristics of ART, the parameter calculation equation of the timing scheme is determined; and (3) select a typical intersection to perform example verification. It combines the principles of the compilation of some urban mixed-road trams. ART operation diagram should meet the specific passenger flow demand and maintain an appropriate departure interval. At the same time, it is also necessary to consider the uncertainty factors when passing through the intersection. Therefore, it is necessary to consider a relatively flexible timetable adjustment algorithm to improve transportation efficiency when preparing the timetable. The minimum green time for ART can be calculated using traditional signal control methods. Finally, the green light time of each phase and the green light scheme of the adjacent intersection are determined [22].

The signal timing scheme of the road traffic interactive systems mainly considers the interval running time, stop time, crossing time, intersection waiting time, etc. Interval running time is the time standard for ART to run between two adjacent stations. The stop time of the middle station of ART depends on the passenger flow, door width, door opening/closing time, and driving organization. The stop time of the first and last stations is usually large, which is determined by way of driving and the way of returning. The crossing time includes the time when ART waits for the authorization signal on the road, the time when ART passes the intersection after receiving the release signal and the time when the tail of ART passes the intersection completely. The road traffic signal control system is modeled and solved by keeping the original signal release phase unchanged and green wave optimization. Finally, the goal of obtaining optimal communication time and reducing the impact on the traffic flow of the main road is achieved.

Under the semiexclusive right-of-way, ART shares the right-of-way with social vehicles. Traffic efficiency is mainly controlled by the permission of crossing. If the train stops frequently before the intersection and increases the number of starts and stops, the running efficiency will be reduced. The priority control system is configured at the intersection. The appropriate priority control strategy is adopted to improve the efficiency of crossing and ensure the efficiency of running on time. Intersection control mode usually has absolute priority and relative priority in two ways.

Absolute priority control ensures the trolley's priority at the intersection. The relative priority approach may not guarantee the priority of ART crossing the intersection. The train sends the priority request information to the ground wired transmission system through the vehicle-ground wireless communication system. The intersection signal priority controller obtains the vehicle priority request information through the ground wired transmission system and sends the corresponding vehicle priority request information to the road traffic signal control machine for priority processing. The on-board control unit sends real-time information such as the position, speed and driving direction of the current vehicle to the intersection signal priority controller. The traffic signal priority controller sends a priority request to the traffic signal controller after information processing. Finally, the traffic signal controller implements a priority strategy.

There is no way to avoid encountering obstacles in the train road interaction system. In this situation, a suitable pedestrian detection algorithm will help ART run efficiently. In addition, the effective detection of pedestrians can not only realize the intelligent driving of trains but also be an important basis for the identification of train accidents. The commonly used pedestrian detection algorithm is machine vision detection based on deep learning.

5.2.4.2 Navigation

ART train positioning and control center is the basic guarantee facility to ensure safe, punctual, reliable and efficient operation of ART. The control center needs to carry out automatic monitoring on the full line of ART to complete the train operation scheduling and emergency handling. Therefore, the positioning of the train is necessary [23].

There are many ways of positioning trams in modern times. At present, the combined positioning method of "satellite positioning + beacon" is widely adopted in China. Because the ART system has no physical track, the positioning equipment installed in the traditional urban rail transit system cannot be applied. Positioning errors will appear when just using satellite positioning. The train will be lost when it passes under the viaduct or underground. Therefore, the positioning of ART adopts the combined positioning mode of "satellite global positioning technology + electronic map," which is mainly based on satellite positioning. It corrects the train position through train speed, line database, and other information.

Satellite positioning terminal equipment is installed on the train. The current position of the train is obtained by differential calculations. The train position is corrected according to the current train speed and line data. The exact train position is sent to the operation scheduling management system of the control center through the vehicle-to-ground wireless communication system. After processing, the location of the train is displayed on the big screen in the dispatching hall.

GPS is the main positioning method. When entering the blind zone, the auxiliary positioning module can be automatically switched. With GPRS as the communication platform, the data collected in real-time (vehicle positioning, vehicle operation, vehicle violation, security, etc.) is transmitted to the control center server. At the same time, the receiving control center server transmits various messages and instructions to realize a two-way communication function. It supports voice and text (LED or LCD) mode intelligent reporting station. In addition, it has functions such as service term, violation prompt, voice prompt, electronic stop sign, etc. The single-schedule scheduling and cluster scheduling instruction information are displayed on the system platform if there is over-speed, stagnant, abnormally open/close, abnormal conditions, road blocking, accidents, faults, disputes, etc.

5.2.4.3 Communication

Train Communication Network (TCN) is a distributed network control system using the computer network as its core. To achieve fast, accurate and reliable communication over the train communication network, the TCN Real-Time Protocols (RTP) need to be followed between the train applications [26]. The RTP can be applied to a variety of train bus communications with functional services, including multifunction vehicle buses and stranded train buses. The TCN system is mainly used as the control, detection and diagnosis system of railway locomotive and vehicle. The communication of ART also involves the communication between trains and vehicles as well as the communication between vehicles and the control center. The overall structure of the vehicle network includes a two-level bus hierarchy, which is combined with Wire Train Bus (WTB) and Multifunction Vehicle Bus (MVB) [27]. The WTB can be used to connect various dynamically marshaled vehicles. The MVB can be used to connect various devices inside a vehicle or a group of vehicle units. The development of the communication system will accelerate the implementation of ART [24].

In general, the train bus or vehicle bus can transmit three types of data, including process data, message data, and monitoring data [28].

(1) Process data

Process data is broadcast data transmitted from the distribution device to multiple user devices. Each process data should contain the following sections: the source buffer for the publisher device and the host buffer for each user device. The newly received data will overwrite the previous old data. Strictly speaking, process data can also be transmitted as contingent data as needed. However, the TCN standard does not support this type of delivery due to the low reliability of the service.

(2) Message data

Message data is target-oriented data sent from the source device to the target device or to all devices on the same bus. Each device needs to provide a pair of queues for all message data it receives or sends. As long as there is space in the queue, the received message data needs to be plugged into the input queue of the target device. Message data is transmitted only when needed. Message data can also be sent periodically. But it is not supported by the TCN standard because of its low reliability.

(3) Monitoring data

There are many types of monitoring data. Monitoring data includes data for monitoring the state of the device, data for detecting quiescent devices, data for the transfer of bus sovereignty, data for the initial operation of the bus, and so on. Some monitoring data is transmitted periodically, while others are transmitted on demand.

At present, the communication methods used in the train system are mainly divided into parallel communication and serial communication. The main communication protocols include Transmission Control Protocol (TCP) communication and User Datagram Protocol (UDP) communication.

5.2.4.4 Serial communication

There are two ways of communication which include parallel communication and serial communication. Serial communication is widely used in the multicomputer system and modern measurement and control system.

(1) Parallel communication

Parallel communication refers to the simultaneous transmission of data bits. One of its advantages is that the transmission speed is fast, the disadvantage is that the transmission line is complex. The communication line cost is high. The parallel transmission is suitable for the near distance and high-speed occasion.

(2) Serial communication

In serial communication, the data are transmitted bit by bit in time-sharing order (e.g., first low, then high). It has the advantages of a few transmission lines and low cost of transmission channels. So serial communication is suitable for long-distance data transmission. Low transmission speed is its disadvantage.

5.2.4.5 TCP communication

TCP is a reliable transmission layer communication protocol [29]. When TCP protocol is used for communication, the connection between the two communication parties should be established at first. Once the connection is established, communication can take place. TCP provides a data confirmation and retransmission mechanism, which ensures that the sent data can reach the communicating party.

However, two programs on the network exchange data through a two-way communication connection, each end of which is called a socket. The socket is the cornerstone of communication and the basic operational unit of network communication that supports TCP/IP protocol. It is an abstract representation of an endpoint during network communication. It contains five types of information necessary for network communication: the protocol used to connect, the IP address of the localhost, the protocol port of the local process, the IP address of the remote host, and the protocol port of the remote process.

The socket is the intermediate software abstraction layer between the application layer and the TCP/IP protocol family. It is a set of interfaces. In design mode, the socket is a facade mode that hides the complex TCP/IP protocol family behind the socket interface.

TCP/IP sockets provide the following three types of sockets.

(1) Sock-Stream

Sock-Stream provides a reliable connection-oriented TCP data transfer service. Data is sent without error or repetition and received in the order in which it was sent. Sock-Stream is equipped with flow control to avoid data overrunning. Data in Sock-Stream is treated as a byte stream with no length limit. The File Transfer Protocol (FTP) uses streaming sockets.

(2) Sock-Dram

Sock-Dram provides a connectionless service UDP. Packets are sent as individual packets, with no guarantee of error-free. Data may be lost or duplicated, and the receiving sequence may be out of order. Network File Systems (NFS) use datagram sockets.

(3) Sock-Raw

The Sock-Raw interface allows direct access to lower-level protocols such as IP and ICMP. It is often used to verify a new protocol implementation or access a new device configured in an existing service.

The server initializes the socket, then binds to the port. The server listens to the port, and waits for the client to connect. At this point, a client initializes a socket and connects it to the server. If the connection between the client and the server is successfully established, the client will send the data request. The server receives the request and processes the request. The server then sends the response data to the client. The client reads the data. Finally, the connection is closed, and one interaction ends.

5.2.4.6 UDP communication

The UDP is a connectionless transport protocol [30]. When using the UDP for communication, there is no need to establish a connection. Data can be directly sent to an IP address in UDP. But there is no guarantee that the other party can receive it. For the UDP-based connectionless socket programming, the concept of server-side and client-side is not particularly strict. The server-side can be called the receiving side, and the client-side is the sending side for sending data.

The server initializes the socket and then binds to the port. If a client initializes a socket at this point, the client sends a data request. The server receives the request, processes it, and sends the response data to the client. The client reads the data and the interaction ends.

After binding, the server does not need to listen to the ports or call to wait for the connection to block. The client does not need to connect with the server in advance.

5.2.4.7 Scheduling management

The traffic dispatching system manages the vehicle according to the effective timetable. The system detects the position of the train through the vehicle message. The system monitors the running condition of the train and the line condition in the entire jurisdiction. It updates the passenger information system of the station in real-time and communicates with the vehicle. The system is connected to the operational control and traffic scheduling system via a trunk wireless device for real-time updates of information (the communication between the operational control center and the intelligent information system in its design can also be done using fiber optics) [25].

ART can provide services in the city orderly and efficiently. ART need not only advanced hardware equipment but also a suitable scheduling management system. Only in this way can ART plays the greatest effect. The scheduling management system combines the characteristics of interactive systems, navigation and positioning system, and communication system. In the dispatching management system, it is not only necessary to apply the related knowledge of the unified research, but also to develop a set of the appropriate management system according to the actual situation of different intersections.

5.2.5 Pedestrian detection algorithms for ART

In recent years, the rapid development of artificial intelligence has led researchers to focus on its application in the field of rail transit. The effective detection of pedestrians cannot only realize the intelligent driving of ART, but also act as an important basis for train accident identification. However, simple video monitoring is far from meeting the requirements. People hope to use intelligent methods to understand the behavior of pedestrians. Use video analysis to ensure that the train can be braked in advance when facing pedestrians on the road during automatic driving. The human motion recognition method based on computer vision requires no sensors and no contact. This breakthrough development has attracted more and more attention from scholars [31].

Pedestrian detection is not only the key to intelligent driving but also the research focus and difficulty of computer vision. In general, determine whether a pedestrian appears in the input image or video sequence is the core to implement pedestrian detection technology. After completing the judgment of the event of "whether there is a pedestrian," the pedestrian position is locked, and the detection is finally completed. This technology is widely used in many areas of intelligent transportation. For example, video surveillance [32,33], human behavior analysis [34], robot development [35], driverless vehicle assistance systems [36–40], etc.

However, there are a lot of factors that affect the recognition performance of pedestrian detection algorithms in the process of vehicle driving. For example, the scene of the light flickering, the change of the climate is uncertain, people's clothing variety, the change of human body posture, etc. Of course, the robustness of the system is particularly important for the application of pedestrian detection technology. Pedestrian detection algorithms are usually required to operate automatically and continuously, so they cannot be so sensitive to factors such as noise, light, and weather. Meanwhile, the processing speed of the system is also very critical for real-time pedestrian detection and tracking system. To obtain good pedestrian recognition results, it is often necessary to solve the two core problems in pedestrian detection technology: feature extraction, classifier, and positioning. The schematic diagram is shown in Fig. 5.4.

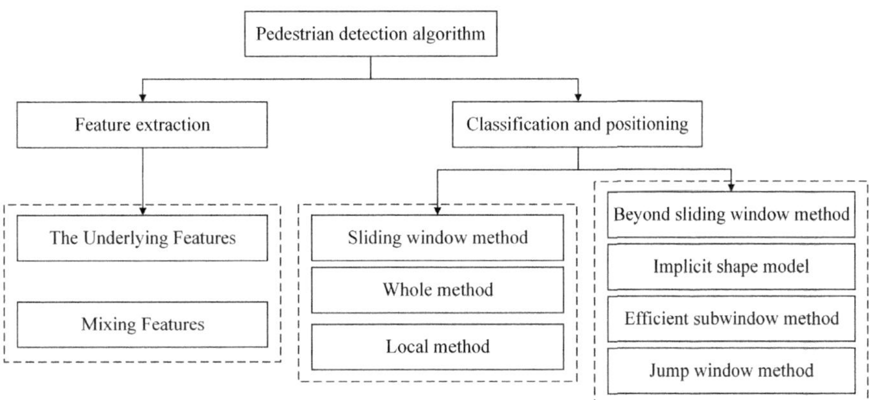

Figure 5.4
The core of the pedestrian detection algorithm.

The pedestrian detection algorithm is widely used in the field of computer applications, such as vehicle assisted driving systems, intelligent video monitoring, robot, aerial image, human-robot interaction system, motion analysis, etc. The pedestrian detection algorithm has been greatly developed in recent years based on computer vision. More expensive devices like laser radar will be replaced by cheap cameras. It means that driverless technology can be landed at a lower cost, which makes the technology more available. Compared with radar data, visual information can give the vehicle a wider field of vision, and unmanned driving will be more reliable. Section 5.4 introduces the pedestrian detection algorithm based on deep learning. At the same time, Section 5.4 also provides a complete example method for pedestrian detection.

5.2.6 Traditional pedestrian detection

Traditional detection algorithms generally use the frame of the sliding window. It mainly includes three steps: the first step is extracting the candidate area of the proposal; the second step is extracting the relevant visual features of the candidate area; the third step is using the classifier to identify. In the early 20th century, many pedestrian detection algorithms were tried based on machine learning, which was summarized in Table 5.2.

The pedestrian detection method of the "Haar wavelet feature + Support Vector Machine (SVM)" proposed by Oren et al. in 1997 may be the earliest target detection method [41]. Based on the Haar wavelet feature, Oren used the SVM as a classifier and then used the Bootstrap method to collect pedestrian negative samples. In 2004, Viola et al. proposed the "Haar + AdaBoost" detection algorithm in combination with the integrated method [42]. Viola et al. used the AdaBoost method as an aid to discriminate and then classify the Haar

Table 5.2: Traditional pedestrian detection algorithms.

Year	Typical method	Specific contents
1997	Haar wavelet template + SVM [41]	Based on the Haar wavelet transform, SVM was used as a classifier, and then the bootstrap method was used to collect negative pedestrian samples
2004	Haar + AdaBoost [42]	The AdaBoost method was used to distinguish and classify the Haar features of the image
2005	Histogram of Oriented Gradient, HOG [43]	Using a linear SVM to classify images
2009	Deformable Part Mode, DPM [44]	The object was described by the relationship between different parts

features of the image. The method has a positive effect on facial recognition. Dalal et al. proposed the pedestrian detection algorithm of the Histogram of Oriented Gradient (HOG) [43].

To classify HOG features, Dalal et al. used a linear SVM to classify images. The HOG operator and VJ profile features were well tested using data from the MIT pedestrian database and the INRIA database. In subsequent research and development, some new human body recognition features had a certain effect on the performance of the pedestrian detection algorithm. In 2009, Felzenszwalb et al. proposed the target detection algorithm for Deformable Part Mode (DPM) [44]. The algorithm took into account the internal structure of the target pedestrian and regarded the pedestrian as a plurality of components. The object was described by the relationship between different parts, and the nonrigid body features were well expressed. The method is applied in the field of pedestrian detection. Pedestrian diversity can be well recognized, thereby improving detection accuracy. The DPM algorithm achieves good results in target detection, but it also has certain drawbacks. The DPM algorithm is complex and the detection efficiency is not high. In addition, the artificially designed pedestrian features are difficult to adapt to the differences in illumination and the complex changes caused by occlusion.

5.2.7 Smart pedestrian detection based on deep learning

Deep Learning is also known as Deep Neural Networks (DNN). Deep learning is a very important branch of machine learning, which has made huge progress in recent years. Deep learning operates mainly by imitating the working mode of the human brain neural network, which is developed from the artificial neural network [45].

5.2.7.1 Convolutional neural networks

Convolutional Neural Networks (CNN) is a deep neural network specially used for processing network-like structural data. The CNN is widely used in a lot of fields, such as natural language processing and image recognition processing [46]. The CNN was proposed by LeCun in 1988 [47]. It is a multilayer supervised learning neural network, which mainly consists of the convolutional layer, pooling layer, and full connection layer.

5.2.7.1.1 Convolution layer

Simply put, the convolutional layer is a key part of neural network construction. Most of the computational tasks of the CNN network model are undertaken by the convolutional layer. In addition, the convolution layer can view the set of multiple filters. Linear product summation is the main process of the convolution operation. Therefore, the convolution operation on the whole image can be regarded as multiplying the original image data of different filtering windows and the filtering matrix of filtering element by element.

Then they are added together. After convolution operation, the original signal features can be enhanced. The noise is reduced and local features are extracted.

Convolutional layer networks typically take the form of filter sliding windows when dealing with high dimensional input data similar to images. Each neuron is connected to a local area of the input data, and the input data is within the filtering window. Since the image data has some local correlation, the local connection characteristics of the convolutional layer can be combined with the local correlation of the image data to extract the local features better. In the convolutional layer, there is an entire set of "weight learnable convolution kernel" filters. The feature map is input to the convolution kernel filter and convoluted with each kernel. The result is processed by the activation function, and the feature map is finally output.

5.2.7.1.2 Pooling layer

Because the image has a local correlation, the pooling layer makes it easy to subsample the image. Firstly, the aggregate statistical operation of the output of the network at the location is replaced by the overall statistical feature of the adjacent output of a certain location. Then the maximum value statistics or the mean statistical strategies are used for the pooling operation. The corresponding pooling layers are also referred to as the mean pooling layer and the maximum pooling layer, respectively. Some pooling layers use the L2 norm strategy or a weighted average statistical strategy based on distance from the center pixel.

However, no matter what pooling strategy is selected, the translational invariance of the pooling layer keeps the input approximately constant when the input data is slightly translated. Translation invariance is a very good property when considering whether a feature appears (wherever it appears). The static nature of the image (that is, the useful features in one area of the image are likely to exist in another area) can achieve the purpose of obtaining the specific position of the confusing features. In addition, the operation of the pooling layer will compress the spatial size of the data body, resulting in the loss of some characteristic information. However, reducing the number of network parameters will reduce the computational resource cost. It is also possible to effectively control overfitting.

5.2.7.1.3 Full connection layer

The full connection layer has a connection structure similar to the Back Propagation Neural Network (BPNN). Although each neural unit in the same layer is independent of each other, it is connected to all units in the previous layer. And the activation value is typically output via softmax, which is the feature extracted from the convolutional neural network. The convolutional and pooled layers on the CNN are usually in the form of alternating connections. In addition, the convolution layer usually operates combination extraction after extracting features. Finally, through the multilayer feature extraction and recombination, it is passed to the full-layer layer to output the description features corresponding to the input deep content.

5.2.7.2 Feature extraction of deep learning

Deep learning can be seen as a large parameter-containing function whose parameters can be obtained by a similar algorithm of gradient descent. To achieve the approximation of complex features by learning deep nonlinear network structures, the following steps are required. Firstly, the network structure with multiple hidden layers is constructed. Then the whole network is trained by a large number of training data. Finally, the deep learning features are extracted by the deep learning network structure. From a mathematical point of view, deep feature performance is far superior to features extracted from shallow models. Therefore, the classification and recognition of features extracted through deep levels are also very effective.

Throughout the deep learning process, the purpose of feature extraction is generally achieved by training a depth model of multiple hidden layers. Compared with the traditional shallow learning and artificial design features, deep learning emphasizes the depth of the model. For example, the AlexNet network has eight layers, the VGG network has 16 layers, and the GoogleNet network has 22 layers [46]. The deep learning network carries out the operation of feature transformation extraction and reorganization layer by layer. The deep learning network maps the pixel features of the image in the original space to a new feature space, so as to make the classification and detection more accurate.

Compared with traditional methods of artificially designing features, deep learning is combined with features learned by big data. And it can describe the intrinsic information of pedestrian data more accurately. It can better solve the difficulty of limiting the development of ART pedestrian detection algorithms. Therefore, deep learning can extract features more stably and perform intelligent pedestrian detection compared with the manually set features.

Existing pedestrian detection algorithms can be directly divided into the traditional detection algorithm and the detection algorithm based on deep learning [48].

In the intelligent driving system, pedestrian detection based on computer vision is an important research branch of human motion analysis based on computer-human vision. Its main research goal is to detect pedestrians under the motion camera quickly and accurately. After the rapid development of deep learning target detection algorithms, people begin to apply this method to the field of intelligent driving pedestrian detection. Pedestrian detection is essentially a dichotomy of image processing: "Does anyone exist?" If someone exists, the output detection target is the confidence of the pedestrian and the location of the pedestrian goal.

Table 5.3: Pedestrian detection algorithms based on deep learning.

Year	Typical method	Specific contents
2014	Regions with CNN features (R−CNN) [49]	Selective Search + feature extraction and classification
2015	Spatial pyramid pooling, SPP [50]	Solve the processing problem of different scale feature maps
2015	SSP-net-ROI pooling [51]	Feature extraction and classification recognition are integrated into a network framework
2015	Region Proposal Networks, RPN [52]	Sharing the parameters of the two network core convolution layers for RPN and Fast R−CNN
2016	YOLO (You Only Look Once) [53]	Using a single-channel end-to-end convolutional neural network
2016	Single Shot Multibox Detector [54]	Adopting an anchor mechanism similar to RPN
2016	Region-Based Fully Convolutional Networks, R−FCN [55]	The faster R−CNN's last full connection layer is replaced by a position-sensitive convolutional network
2017	Anchor + YOLOv2 [56]	Optimize YOLO + combine anchor mechanism

Pedestrian detection algorithms based on deep learning in recent years are summarized in Table 5.3.

At present, pedestrian detection algorithms based on deep learning are generally divided into deep learning pedestrian detection algorithms based on region proposal and end-to-end deep learning pedestrian detection algorithms without region proposal.

5.2.7.3 Detection algorithm based on region proposal

In 2014, Girshick et al. proposed a regional algorithm with CNN features, referred to as the R−CNN [49]. The method firstly uses the selective search for regional proposals and then uses CNN for the feature extraction and classification. This method also proves the validity of the feature extracted by convolutional neural networks. R−CNN is the pioneering of neural networks based on regional proposals. R−CNN is the first truly industrial-grade solution. But the R−CNN algorithm also has some drawbacks. The training process of the R−CNN algorithm is too complicated. The training and test calculations of R−CNN are repeated. The space and time of training are too high. In addition, the whole connection layer of CNN requires the input image having a specific size. Most of the actual picture sizes are different, especially the Regions of Interests (ROI) obtained by selective search need to be unified after cutting. However, the operation of clipping will result in incomplete target, and deformation will result in serious deformation of the target.

The SPP-net algorithm proposed by He et al. in 2015 mainly focused on Spatial Pyramid Pooling (SPP) for feature maps [50]. This method replaces the operations of the clipping/deformation of the original image. This method removes the fixed-size constraint and solves the problem of how to deal with the feature maps of different scales. In 2015, Girshick proposed a single-layer SPP based on the SSP-net-inductive region pooling (ROI pooling) [51]. This method integrates feature extraction, classification and recognition into a network framework. This method proposes a Fast Regions with Convolutional Neural Network features (Fast R–CNN), which shares the feature extraction through CNN. This method greatly reduces the complexity of the training process. In addition, the detection speed and accuracy of the model are improved again. However, due to the inefficiency of selective search to generate the proposal area, the detection speed is still not fast enough.

In 2015, Ren et al. realized feature extraction through further sharing of CNN and put forward Region Proposal Networks (RPN) to generate ROIs [52]. This method is combined with the Fast R–CNN to design Faster Regions with Convolutional Neural Network features (Faster R–CNN). This method can obtain a unified detection network. By sharing the parameters of the two network core convolution layers of RPN and Fast R–CNN, further acceleration is achieved. The end-to-end detection mode is truly realized. However, since the middle part of the ROI pooling is composed of a fully connected layer, the feature layer after the previous ROI pooling is mapped into two parts. One is object classification and the other is coordinate regression.

However, in the subsequent research, more and more base CNNs have proved that not all links are required, such as GoogleNet [57], ResNet [58], and other full-convolution networks. The effect is better and can be adapt to different scales of pictures. In 2016, Dai et al. proposed the algorithm of Region-based Fully Convolutional Networks (R–FCN) [55]. This method replaces the final fully connected layer of the Faster R–CNN with a location-sensitive convolutional network, allowing all calculations to be shared. This method of proposal greatly improves the accuracy of pedestrian detection. But the number of network layers is too big to make the detection speed fast. And it cannot achieve vehicle detection.

The neural network based on region proposal is a target detection algorithm that combines region proposal and convolution neural network. In the algorithm, Region of Interest (ROI) is usually obtained by a selection search or regional proposal network. After that, the CNN is used to classify each proposal region to obtain classification category and confidence. At present, the main regional proposed neural network algorithms mainly include the R–CNN, SPP-net, Fast R–CNN, Faster R–CNN, R–FCN, etc.

5.2.7.4 End-to-end detection algorithm based on deep learning

In 2016, Redmon et al. proposed the real-time target detection network: You Only Look Once (YOLO) [53]. The method uses a single-channel end-to-end convolutional neural network, and the object detection problems are solved as a regression problem. Finally, the original image is input. Once detects, the output results of the object category, corresponding confidence probability and location information are obtained. Since the training and inspection of the YOLO are all run in single-channel network architecture, the detection algorithms that differ from regional proposals seek regional proposal processes clearly. This operation greatly reduces the training complexity and greatly improves the detection speed. However, this method also has some drawbacks. For instance, the YOLO forces the original image into fixed-size regions directly. So class prediction on the mesh region is difficult to accurately identify multiple objects falling within the same region.

Liu et al. proposed a Single Shot multibox Detector (SSD) target detection network [54]. The method adopts an anchor mechanism similar to RPN, which is recognized on the feature map of different levels to cover more scope. The method predicts the category prediction probability based on the anchor box to achieve the purpose of trying to improve the detection accuracy of small targets. The SSD is a master of the single-channel network target detection model, which is close to the recognition accuracy of the regional proposal network. The SSD is faster than the detection network based on the regional proposal with an order of magnitude. In 2016, Redmon et al. carried out a series of optimizations to the YOLO and proposed the YOLOv2 algorithm in combination with the anchor mechanism, which greatly improved the efficiency of network execution [56].

The end-to-end deep learning pedestrian detection method uses a single-channel network architecture without a regional proposal. The detection method unifies pedestrian positioning and pedestrian identification into a single network. The method outputs the position information of the pedestrian and the confidence of the pedestrian at one time. At present, the commonly used methods are the YOLO, SSD, YOLOv2, etc.

5.2.7.5 ART pedestrian detection application

In this section, a set of pedestrian detection methods for ART is introduced. The intelligent algorithm includes four parts: pedestrian keyframe extraction, pedestrian contour extraction, behavior characteristic parameter extraction, and BP neural network classification prediction. Pedestrian posture data are obtained from the Weiziman database.

5.2.7.6 Pedestrian keyframe extraction HOG + SVM

5.2.7.6.1 HOG feature extraction

At present, HOG is widely used in pedestrian detection of the still image. The pedestrian detection method adopted in this chapter is the HOG feature + SVM classifier method.

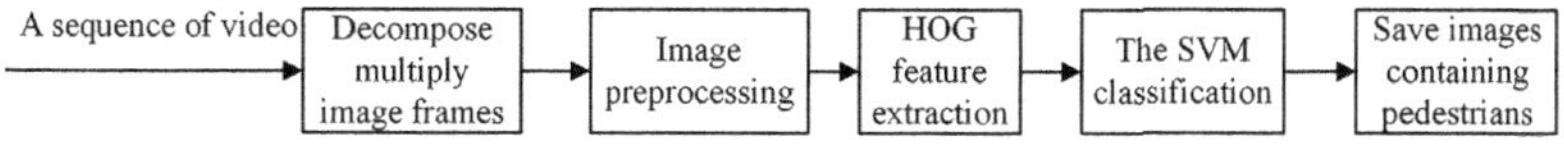

Figure 5.5
The flow chart of Histogram of Oriented Gradient (HOG) feature + Support Vector Machine (SVM) classification.

A large number of experiments have proved that this method has good performance in time, speed and effect. Fig. 5.5 is the flow chart of the method in this section.

HOG is a feature descriptor used for object detection, which is widely used in the field of computer vision and image processing [59]. This feature is obtained by calculating and counting the gradient histogram of the local area of the image. The specific implementation method is to first divide the image into small connected cell units. Then, the histogram of the direction gradient or edge of each pixel in the cell—cell is extracted. Then these histograms are combined in a certain way to form a feature descriptor of the image, which can be used for object detection in the next step [60].

HOG feature extraction mainly includes the following steps:

(1) Image grayscale processing and standardization of Gamma space;
(2) The image gradient calculation;
(3) The histogram of gradient direction corresponding to each cell unit is computed;
(4) In addition, cell—cells are combined into blocks, in which the histogram of gradient direction is normalized.

5.2.7.6.2 SVM classifier classification

The calculation process of the classifier is shown in Fig. 5.6:

Firstly, the training module trains the pedestrian samples and nonpedestrian samples in the database to obtain a 3780-dimensional coefficient m and a threshold τ, which are the parameters of the classification hyperplane obtained by the SVM classifier. Then, multiply and accumulate the feature descriptors H and m of each detection window. The cumulative result is compared with the threshold τ. After the comparator, if $y > 0$ is output, there will

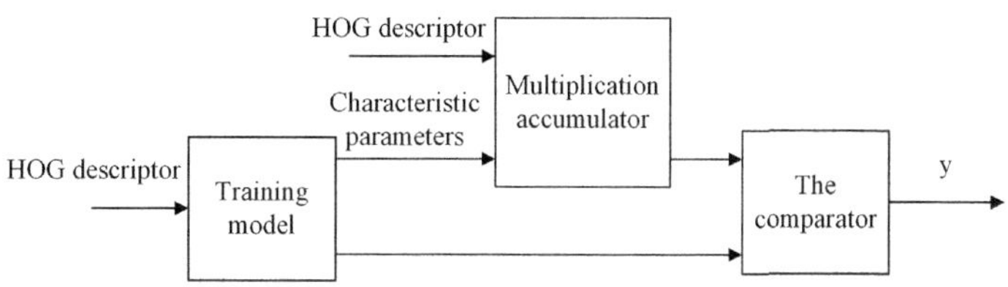

Figure 5.6
The flow chart of the Support Vector Machine (SVM) classification.

be pedestrians in the window; otherwise, there will be no pedestrians in the window. The calculation equation is shown as follows [61]:

$$y = \sum_{i=0}^{3779} m_i^T \times H_i + \tau \tag{5.1}$$

For each detection window, 3780 cycles are needed to complete the discrimination. An $M*N$ resolution image requires $3780 \times \left[\frac{(M-64)}{8}+1\right] \times \left[\frac{(N-128)}{8}+1\right]$ cycles. Therefore, this method requires a higher amount of calculation. To reduce the need for computer storage space, it is able to store the feature coefficient m separately in pixel blocks.

This section mainly introduces the HOG + SVM method used for pedestrian detection. The above method is used to screen images and extract the image frames containing pedestrians for the convenience of subsequent processing.

5.2.7.7 Edge detection and background subtraction fusion for pedestrian contour extraction

The image background is obtained by background subtraction. The rough contour of the pedestrian is extracted by the method of background difference and Canny operator edge detection.

The background subtraction method is simple and easy to understand. It has been widely used in object contour extraction. However, the method still has many shortcomings, such as the absence of contour image in the target image when the gray value of the target image is similar to that of the background image. The simple edge detection method has a lot of interference from the nontarget region under complex background [62]. In this chapter, the foreground image processed by background subtraction is extracted by the Canny operator which combines background subtraction with edge extraction. Then the foreground image processed by background subtraction and the edge image obtained by edge detection is processed by logical "or" operation. Finally, minimize the impact of noise. Fig. 5.7 is the flow chart of pedestrian contour extraction [63].

If the gray value of the target is close to the background in background difference, it is difficult to extract the complete contour of the target. Therefore, the two results obtained by background difference and edge detection are carried out logic or operation to obtain the final detection target.

After background difference and edge detection, there is still a lot of noise in the image, which needs to be processed in the later stage morphologically. The methods adopted in this chapter are median filtering, expansion, and corrosion.

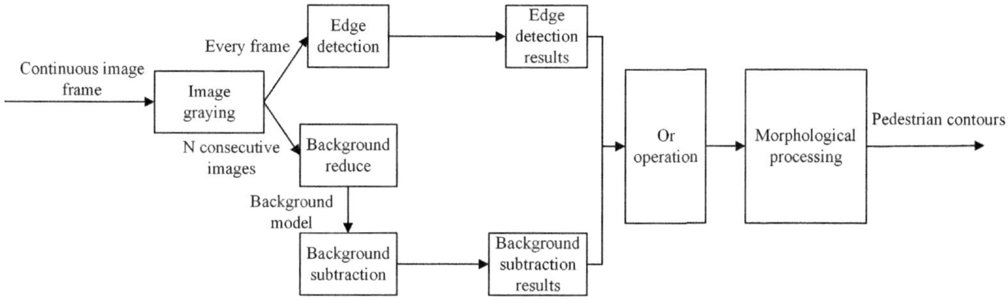

Figure 5.7
The flow chart of pedestrian contour extraction.

(1) Median filtering

Median filtering is a nonlinear method in noise processing. It has a good ability to suppress impulse noise and good edge retention. The basic principle is that the pixel value of each pixel is replaced by the median pixel value of each pixel in the neighborhood of the point. The octet neighborhood method is adopted in this chapter. The method can restrain impulse noise.

(2) Expansion

Expansion is the process of expanding the boundary of the studied target outward, that is, all the background points connected to the target are regarded as points on the target. The effect is to enlarge the target image by one circle. It can be widely used to modify the cavity in the extracted target contour. This method can be expressed as a set: assuming that M and N are both subsets of E. The expansion of M with N can be described as follows [64]:

$$M \oplus N = \{z \in E | z = m + n, m \in M \cap n \in N\} \tag{5.2}$$

(3) Corrosion

Corrosion operation and expansion operation are opposite processes, which make the studied target boundary shrink inward. Their effect is to make the target image shrink one circle. Two connected objects can be separated into image processing. If M and N are both subsets of E, corrosion of M with N can be described as follows [64]:

$$M * N = \{z \in E | z + n \in M, n \in N\} \tag{5.3}$$

Generally, the difference between light and dark in the contour map can be seen directly by the naked eye. This is due to changes in light, which creates shadows in the image. In future research, the illumination change of the image can be studied and discussed. A solution to eliminating the shadow in the image will be proposed.

The above method is applied to extract the pedestrian contour from the pedestrian posture database.

Since the pedestrian posture always remains standing, the background model cannot be constructed by background reduction when obtaining the contour map of the pedestrian waving gesture. However, the background in this video of the Weiziman database is simple, so this paper adopts the method of background segmentation and edge detection to extract the contour of pedestrian waving posture. The background segmentation method is based on the idea that the Red-Green-Blue (RGB) image is transformed into a gray image and the threshold is set artificially or based on the statistical idea. Reset the pixel value of points with a pixel value below this threshold to 0. The pixel value of points with a pixel value above this threshold is equal to 255. Finally, the contour map is obtained with good effect, which can ensure the smooth completion of subsequent processes.

This section mainly analyzes the commonly used edge detection operators in image edge detection algorithms. The background of the image is extracted by the method of background reduction. The general contour of the pedestrian is obtained by the method of background subtraction and edge detection. The complete contour of the pedestrian is obtained by the method of median noise reduction, expansion, and corrosion in morphology, so as to facilitate the subsequent feature extraction.

5.2.7.8 Establishment of pedestrian pose behavior parameter model and extraction of characteristic parameters

5.2.7.8.1 Pedestrian posture behavior parameter calculation

In this chapter, the proposed pedestrian posture identification method is used to detect and extract the pedestrian body area by means of background cutting and edge detection. Then the contour of the pedestrian in the image is obtained by removing the noise by means of morphological image processing. Then the key parameters of the contour features for the pedestrian are extracted. The recognition of pedestrian attitude is realized by the method of feature parameter fusion. The selection of appropriate features is the key to identify.

The features commonly used for pedestrian posture recognition are given as

(1) the area within the pedestrian profile area Ar;
(2) the width of the pedestrian profile W; and
(3) the height of pedestrian profile H.

The features of the aforementioned three pedestrian contours are the absolute features of the pedestrian contour. For pedestrian targets far from the camera device, these feature parameters will be reduced by two times. Therefore, when actually using these features, pedestrians must be normalized firstly as

(4) The complexity of the shape of the pedestrian contour area e.

Shape complexity can be expressed by the following equation [65]:

$$e = \frac{L^2}{A_r} \tag{5.4}$$

where L represents the perimeter of the graph, and Ar is the area of the graph.

(5) Rectangle g of pedestrian contour area

The rectangularity is the ratio of the area inside the pedestrian profile to the minimum area outside the pedestrian profile. The calculation method is shown as follows [65]:

$$g = \frac{A_r}{WH} \tag{5.5}$$

5.2.7.8.2 Feature selection

Stellate bone feature description is a good description of pedestrian posture. For example, there is a significant difference in the amplitude of arm and leg swing for walking and running postures. The difference can be shown in the stellate bone descriptor. Finally, the characteristic parameters of pedestrian posture in this chapter are selected as the prototype.

A total of six feature points are selected for the stellate bone descriptor. They are the centroid of the pedestrian, the vertex of the pedestrian's limbs and the vertex of the pedestrian's head.

This chapter firstly introduces the commonly used pedestrian posture identification methods. Then a pedestrian posture identification method is proposed based on contour feature parameters. After consulting a large number of literature, the contour feature parameters used for pedestrian posture identification are summarized. Finally, a parameter model of human posture behavior oriented to visual identification is formed.

5.2.7.9 Pedestrian posture prediction

Recognition of pedestrian behavior and posture is an important aspect of image recognition. This chapter mainly focuses on the train pedestrian posture discrimination. According to the actual situation, the common pedestrian posture mainly includes running, walking and waving. The calculation process of pedestrian posture recognition is to input the monitoring image video into the discrimination system. First, subtract video to process the image frame. The pedestrian posture features in the image are parameterized through the operation of pedestrian detection, background difference, contour extraction, and feature extraction. The extracted characteristic parameters of pedestrian posture can be regarded as the mode of this posture and compared with the known reference mode. The best matching reference pattern is the recognition result of this pedestrian posture. The pedestrian posture identification process is shown in Fig. 5.8.

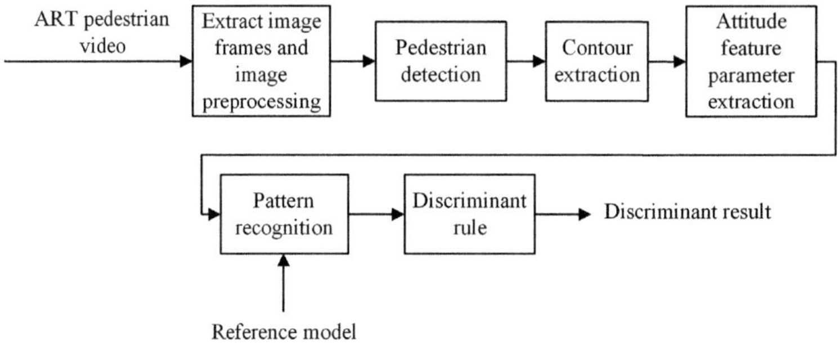

Figure 5.8
Posture recognition process.

In this chapter, postures that need to be identified are running, walking, and waving. The feature signals of each posture are 13 dimensions. For each posture, 500 groups of feature data are extracted respectively. Then the BP neural network can classify and identify the three postures according to the feature data.

A large number of experiments have proved the necessity of bounded hidden layer node function. Therefore, the BP neural network adopts a three-layer structure in this chapter. The hidden layer function is set as the sigmoid function, while the nonlinear function is set as the output node. The number of neuron nodes in the input layer is set. The neuron in the input layer is the link between the external signal and the BP neural network. Whether its value is representative affects the accuracy of the classification result directly. The input of the input layer in this article is the 13 parameters of pedestrian contours: the six distance between the centroid and the leftmost boundary point of the pedestrian contour, the rightmost boundary point of the pedestrian contour, the uppermost boundary point of the pedestrian left contour, the lowest boundary point of the pedestrian left contour, the uppermost boundary point of the pedestrian right contour, the lowermost boundary point of the pedestrian right contour; the angle between adjacent connected line segments; the ratio of pedestrian contour width to height. Therefore, the number of neuron nodes in the input layer of this artificial network is 13.

By determining the number of neuron nodes in the output layer, this network mainly realizes the classification and identification of pedestrian posture. It means the classification of the posture according to the input posture characteristic parameters is realized. There are three types, namely running, walking and waving. The number of neuron nodes in the output layer is equal to 3. The determination of the number of neuron nodes in the hidden layer is the key to the establishment of neuron structure. Due to its importance, many scholars have done researches on it. A large number of experiments show that if the number of hidden layer neuron nodes is set too big, it may result in too long training time and low efficiency of the network, even if the neural network can learn well and eventually achieve high accuracy. There may also be overfitting of the network to affect the generalization ability of the network. If the number of neuron nodes in the hidden layer is too small, the network cannot be effectively trained and the accuracy of identification will be affected. Based on a large number of tests, predecessors put forward many empirical equations are given as follows [66]:

$$t < n - 1 \tag{5.6}$$

$$t < \sqrt{n + k} + i \tag{5.7}$$

Some scholars also put forward an empirical equation to calculate the number of nodes in the hidden layer directly [66]:

$$t = 2\sqrt{n + k} + 1 \tag{5.8}$$

In the aforementioned equations, t represents the number of neurons in the hidden layer, n represents the number of neurons in the input layer, and I represents an integer between 0 and 9. In the actual selection process, the most appropriate number of nodes can be selected by trial and error method under the aforementioned conditions. The basic principle of selection is to select as few neuron nodes as possible under the condition of satisfying the accuracy requirement. The number of hidden layer neuron nodes in this chapter is $t = 10$.

The training process of the neural networks requires training data firstly. In this chapter, 1500 sets of training data are extracted. Among them, 1200 sets of data are used as training data of the BP neural network and 300 sets of data are used as the test data of the BP neural network. After the training of 1200 sets of data, the neural network system classifies the pedestrian posture corresponding to 300 sets of test data to achieve the function of pedestrian posture identification.

When the number of hidden layer neurons is $t = 10$ and the number of iterations of the neural network training process is $I = 20$, the classification and recognition results of the neural network are shown in Table 5.4.

According to the above results, it can be seen that the BP neural network achieves a high accuracy in the classification of pedestrian posture recognition. So the BP neural network can be applied to the recognition of pedestrian posture.

Table 5.4: The pedestrian posture recognition accuracy of BP neural network.

Pedestrian attitude type	Running	Walking	Waving
Recognition accuracy	0.9123	0.9684	1.0000

5.2.8 Conclusion and outlook

This chapter introduces the composition of hardware equipment and software systems of ART. The pedestrian detection algorithm in the process of unmanned driving is also introduced. The hardware and software complement the shortcoming of each other. Hardware provides a material basis for software. The software cannot play its role without the support of corresponding hardware. Only supported by the software, the hardware can provide more powerful function. Therefore, the software and hardware constitute a complete ART system. The main contents of this chapter can be summarized as follows:

(1) The hardware equipment

The hardware of modern ART can be divided into two parts. One is the main component of the vehicle, including the train body, independent wheel sets, traction devices, end-connection equipment, electronic control systems, auxiliary systems, air conditioning, and ventilation systems. These hardware devices constitute the basic model of the train. In addition, the interior equipment of ART also includes a vehicle-mounted multimedia terminal system and vehicle-mounted intelligent terminal system.

To realize accurate and fully automatic driving with a diver on the virtual track, ART needs to integrate various sensors. Abundant sensors are the second part of ART hardware. In general, sensors need to involve hearing, vision, perception and other aspects. In addition, the radar sensor and positioning sensor enables the train to move with higher accuracy than that only uses the GPS.

(2) The software system

Only with the hardware equipment, ART cannot meet the requirements of urban rail usage. ARTs should face not only other trains but also complicated terrain and traffic conditions when more than one ART is on the street. So the design of software system is very important. To realize efficient and safe transportation, the following software systems are needed: road traffic interaction system, navigation system, communication system, and scheduling management system.

The road traffic interaction system involves the assignment of two road rights at the intersection of street lights. It requires the use of pedestrian detection algorithms for reasonable obstacle avoidance and special crossing signal timing schemes, so that ART can adapt to the normal operation of road traffic. The navigation system includes automatic monitoring of ART. The navigation system can ensure the safe, punctual, reliable and efficient operation of ART. At the same time, the train dispatching and emergency treatment are completed. The communication system includes communication between the vehicles and communication between the vehicle and the control center. The scheduling management system is equivalent to the integration of the interactive system, the navigation system, and the communication system. The cross-use of the aforementioned four systems ensures the efficiency and safety of ART.

(3) Pedestrian detection algorithm in intelligent driving

The pedestrian detection algorithm is an important part of the intelligent driving system of ART. At present, pedestrian detection algorithms mostly based on deep learning. According to different stages of development, the detection algorithms based on deep learning can be divided into two parts. One is the detection algorithms based on region proposal, and the other is the end-to-end detection algorithms based on deep learning. The development of vision-based pedestrian detection algorithms can effectively reduce the cost of vehicle sensors. The high price of sensor devices such as laser radar can be replaced by cameras with image detection technology. The costs are reduced in this way. The era of future unmanned driving is worth looking forward to. In addition, an intelligent prediction method for pedestrian attitude-behavior parameter fusion of ART is introduced.

First, the pedestrian video is divided into a continuous image frame. The image is detected by HOG feature classification. Then, the image containing pedestrian is saved.

Secondly, the background subtraction method is used to get the image background. The approximate contour of the traveler is extracted by the combination of background difference and canny operator edge detection. After that, the contour boundary is clarified by the morphological method. In this chapter, a model of human posture behavior parameters oriented to visual identification is established. The model includes 13 main feature parameters for pedestrian posture identification.

Finally, the BPNN is used to fuse the feature parameters and posture prediction. The pedestrian posture database adopts the Weiziman database and selects 500 sets of pedestrian data. 1200 sets of data are randomly selected to train BP neural network, and the remaining 300 sets of data are tested on neural networks. This algorithm can satisfy the requirement of natural interaction for the correct rate of pedestrian posture identification and can be widely used in train intelligent monitoring.

In the future, it is also possible to increase the capacity of single train passengers using technologies such as marshaling or wireless reconnection. The improvement of ART vehicle interior service equipment can increase the passenger experience. In the future, the advantages of ART in reducing the consumption of nonrenewable energy through green electric drive will gradually become prominent with the maturity of battery and motor technology. Shorter charging time and longer driving duration make the operation of ARTs more reliable. In densely populated cities, the characteristics of ART are more suitable for urban traffic. ARTs will show greater potential in passenger capacity. With the development of science and technology, computer vision will be more widely applied to ART. The realization of this technology will greatly reduce the hardware cost of ART, which is very helpful for its popularization. The traffic efficiency of the city will be improved again due to the popularization of artificial intelligence technologies. In the future, people will have more choices to travel.

References

[1] G. Wilson, London United Tramways: A History 1894–1933, Routledge, 2013.
[2] John Cristian Borges Gamboa, Deep Learning for Time-Series Analysis, 2017 arXiv preprint arXiv:170101887.
[3] S. Zhao, Research on key technical scheme of ART signal system, Railway Commun. Signal Eng. Technol. 16 (2019) 56–58+61.
[4] M. Tanabe, H. Wakui, N. Matsumoto, et al., Computational model of a Shinkansen train running on the railway structure and the industrial applications, J. Mater. Process. Technol. 140 (2003) 705–710.
[5] J. Lichtblau, B. Sanftl, C. Kraus, et al., Influence of train interior on train-onboard communication, in: 22nd International Microwave and Radar Conference (MIKON), 2018, pp. 605–606.
[6] H. Niu, X. Wu, Rayafei, et al., Dynamic characteristic test bed and test system development of the electro-hydraulic servo steering system of ART, Chin. Hydraul. Pneum. 7 (2019) 120–127.
[7] F. Jing, G. Xu, Y. Xu, et al., Research on optimal control strategy of coupler force in urban rail transit, in: International Symposium for Intelligent Transportation and Smart City, 2017, pp. 163–181.
[8] G. Skotnikov, M. Jileykin, K. Ai, Increasing the stability of the articulated lorry at braking by locking the fifth wheel coupling, in: IOP Conference Series: Materials Science and Engineering, vol. 315, 2018, 012027.
[9] H. Meng, H. Wang, G. Ming, et al., Experimental and simulation study on a novel uniform air/smoke exhaust device in subway, in: IOP Conference Series: Earth and Environmental Science, vol. 238, 2019, 012022.
[10] P. Xie, L. Du, B. Zhou, et al., Design and implementation of vehicle multimedia system automated test, in: IOP Conference Series: Materials Science and Engineering, vol. 490, 2019, 072048.
[11] K. Hubert, P.A. Zuber, The IEC/IEEE train communication network, IEEE Micro 21 (2001) 81–92.
[12] B. Hofmann-Wellenhof, H. Lichtenegger, E. Wasle, GNSS–Global Navigation Satellite Systems: GPS, GLONASS, Galileo, and More, Springer Science & Business Media, 2007.
[13] T. Caselitz, B. Steder, M. Ruhnke, et al. Monocular camera localization in 3d lidar maps. In: 2016 IEEE/RSJ International Conference on Intelligent Robots and Systems (IROS). 1926-1931, 2016.
[14] L. Cao, C. Wang, J. Li, Robust depth-based object tracking from a moving binocular camera, Signal Process. 112 (2015) 154–161.
[15] S. Pollard, M. Pilu, S. Hayes, et al., View synthesis by trinocular edge matching and transfer, Image Vis. Comput. 18 (2000) 749–757.
[16] S.E. Reutebuch, H.-E. Andersen, R.J. Mcgaughey, Light detection and ranging (LIDAR): an emerging tool for multiple resource inventory, J. For. 103 (2005) 286–292.
[17] M.E. Russell, A. Crain, A. Curran, et al., Millimeter wave radar sensor for automotive intelligent Cruise control, in: IEEE MTT-S International Microwave Symposium Digest, vol. 3, 1997, pp. 1257–1260.
[18] F. Alonge, F.M. Marco Branciforte, A novel method of distance measurement based on pulse position modulation and synchronization of chaotic signals using ultrasonic radar systems, IEEE Trans. Instrument. Meas. 58 (2008) 318–329.
[19] A. Eskandarian, Handbook of Intelligent Vehicles, Springer, 2012.
[20] S. Park, J. Deyst, J. How, A new nonlinear guidance logic for trajectory tracking, in: AIAA Guidance, Navigation, and Control Conference and Exhibit, 2004, 4900.

[21] M. Hartong, R. Goel, D. Wijesekera, Communications security concerns in communications based train control, WIT Trans. Built Environ. 88 (2006).

[22] A. Sładkowski, W. Pamuła, Intelligent Transportation Systems-Problems and Perspectives, Springer, 2016.

[23] K. Kim, S.-H. Kong, S.-Y. Jeon, Slip and slide detection and adaptive information sharing algorithms for high-speed train navigation systems, IEEE Trans. Intell. Transport. Syst. 16 (2015) 3193–3203.

[24] V. Vahidi, E. Saberinia, OFDM high speed train communication systems in 5G cellular networks, in: 15th IEEE Annual Consumer Communications & Networking Conference (CCNC), 2018, pp. 1–6.

[25] M. Samà, C. Meloni, A. D'ariano, et al., A multi-criteria decision support methodology for real-time train scheduling, J. Rail Transport Plann. Manag. 5 (2015) 146–162.

[26] J.C. Moreno, E. Laloya, J. Navarro, A link-layer slave device design of the MVB-TCN bus (IEC 61375 and IEEE 1473-T), IEEE Trans. Veh. Technol. 56 (2007) 3457–3468.

[27] J.C. Moreno, E.J. Laloya, J. Navarro, Line redundancy in MVB-TCN devices: a control unit design, in: MELECON 2006-2006 IEEE Mediterranean Electrotechnical Conference, 2006, pp. 789–794.

[28] T. Chen, Q. Zeng, Train Microcomputer and Network Control Technology and Application, Science Press, Beijing, 2012.

[29] Y. Ahmed Yahya, A.T. Al-Hammouri, Transmission control protocol global synchronization problem in wide area monitoring and control systems, UHD J. Sci. Technol. 1 (2017) 7–12.

[30] N.M. Garcia, F. Gil, B. Matos, et al., Keyed user datagram protocol: concepts and operation of an Almost reliable connectionless transport protocol, IEEE Access 7 (2019) 18951–18963.

[31] S. Zhang, R. Benenson, O. Mohamed, et al., Towards reaching human performance in pedestrian detection, IEEE Trans. Pattern Anal. Mach. Intell. 40 (2017) 973–986.

[32] A. Pentland, Looking at people: sensing for ubiquitous and wearable computing, IEEE Trans. Pattern Anal. Mach. Intell. 22 (2000) 107–119.

[33] E. Zhang, F. Chen, A fast and robust people counting method in video surveillance, in: International Conference on Computational Intelligence and Security (CIS 2007), 2007, pp. 339–343.

[34] J.K. Aggarwal, M.S. Ryoo, Human activity analysis: a review, ACM Comput. Surv. 43 (16) (2011).

[35] T. Germa, F. Lerasle, N. Ouadah, et al., Vision and RFID data fusion for tracking people in crowds by a mobile robot, Comput. Vis. Image Understand. 114 (2010) 641–651.

[36] J. Ge, Y. Luo, G. Tei, Real-time pedestrian detection and tracking at nighttime for driver-assistance systems, IEEE Trans. Intell. Transport. Syst. 10 (2009) 283–298.

[37] C. Tomiuc, S. Nedevschi, M. Michael Meinecke, Pedestrian detection and classification based on 2d and 3d information for driving assistance systems, IEEE Int. Conf. Intell. Comput. Commun. Process. (2007) 133–139, 2007.

[38] P. Dollar, C. Wojek, B. Schiele, et al., Pedestrian detection: an evaluation of the state of the art, IEEE Trans. Pattern Anal. Mach. Intell. 34 (2011) 743–761.

[39] D. Geronimo, A.M. Lopez, A.D. Sappa, et al., Survey of pedestrian detection for advanced driver assistance systems, IEEE Trans. Pattern Anal. Mach. Intell. 32 (2010) 1239–1258.

[40] K.-Y. Byun, B.-S. Kim, H.-K. Kim, et al., An effective pedestrian detection method for driver assistance system, in: IEEE International Conference on Consumer Electronics (ICCE), 2012, pp. 229–230.

[41] M. Oren, C. Papageorgiou, P. Sinha, et al., Pedestrian detection using wavelet templates, in: Proceedings of the IEEE Conference on Computer Vision and Pattern Recognition, vol. 97, 1997, pp. 193–199.

[42] V. Paul, J. Michael, Jones, Robust real-time face detection, Int. J. Comput. Vis. 57 (2004) 137–154.

[43] N. Dalal, B. Triggs, Histograms of oriented gradients for human detection, in: IEEE Computer Society Conference on Computer Vision and Pattern Recognition, 2005, pp. 886–893.

[44] P.F. Felzenszwalb, R.B. Girshick, D. Mcallester, et al., Object detection with discriminatively trained part-based models, IEEE Trans. Pattern Anal. Mach. Intell. 32 (2009) 1627–1645.

[45] Y. Lecun, Y. Bengio, G. Hinton, Deep learning, Nature. 521 (2015) 436.

[46] A. Krizhevsky, I. Sutskever, G.E. Hinton, Imagenet classification with deep convolutional neural networks, in: Advances in Neural Information Processing Systems, 2012, pp. 1097–1105.

[47] Y. Lecun, Y. Bengio, Convolutional networks for images, speech, and time series, in: The Handbook of Brain Theory and Neural Networks, vol. 3361, 1995, pp. 255–258.
[48] S. Zhang, R. Benenson, O. Mohamed, et al., How far are we from solving pedestrian detection?, in: Proceedings of the IEEE Conference on Computer Vision and Pattern Recognition, 2016, pp. 1259–1267.
[49] R. Girshick, J. Donahue, T. Darrell, et al., Rich feature hierarchies for accurate object detection and semantic segmentation, in: Proceedings of the IEEE Conference on Computer Vision and Pattern Recognition, 2014, pp. 580–587.
[50] K. He, X. Zhang, S. Ren, et al., Spatial pyramid pooling in deep convolutional networks for visual recognition, IEEE Trans. Pattern Anal. Mach. Intell. 37 (2015) 1904–1916.
[51] R.-C.N.N. Ross Girshick Fast, in: Proceedings of the IEEE International Conference on Computer Vision, 2015, pp. 1440–1448.
[52] S. Ren, K. He, R. Girshick, et al., Faster R-CNN: towards real-time object detection with region proposal networks, in: Advances in Neural Information Processing Systems, 2015, pp. 91–99.
[53] R. Joseph, S. Divvala, R. Girshick, et al., You only look once: unified, real-time object detection, in: Proceedings of the IEEE Conference on Computer Vision and Pattern Recognition, 2016, pp. 779–788.
[54] W. Liu, D. Anguelov, D. Erhan, et al., Ssd: single shot multibox detector, in: European Conference on Computer Vision, 2016, pp. 21–37.
[55] J. Dai, Y. Li, K. He, et al., R-FCN: Object detection via region-based fully convolutional networks, in: Advances in Neural Information Processing Systems, 2016, pp. 379–387.
[56] R. Joseph, F. Ali, YOLO9000: better, faster, stronger, in: Proceedings of the IEEE Conference on Computer Vision and Pattern Recognition, 2017, pp. 7263–7271.
[57] C. Szegedy, W. Liu, Y. Jia, et al., Going deeper with convolutions, in: Proceedings of the IEEE Conference on Computer Vision and Pattern Recognition, 2015, pp. 1–9.
[58] K. He, X. Zhang, S. Ren, et al., Deep residual learning for image recognition, in: Proceedings of the IEEE Conference on Computer Vision and Pattern Recognition, 2016, pp. 770–778.
[59] S. Zhang, De Cheng, Y. Gong, et al., Pedestrian search in surveillance videos by learning discriminative deep features, Neurocomputing 283 (2018) 120–128.
[60] H.C. Song, G.H. Lee, D.-S. Shim, et al., Visual distinctiveness detection of pedestrian based on statistically weighting PLSA for intelligent systems, Int. J. Contr. Autom. Syst. 16 (2018) 815–822.
[61] D.F. Llorca, R. Arroyo, M.-A. Sotelo, Vehicle logo recognition in traffic images using HOG features and SVM, in: 16th International IEEE Conference on Intelligent Transportation Systems, 2013, pp. 2229–2234.
[62] S. Yuan, X. Dong, X. Liu, et al., Edge detection based on computational ghost imaging with structured illuminations, Optic Commun. 410 (2018) 350–355.
[63] J.W. Davis, V. Sharma, Fusion-based background-subtraction using contour saliency, in: IEEE Computer Society Conference on Computer Vision and Pattern Recognition, 2005, pp. 11–19.
[64] J.W. Davis, V. Sharma, Background-subtraction using contour-based fusion of thermal and visible imagery, Comput. Vis. Image Understand. 106 (2007) 162–182.
[65] T. Osaragi, Modeling of pedestrian behavior and its applications to spatial evaluation, in: Proceedings of the Third International Joint Conference on Autonomous Agents and Multiagent Systems, vol. 2, 2004, pp. 836–843.
[66] S. Nakamura, T. Kawasaki, A method for predicting the lateral girder response of footbridges induced by pedestrians, J. Constr. Steel Res. 65 (2009) 1705–1711.

CHAPTER 6

Rail transit inspection robot systems

6.1 Overview of rail transit inspection robots

6.1.1 Background of rail transit inspection robots

With the improvement of social productivity and the continuous development of urbanization, the sustained and rapid urban growth of the population has brought opportunities for development, as well as challenges to transportation. Urban rail transit has become an effective mode of transportation for solving urban traffic congestion problems because of its large capacity, fast speed, low energy consumption, high punctuality, high transportation efficiency, and low carbon environmental protection. It plays an indispensable role in major cities. The increasingly important role of rail transit in urban traffic is increasing, ensuring people's normal commuting and travel.

Rail transit is in a period of rapid development and rapid growth. Once trains are delayed, the speed of trains is decelerated, sections are out of service, and passengers are stranded. These consequences are quite serious. Due to the dominant position of rail transit in urban traffic, there will be a lot of negative effects once the rail transit network is interrupted. For example, some passengers will give up rail and ground transportation. This not only leads to ground traffic congestion but also affects the development of the rail transit system. As these negative effects increase, the efficiency of the public transport system throughout the city would be significantly reduced [1].

The main problem of rail transit is to lay electrified equipment along the track, such as transmission line, temperature monitoring of the transformer, data collection of various instruments, foreign matter intrusion of the track, normal operation of the switch, and the rail transit vehicle. Power lines, terminals, and electrical components are often abnormally heated due to short circuits, increased contact resistance, and the like. This will cause many problems such as electrical fire accidents in trains, damage to trains or track facilities, serious casualties, and loss of state property [2].

How to take effective measures to improve equipment quality and ensure safety is a major security issue that is urgently needed to be addressed in the rail transit. The difficulty in rail transit detection is the difficulty in control. From a technical point of view, vibration can cause damage to joints, insulation, or failure to load. Short-circuit and improper installation process will result in increased contact resistance. Excessive using of various

Robot Systems for Rail Transit Applications. https://doi.org/10.1016/B978-0-12-822968-2.00006-1

electrical devices will cause overheating in some places and, in severe cases, a fire. The main reason is that the detection means is limited, and the fault cannot be located in time. The dynamic monitoring of the equipment in the train operation is carried out by the train attendants that are individually combated and highly dispersed. The operations of train attendants depend largely on their consciousness, so the problem can be found at the beginning. Then, the accident can be easily triggered [3]. There are many reasons for the electrical fire accidents of trains.

In view of nature of the occurrence, there are problems caused by the quality and improper use of electrical appliances. There are quality problems in the design and installation process of the manufacturers. There are also problems with inadequate maintenance, as well as improper management in use. From the category of occurrence, there are problems with the equipment itself, as well as personnel awareness, training, skills, operating standards, etc. From the perspective of management, there are both management problems and problems at the operational level. From the types of problems, there are both technical problems and management loopholes. From technical reasons, there are short-circuit problems, overload problems, leakage current problems, and arcing problems.

At present, the main method of inspection in rail transit is with a workforce, but the working environment of rail transit equipment is complex, and the conditions are poor [4]. In the case of bad weather, the staff who conduct inspections have personal safety risks, and the inspection efficiency is low. The inspection results depend on the business proficiency and subjective initiative of the inspection personnel. The rail transit inspection robot is developed because it is easy to cause negligence under the long-term labor of mechanical repetition. The rail transit inspection robot will become a new means of inspection of rail transit lines and equipment in the station. Manual inspection will be gradually phased out. The inspection robot can use various sensors equipped by itself to detect the rail transit lines and rail transit vehicle equipment through various advanced technologies in a fully autonomous manner or in cooperation with human inspectors. The use of inspection robots can detect working conditions of rail transit equipment under any severe conditions, and ensure the orderly operation of rail transit. This is very important for the future realization of the full automation of rail transit. It is the future development trend.

The inspection robot is an autonomous mobile robot. It is an intelligent robot that integrates the latest positioning and navigation technology, path-planning technology and information collection technology. It can sense the operating environment and make correct reaction behaviors. It can obtain its current position and gesture information in real-time, and then use the path-planning algorithm to obtain the path through the on-board computer. Therefore, the inspection tasks can be completed on the premise of ensuring the safety of the robot itself and the environment. The realization of this capability comes from the sensors equipped by the robot, such as cameras, inertial navigation devices, laser radars, encoders, and range finders [5]. It is necessary to process information obtained by these sensors in a specific way to perceive the surrounding environment. It is also important to accurately calculate its own position, gesture and make correct control decisions.

The inspection robot is an intelligent robot that performs inspection work in a certain order according to the preset target. It can complete environmental monitoring, safety monitoring, equipment inspection, data reading, and icon recognition. Therefore, inspection robots can be widely used in industry, infrastructure operation and maintenance, and life. They are especially useful in the work environments which are harmful to the human body such as rail transit, nuclear power plants, reservoirs, chemical plants, and underground pipelines. Inspection robots can replace human workers to solve problems with high danger, high labor costs, low operating efficiency, and boring content. The effective navigation control method is an important part of the inspection function of the autonomous mobile robot. The key lies in the realization of its own positioning and perception of the surrounding environment.

At present, the inspection robot navigation technology of fixed trajectory has been mature and applied. However, in the environment of the unknown, large scenes, weak texture, and repeated texture, it is still a technical difficulty for the inspection robot to determine its own position and gesture, sense the surrounding environment, and construct a three-dimensional (3D) map. With the rapid development of computer communication technology and network technology, wireless image transmission breaks the disadvantages of traditional coaxial cable and optical fiber image surveillance due to hardware connection, and has greater flexibility and convenience. Therefore, an image surveillance system based on wireless network emerges. Wireless image surveillance has become the main method used in the surveillance field.

In the past, wired image surveillance systems often faced many problems such as the need to lay many ground and underground lines, high costs, and long construction periods. The rail transit inspection robot is mainly used for outdoor inspection, assisting or replacing the workers to complete equipment inspection tasks. It is required that the system can directly carry the infrared camera and the visible light Charge Coupled Device (CCD) to detect the field devices. Using pattern recognition and image processing technology to complete real-time monitoring of rail transit equipment under the condition of no or few people [6]. At the same time, according to the task operation or preset task of the base station operator, the operator is assisted to implement and complete the image detection of the outdoor rail transit device and the infrared detection of the primary device.

The rail transit inspection robot can record, store feedback device information, and provide abnormal alarms in time according to site conditions. At the same time, it can provide staff with relatively complete data reports on rail transit equipment, which provides great convenience for staff to diagnose the safety hazards of equipment and analyze equipment failures. The rail transit inspection robot can complete the detection of the equipment more effectively. At the same time, it can also provide more comprehensive processed equipment working status data to the maintenance personnel, and form a historical curve by analyzing the data, which greatly improves the predictability of equipment. It also provides more effective reference data for device status detection or device status assessment, further reducing the equipment accident rate. In summary, the following conclusions can be drawn: using the rail transit inspection robot to inspect the equipment, the maintenance personnel can remotely control the inspection robot at the base station and thus stay away from the live equipment, which significantly improves the safety and reliability of the inspection work. The inspection robot is a kind of on-site supervision and testing tool that applies innovative technology. It promotes the promotion and application of intelligent and unmanned inspection. At the same time, the real-time performance of rail transit equipment status detection and the automation level of rail transit have also been significantly improved [7].

The design of the inspection robot includes two parts: the hardware structure of the inspection robot and the host computer system [8]. The hardware structure mainly refers to the inspection robot body. The inspection robot is mainly composed of mechanical system, hardware system, and software system. The mechanical system mainly includes the inspection robot body, the motion mechanism, and the pan/tilt equipment. The hardware system includes the central controller for realizing the robot control, the motor driver and other driving devices, the laser sensor, the ultrasonic sensor, the infrared camera, the visible light camera, and the Radio Frequency Identification (RFID) sensors. The software system mainly realizes the intelligent control of the robot and the intelligent recognition of the target. The inspection robot uses navigation methods such as visual navigation and magnetic navigation to realize the autonomous driving of the robot. The pan/tilt device is equipped with a visible light camera and an infrared camera to realize intelligent identification and detection of the rail transit equipment. In the design of the inspection robot system, mechanical system, autonomous navigation and positioning, path-planning and instrumentation detection are the key technologies for the inspection robot to complete the inspection work. The structure of the inspection robot is shown in Fig. 6.1.

6.1.2 Development progress of inspection robots

At present, the inspection work in most of the infrastructure is mainly done through the available workforce. The staff regularly inspects the equipment and records the readings of

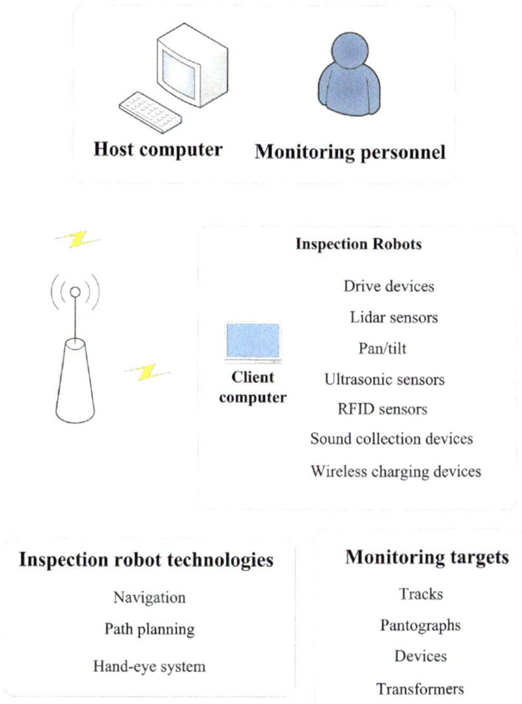

Figure 6.1
Inspection robot structure.

various instruments. This method has a large workload and low efficiency. There is inevitably a problem of inaccurate data recording caused by manual errors. High-risk and toxic environments such as nuclear power plants and chemical plants may cause safety hazards to workers. To reduce the demand for a workforce and improve the accuracy and efficiency of inspections, inspection robots have been gradually developed.

In the 1950s, with the maturity of automation technology, mobile robots began to emerge. In the following years, mobile robot technology began to develop rapidly and was used in an extensive range of fields. In the 1960s, the Stanford Research Institute of the United States developed an intelligent robot named Shakey, which has the ability to perceive the surrounding environment and use algorithms to independently complete path-planning [9]. With the rapid development of computer technology, mobile robot technology has been further developed. The Japanese power company has developed a mobile robot for the substation, which is equipped with an infrared camera, visible light camera, sound collection device, and other sensors. It is the prototype of the inspection robot. With the development of mobile robots, inspection robots have gradually replaced humans for inspection work, forming a new research direction.

In the 1980s, Japan began to study the use of robots for live maintenance of high-voltage overhead transmission lines, using a variety of sensors to detect electrical equipment in the substation and to collect parameters related to transmission equipment [10]. The application of robots realizes a high degree of automation of inspection tasks and reduces the workload of the staff. The inspection robot developed by Quebec in Canada has a supporting management system, which can be remotely controlled to realize equipment detection [11,12]. An inspection robot developed in Brazil is used to detect the substation equipment for abnormally high temperatures, and the real-time detection of the surface temperature of the equipment is realized by carrying various sensors. A lot of research results in the fields of robot motion control, sensors, and positioning and navigation have been obtained. More and more research institutions, technology companies and university research groups have taken the technologies related to inspection robots as research and development topics.

The Quebec Institute of Hydropower Research in Canada developed an inspection robot for wire deicing. It uses pulleys to move on high-voltage transmission lines. It has been continuously updated. On-board visual sensors have been added to monitor high-voltage transmission lines and information is transmitted back in real-time. The inspection robot can be controlled by the base station operator to perform high-altitude operations and realize basic functions such as maintenance and reinforcement.

A track-type inspection robot was jointly developed by Japan's Shikoku Electric Power Co., Ltd. and Toshiba Corporation to inspect power equipment in the substation. The first inspection robot in China was developed by the scientific research personnel of the State Grid Corporation of China. The inspection robot was used for the inspection of power grid equipment and instruments, and began to replace the manual inspection work [8]. The research work can be traced back to the 1990s. The inspection robot is equipped with a variety of sensors such as ultrasonic sensors, laser sensors, and cameras, and its navigation method is magnetic navigation. Since then, the State Grid Corporation of China successfully developed multiple generations of inspection robots whose performance tends to be stable. They are more suitable for inspection work in outdoor complex environments. On this basis, Chongqing Electric Power Company and Zhejiang Guozi Robotics Co., Ltd. have successively developed inspection robots and put them into operation at local substations. The operation of these inspection robots is in good condition and can completely replace the manual monitoring of the temperature of the equipment and the readings of the equipment. The inspection work began to enter the era of robots.

In 2010, Shanghai University developed an inspection robot [13]. The robot patrols the high-voltage line using two arms hanging on the same side, with independent inspection and obstacle avoidance. In 2018, Chinese Chang'e-4 lunar explorer was successfully sent into space and landed safely on the surface of the moon [14]. The Yutu-2 lunar rover was equipped with advanced sensors for data collection on the moon, which is a milestone for mobile robots in China.

With the implementation of "A Country with Strong Transportation Network" in China, the demand for the intelligent robot market and technology has been greatly stimulated, and other industries have gradually transformed in an intelligent direction. In the recent trend of higher labor costs, the use of robots instead of using workers can effectively reduce costs. For traditional substation inspections, the use of inspection robots can solve the drawback of overreliance on a workforce. As an important aid tool, the inspection robots have developed rapidly and gradually penetrated all walks of life. The application of inspection robots to complete inspection tasks is the mainstream trend in the development of rail transit inspections. Under the current situation that the labor cost is getting higher and higher, the use of inspection robots to complete the cumbersome, heavy, and dangerous rail transit inspection work is a trend of social development and a result of scientific and technological progress.

The main inspection robots include wheeled autonomous mobile inspection robots, orbital inspection robots, suspended inspection robots, disaster search and rescue inspection robots, and pipeline inspection robots [15]. The applications of the inspection robots in the railway mainly contain two aspects, including inspection for traction substation and inspection along the railway. The applications are explained in Fig. 6.2.

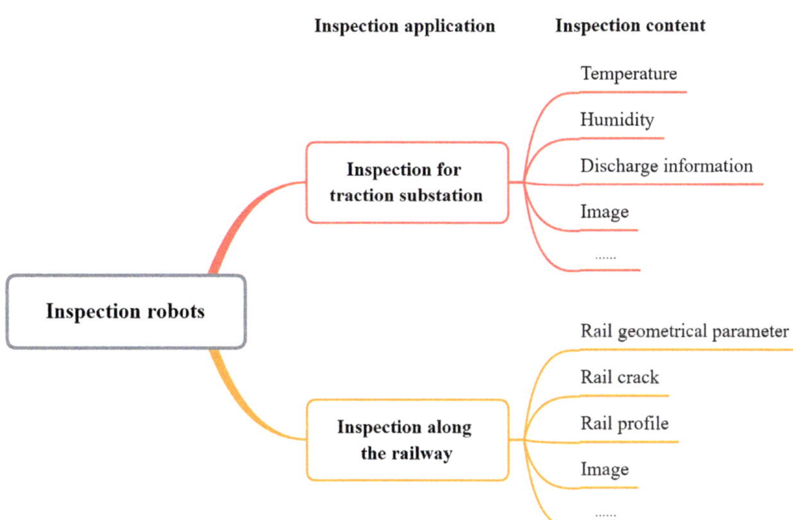

Figure 6.2
The railway applications of the inspection robots.

6.2 Main components of rail transit inspection robots

The main components of rail transit inspection robots are mainly embodied in the design of the motion mechanism of the robot. Considering the complex environment of rail transit, the inspection robot needs to have flexible motion characteristics to ensure that it can cover the rail transit along the line and every corner of the station as completely as possible. The autonomous navigation technology embodies the function of "patrol" for the inspection robot, which is the basic guarantee for completing the inspection task. Only when it moves smoothly to the position of the device to be tested can the detection work be realized. Intelligent detection technology is an important technical means to achieve the "inspection" of inspection work. Only when the detection of equipment is accurately realized, can intelligent inspection replace manual inspection. In summary, according to the current development status of inspection technology, the main components required for the intelligent inspection robot system mainly include the following: driving wheel, laser sensing device, pan/tilt, ultrasonic sensor, RFID sensor, sound collection device, and wireless recharging device [16]. These components are shown in Fig. 6.3.

6.2.1 Driving devices

The driving device serves as the basis of the mobile robot operation and controls the speed, acceleration, and steering. It can ensure the efficiency and safety of the robot in carrying tasks. According to the different layout, there are four main types of driving wheels: differential wheel, single steering wheel, double steering wheel, and full steering wheel. Among them, the inspection robot mainly adopts the differential wheel.

(1) Differential wheel

Differential wheel layout method symmetrically arranges two driving wheels in the middle position of the mobile robot, and installs universal wheels at the front and rear positions to

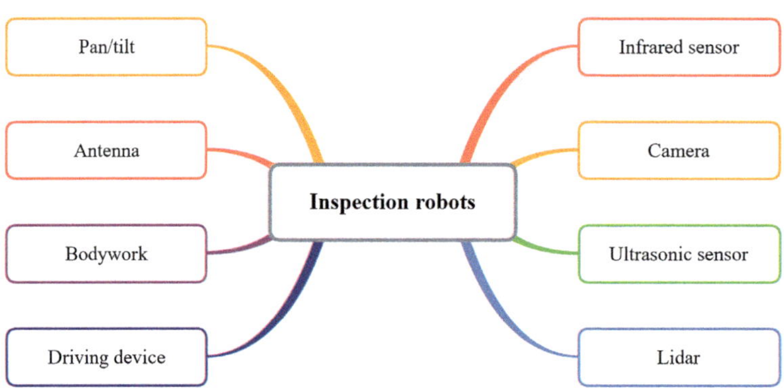

Figure 6.3
Main components of the inspection robots.

ensure the robot's linear travel and steering balance. The main advantage of this layout is that the control method is simple, which is convenient for the mobile robot to turn, and the four-wheel landing method is conducive to the inspection robot's movement in the outdoor complex environment [17].

(2) Single steering wheel

Single steering wheel layout uses only one driving wheel and two driven wheels. Two driven wheels are used to stabilize the robot's travel and ensure the robot's steering ability. Compared with the former layout, this layout method is slightly worse in off-roading capability, and it is easier to lose balance in the complex environment of rail transit. Once the only driving wheel idles, the entire mobile robot will lose its ability to move. The steering action of this layout is easy to implement, but the steering radius is longer.

(3) Double steering wheel

Double steering wheel layout uses two steering wheels for the mobile robot, and the other two universal wheels to assist the support. Each steering wheel is controlled by two motors. Through the steering motor to control the steering angle of the steering wheel, the mobile robot can realize back and forth travel, turning, oblique travel, lateral travel, and rotation in the plane.

(4) Full steering wheel

Full steering wheel layout installs four steering wheels on the mobile robot. Each wheel is controlled by two motors, and the steering of the steering wheel is controlled by the motors. Compared with the first layout, the robot using this layout can realize the lateral and oblique movement, as well as in-situ rotation. Its movement is also more flexible. However, with the improvement of functions, such wheels are expensive and the control is complicated. It is usually used for heavy-load mobile robots.

6.2.2 Sensors

6.2.2.1 LiDAR sensor

Light Detection and Ranging (LiDAR) is a distance sensor, consisting of a light detection device and a radar. The point cloud information detected by the LiDAR has a large amount of target information. The parameters such as the distance, orientation, height, speed, attitude, and even shape of the surrounding targets detected by emitting a laser beam are mainly used for high-precision 3D modeling. Using LiDAR for high-precision 3D modeling can improve the accuracy and effectiveness of the map. LiDAR has high resolution, strong antiinterference ability, and strong low-altitude detection ability. It can work continuously and is very suitable for the outdoor work of inspection robots. Its shortcomings are also obvious. The LiDAR is greatly affected by the weather when it is working. The propagation distance of the laser beam will be greatly reduced in rainy and foggy weather, and other sensors need to work with LiDAR together [18]. At present, the highest accuracy of 3D modeling is obtained by 3D laser imaging radar. The 3D laser imaging radar uses a direct ranging method to obtain point cloud data of the surrounding environment, then uses related algorithms to establish the 3D model of the surrounding environment.

With the popularization of LiDAR for civil use, there are many types of LiDARs that meet different scenarios. The LiDAR can be divided into two types according to the detection method: direct detection LiDAR and coherent detection LiDAR. The direct detection LiDAR is used in mobile robotics, unmanned driving, mapping, and other fields. Compared with the direct detection LiDAR, the coherent detection LiDAR is more complex, has better antiinterference ability, and the obtained map is more detailed, but the technology is not mature enough.

At present, there are two main ranging methods, Time of Flight (ToF) method, and triangulation method. The ToF can be divided into two types, pulse ranging method, and phase ranging method.

(1) Laser pulse ranging

The target distance is obtained by multiplying the flying time of the laser pulse between the target and the pulse transmitter by the propagation speed of the laser pulse. The fast measurement speed and strong antiinterference ability are the main advantages, but the technology is difficult to use.

(2) Laser phase ranging

The target distance is obtained by using the phase difference of the laser signal between the radar and the target. According to the principle, it can be known that the laser phase ranging technology takes longer than the laser pulse ranging technology, but the resolution of laser pulse ranging can be very high, and it is very common to use this technology on robots.

(3) Triangulation method

After the laser transmitter emits the laser light, the laser light is reflected by the target in the surrounding environment and received by the linear CCD. Because there is a certain distance between the laser transmitter and the detector, the LiDAR can deduce the distance of the measured object according to the optical path and the imaging positions of three different objects on the CCD through triangular equation.

6.2.2.2 Ultrasonic sensor

6.2.2.2.1 Ultrasonic ranging principle

Ultrasound is defined as sound waves with a vibration frequency exceeding 20 kHz, which cannot be heard by the human ear. Ultrasound can be used to transmit information. In sonar systems, it is widely used in underwater environment detection due to its excellent underwater propagation ability. In the field of robots, ultrasonic sensors are mainly used to detect obstacles in the environment around the robot [19]. Ultrasound can work in any light environment for 24 h. In the modern environment, when the target is glass, ultrasonic sensors can work normally, but laser sensors cannot detect glass. The attenuation rate of ultrasound propagating in the air is very fast, and the attenuation rate after reflection is also very fast. Therefore, ultrasonic waves are mainly used for short-range and high-precision obstacle detection of the robot. The ultrasonic ranging method obtains the distance of the target object by calculating the time from when the ultrasonic wave is transmitted to when the return signal is received. Assuming that the time elapsed from the transmission of the ultrasonic pulse to the reception is t, and the propagation speed of the ultrasonic wave in the air is c, the measured distance s of the target obstacle can be expressed as Eq. (6.1) [20]:

$$s = \frac{c \cdot t}{2} \tag{6.1}$$

(1) Temperature compensation

The propagation speed of ultrasound is mainly affected by temperature, medium, and air pressure. First, the medium has the most influencer. The speed of ultrasound in solids is greater than in liquids, and the speed of ultrasound in liquids is greater than in air. Second, the speed is also affected by temperature. It is concluded that the relationship between the speed of ultrasound in air and temperature can be expressed as Eq. (6.2) [20]:

$$c = 331.4 \times \sqrt{1 + T/273} = 331.4 + 0.607T \ (\mathrm{m/s}) \tag{6.2}$$

where T is the temperature.

(2) Selection of transmission frequency

Ultrasonic waves are sound waves with a frequency greater than 20 kHz, which causes a certain attenuation when propagating in the air, and the higher the frequency of the ultrasonic waves, the faster the attenuation. Therefore, the selection of the transmission frequency not only affects the diffusion and absorption loss of the ultrasonic wave in the air, but also has a certain influence on the obstacle reflection loss, background noise and other interference factors, so it is necessary to select a suitable transmission frequency. The selection of the ultrasonic working frequency mainly considers the requirements for ranging. When strong ranging capability is required, the consumption of ultrasonic waves in the propagation process should be minimized. Therefore, within the allowable range of working frequency, low operating frequency should be selected. When it is necessary to achieve a more accurate measurement of the obstacle surface, higher operating frequency should be selected to achieve higher resolution.

6.2.2.2.2 Determination of the position of obstacles based on ultrasonic ranging

When the inspection robot encounters obstacles during the movement, it should first determine the position of the obstacles, so that the obstacles can be avoided effectively. Therefore, the determination of the position of obstacle is crucial for the robot obstacle avoidance system [21]. According to the number of ultrasonic waves that detect obstacles, the obstacle positioning is divided into single ultrasonic positioning, two ultrasonic positioning, and multiple ultrasonic positioning.

(1) Single ultrasonic wave to determine obstacles

Assuming that an obstacle in the space is directly facing an ultrasonic transmitting module and can be completely covered by the ultrasonic transmitting module without reflecting other ultrasonic waves, then the obstacle is detected by only one ultrasonic module. Assume that the coordinate of the ultrasonic module is (x_1, y_1), and the emitted ultrasonic wave is reflected at a certain point O on the obstacle. The measured distance is d, and the coordinate $(x_1, y_1 + d)$ of the obstacle O is easily obtained.

(2) Two ultrasonic waves to determine obstacles

When an obstacle in space can be detected by two ultrasonic transmitting module, information fusion techniques are needed. In this case, triangulation information fusion can be used to achieve the positioning of obstacles. It is assumed that the ultrasonic waves emitted by the ultrasonic modules $M(-s, 0)$ and $N(s, 0)$ are reflected back by the obstacle point $S(x, y)$. Using the midpoint of MN as the origin of the coordinates, a coordinate system with the line MN as its x-axis can be established. Assume that the length of SM is l_1, the length of SN is l_2, and the angle between SM and SN is θ. According to the Pythagorean theorem, the following equation can be obtained [22]:

$$(x - s)^2 + y^2 = l_1^2 \tag{6.3}$$

$$(x + s)^2 + y^2 = l_2^2 \tag{6.4}$$

According to Eqs. (6.3) and (6.4), the coordinates of point S can be expressed by the following equations [22]:

$$x = \frac{\left|l_1^2 - l_2^2\right|}{4s}, \; y = \sqrt{l_2^2 - \frac{\left(l_1^2 - l_2^2 - 4s^2\right)^2}{4s}} \tag{6.5}$$

6.2.2.3 RFID sensor

6.2.2.3.1 RFID composition

RFID technology mainly uses wireless radio frequency signals to transmit information about targets. The RFID tag is used to mark the object to be positioned, and then the target positioning can be achieved by performing data processing on the strength of the signal, the time difference between the transmission and reception, the angle and the phase [23]. RFID technology itself has the characteristics of automatic identification, no need for direct contact, Nonline of Sight (NLOS) recognition, and low price. It has been gradually integrated into people's daily lives and is used to locate books, people, goods, etc. in schools, warehouses, logistics, etc., providing a more convenient living, learning, and working environment for people. Typically, a complete RFID system primarily consists of computers, RFID readers, RF antennas, and RFID tags.

(1) RFID reader

The reader is the core component of the RFID system. The RFID reader is mainly composed of a data processing module. If a radio frequency transmitter and a radio frequency receiver are built-in, the wireless signal can be transmitted and received, and the information in the signal can be processed. The reader can not only connect to the computer through the network interface and USB interface, but also communicate with the tag through the antenna. The reader can also control the frequency and power of the transmitting signal of the antenna through the wired interface, and can read the ID number of the tag and the strength and phase of the signal from the received wireless signal, thereby realizing the function of target recognition [24].

(2) RFID tags

The RFID tags are mainly based on microchips integrated with antennas and have a worldwide identification code, so they can be used directly to mark targets. At the same time, the information we need can be written into the tag by writing the internal chip, and the tag can be used multiple times. Besides, RFID tags can be specifically divided into active tag, passive tag, and semiactive tag based on whether a power supply circuit is provided inside the tag. Among them, the active tag itself has long reading distance and internal storage and provides energy through its own internal power supply to continuously transmit wireless signals to the environment, but it is susceptible to the size of the power supply. The passive tag mainly passively responds to the wireless signal transmitted by the reader, and then extracts energy from the obtained wireless signal to wake up the tag and backscatter the information carried by its chip, thereby having better environmental applicability. However, because there is no power inside, the distance of backscattering is limited. The semiactive tag carries a power supply inside, but only enables the chip to work to receive signals. Therefore, the antenna only needs to transmit wireless signals, which is more efficient. Besides, the RFID tags, like antennas, have different operating frequencies, and different working frequency bands are suitable for different application scenarios.

(3) RFID antenna

Linearly polarized antennas (dipole antennas) and circularly polarized antennas are two antennas currently in common use. The linear polarization (or plane polarization) of a linearly polarized antenna propagates electromagnetic waves perpendicularly or horizontally along a given plane. Therefore, to achieve a better reading range and rate, linearly polarized antennas and tags typically require a fixed position. This antenna provides the best performance when the direction of the label is known. The circularly polarized antenna propagates electromagnetic waves in a circular spiral pattern along two planes. Due to the continuous transmission of radio waves from the antenna, the electromagnetic waves can cover the tag in any direction in the environment. If the direction of the label is unknown or not fixed, circular polarization is the best choice.

6.2.2.3.2 How an RFID sensor works

First, run the driver for the RFID reader on the computer or data controller side. At the same time, the parameters of the driver can be modified to control the working mode, working channel, frequency of the wireless signal transmitted by the antenna, signal transmission power, and the sampling time of the reader. The reader then controls the antenna to transmit a wireless signal according to the settings. Within the radiation range of the antenna, the tag marked on the object to be located begins to work by receiving a wireless signal transmitted by the antenna which carries information such as the clock, data, and energy.

At this point, the tag in the environment is woken up. The wireless signal received is demodulated by the control module. And the information stored is backscattered by the wireless signal through the built-in antenna. Subsequently, the antenna connected to the reader obtains a wireless signal reflected by the tag, and the signal carries information of the tag itself such as signal strength, phase, time, and the like. At the same time, the signal is sent to the data processing module in the reader for demodulation processing. Finally, the reader sends the collected signal strength, phase, time and other information to the computer for processing and judgment. Besides, if the RF antenna is installed inside the reader, there is no need to connect an external antenna, and the wireless signal can be directly transmitted and received through the internal antenna. But by doing this, the reading range of the reader will be greatly limited [25].

Since the RFID is mainly based on wireless signals for related data transmission, it has the following advantages. Within the radiation range of the antenna's electromagnetic field, information about the RFID tag can be directly identified by the RFID reader. It can avoid the restrictions of direct contact and the line-of-sight, and can penetrate nonmetallic objects. Since the wireless signal is transmitted at a fast speed, the reader can recognize multiple RFID tags in a short time interval, so that it can be applied in a complicated environment. With the development of semiconductor technology, the capacity of the RFID tag chip has been greatly expanded, making it more widely used. The content of the tag chip can be repeatedly written and modified, so the tag can be reused and resources can be saved. The ID number of the RFID tag uses the electronic product code, which is globally unique, so it has high security and can be used to mark special targets in special scenes. Since the RFID tag itself is based on a microchip, it has the advantages of small size and convenient carrying [26].

However, the RFID technology also has many disadvantages. Among them, a more complicated method is needed to perform positioning because the RFID technology is mainly based on the signal strength, transmission and reception time, phase, angle and other information of wireless signals for positioning, marking and other work, and it cannot directly provide the location information of the target. When using RFID technology, its signal strength information is often used for data analysis. Therefore, it is necessary to establish a signal strength model and train it multiple times, which requires a long preparation time before use. At the same time, the radio frequency signal is greatly affected by factors such as metal and water in the environment, which is likely to cause information interference, resulting in less accurate positioning.

Besides, in the indoor environment, the radio frequency signal is prone to multipath effect, which causes the signal received by the reader to have large errors and more interference, and makes the positioning accuracy worse. Besides, the method based on the difference in arrival time and angle of the RFID signal requires high-precision hardware devices, and requires more difficult operations and analysis methods for positioning or detection. At the same time, the phase information of RFID can also be used for positioning, but there will be inevitable full-cycle blur, which can cause a large ranging error. Finally, the standards of RFID in different countries are not the same. Each country has its own RFID frequency band, which makes the development and promotion of RFID technology more difficult.

6.2.2.3.3 Principle of RFID measurement

When the RFID tag in the environment senses a radio wave, the internal circuit is activated. It then transmits the information it carries to its antenna, and the reader interprets the signals after receiving them through its antenna. Phase information and signal strength information of the RFID can be obtained after interpretation. The information can be used to estimate the distance between the tag and the reader, but the accuracy of the estimation is not high due to the environmental influences.

(1) Phase information

The main principle of phase ranging is to estimate the distance between the receiver and the transmitter by multiplying the signal round-trip time by the speed of light. The round-trip time of the signal can be obtained by the ratio of the phase to the angular frequency of the signal. Therefore, the distance between the transmitter and the receiver can be expressed as [27]

$$2d = \frac{\theta}{\omega} \cdot c \tag{6.6}$$

where d is the distance between the transmitter and the receiver, θ is the phase of the signal, c is the speed of light, and w is the angular frequency of the signal. And from the equation $\omega = 2\pi f$, the following equation can be obtained [27]:

$$d = \frac{c \cdot \theta}{4\pi f} \tag{6.7}$$

where f represents the frequency of the signal. The relationship between phase information and distance lays the foundation for measuring the radial velocity.

(2) Signal strength information

In an indoor environment, signal strength is related to distance. The farther the distance between the receiver and the transmitter, the weaker the signal strength obtained. The change of signal strength with distance can be represented by a signal attenuation model [28]:

$$d = 10^{[p(d_0) - p(d)]/10n} \tag{6.8}$$

where d_0 represents the reference distance and n represents the path attenuation index. The distance between the tag and the reader can be roughly estimated by Eq. (6.8). However, in the process of propagation, the signal is easily affected by the metal and water in the environment, and the distance accuracy estimated by this method is also low. The attenuation characteristics of signal strength can be used to establish an RFID sensor model to reduce the error caused by the environment on the estimated distance.

6.2.2.4 Sound collection device

The sound collection of the inspection robot refers to the collection and processing of abnormal sounds heard by the inspection robot during the inspection process. By identifying the abnormal sound, and performing feature extraction and feature matching on the sound in the controller of the lower computer, it is determined whether the inspection environment is abnormal. The inspection robot sends the position information of the place where the abnormal sound is heard during the inspection process to the upper computer, then the upper computer will inform the inspector after receiving the information.

The hardware for monitoring sound uses an electret microphone as the sound monitoring unit, which is a capacitive "sound-electricity" converter. The electret material is a substance that is formed by the polarization of the high molecular polymer, which maintains the state of the electrode for a long period of time. Microphones made of electret materials have the characteristics of small size, high fidelity, high sensitivity, wide frequency range, and low price. The left side of the electret film is coated with metal. There is a very small gap between the electret film and the metal back electrode. The electret and air are used as insulators, and the back electrode and the metal on the electret form the two poles of the capacitor. When the vibration of the sound wave causes the distance between the two poles to change, the capacitance of the capacitor will also be changed accordingly. It is known from equation $Q = UC$ that when the amount of charge Q does not change, if the capacitance C changes, the voltage U will change accordingly, thereby achieving sound-electricity conversion [29].

Electret microphones usually have very small capacitors, and the output electrical signal is extremely weak. Because the output impedance value is very high, which can be as high as hundreds of megaohms or more, the electret microphone cannot be directly connected to the input circuit of the audio amplifier. To solve this problem, a junction field-effect transistor (JFET) is introduced into the output electrode of the microphone to perform impedance transformation. When the voltage of the capacitor changes, it will cause the drain current of the JFET to change, and then the impedance transformation is realized. The JFET must operate with an external DC voltage, so the electret microphone is an active device.

6.2.3 Pan/tilt

Pan/tilt integrates the infrared thermal imager and the binocular vision camera. During the inspection process, the two sensors can capture the equipment and equipment in the monitoring environment in all directions. Therefore, the pan/tilt is required to rotate horizontally in all directions, and 90° rotation in the vertical direction. A wiper is installed on the gimbal to ensure that the robot inspection work can also operate normally in rainy weather conditions.

6.2.3.1 Infrared thermal imager sensor

When British scientist William Herschel studied the solar thermal effect, he split the sunlight into seven different monochromatic lights. When he moved the thermometer outside of the red light, he found that the temperature had risen, so he discovered infrared. Subsequent experiments by many scholars have confirmed the existence of infrared. Any object in nature with a temperature greater than absolute zero will always radiate infrared. The amount of energy radiated depends on the chemical composition, physical structure, volume shape, and other factors of the radiating object [30]. Since infrared cannot be directly observed by people and there is a one-to-one correspondence between temperature and infrared radiation, so infrared thermal imaging technology has been formed.

Through the infrared thermal imaging technology, the surface temperature of the object can be obtained in a noncontact manner. The professional thermal analysis software supporting the infrared thermal imaging system is used to monitor the image. Objects with abnormal temperatures can be warned in time. Moreover, using infrared thermal imaging technology, infrared images can be obtained at long distances without contacting the measured object. The infrared image and visible light image are different. The pixel value of the infrared image represents the temperature of the surface of the measured object. Therefore, infrared thermal imaging technology is widely used in unmanned inspection, security monitoring, night vision imaging, and other fields [31].

Infrared is part of electromagnetic waves and forms a continuous and complete electromagnetic spectrum with gamma rays, X-rays, ultraviolet rays, visible rays, and radio waves. Infrared has a wavelength range of 0.76—1000 μm, extending from the red-light edge of the visible light region to the radio wave boundary. In the field of infrared detection research, the infrared spectrum is generally divided into four bands, namely, the near-infrared band (0.76—3 μm), the midinfrared band (3—6 μm), the far-infrared band (6—15 μm), and the extrafar-infrared band (15—1000 μm). The first three infrared spectral bands each contain an atmospheric window of 2—2.5 μm, 3—5 μm, and 8—14 μm [32]. In these bands, the attenuation of infrared radiation energy is small, and the transmittance of the infrared is small. Therefore, most infrared detection systems use these infrared bands that correspond to the atmospheric window.

The imager for infrared thermal imaging was invented in the United States in the 1960s. After that, researchers around the world researched infrared technology. The infrared thermal imager can capture the infrared radiation radiated by the object, which is then converted into an electrical signal by the infrared device to form a visible infrared image. The main functions of the infrared thermal imager are the following: (1) extension of human observation range, (2) detection of fast-moving objects, and (3) monitoring of thermal motion. The infrared thermal imager can filter the interference in the air environment, has strong antiinterference ability. It can detect infrared signals and preprocess the obtained images. The infrared thermal imagers can be divided into two types from the perspective of the working principle: photoelectric detectors and thermal detectors. The principle of the photoelectric detectors is to use the photoelectric effect of the material to convert optical information into electrical information. The principle of the thermal detectors is to directly detect the infrared radiation of the target object, convert the infrared radiation into a measurable physical quantity, and then transmit it to the host computer.

6.2.3.2 Binocular vision sensor

In nature, sunlight hits an object, and the surface of the object will absorb part of the sunlight, and the reflected sunlight is captured by human eyes. Binocular stereo vision imitates the imaging principle of human eyes. Two cameras that collect light are installed on the camera to obtain respective images, and then a 3D model of the image is constructed using the principle of triangulation [33]. A complete binocular stereo vision system is divided into five parts: (1) camera calibration, (2) image acquisition and preprocessing, (3) feature extraction, (4) stereo reconstruction, and (5) 3D reconstruction.

(1) Camera calibration

Camera calibration is the primary problem in binocular stereo vision systems and is the foundation and prerequisite [34]. Camera calibration needs to obtain the basic parameters and distortion coefficients of the camera, and to establish the geometric model of the camera imaging. It is necessary to obtain the imaging characteristics of the binocular camera, the conversion equation of the binocular camera coordinate system and the world coordinate system, and establish a point-to-point mapping relationship. The main calibration methods are given as follows:

(a) Traditional photogrammetry calibration methods

The method expresses the conversion between each image and object with a constraint relationship of not less than 17 parameters, which can achieve good precision, but it brings a lot of calculation due to too many parameters.

(b) Two-step calibration

The method is currently the most widely used calibration method. The calibration algorithm proposed by Zhang is simple to operate, and only needs to take multiple photos of the calibration plate to obtain multiple photos to calibrate the camera [35].

(c) Perspective transformation matrix calibration method

The method is relatively simple to operate. It only needs the position of the 3D space of the calibration point and the image coordinates corresponding to the calibration point to perform camera calibration. Because it is simple, the calibration calculation can be performed in real-time.

(d) Double plane calibration method

The method requires more accurate external parameters, which are used for dual-camera calibration. The calibration has at least 10 3D coordinates of the calibration point. This calibration method is affected by the nonlinear distortion of the lens.

(e) Direct linear transformation calibration.

The algorithm for solving linear equations is used to obtain the parameters of the camera model. The problem of nonlinear distortion of the lens is not considered in the algorithm. The calibration accuracy needs to be improved using a nonlinear optimization algorithm.

(2) Image acquisition and preprocessing

Image acquisition and preprocessing are important parts of the binocular stereo vision system, which determine the quality of the images and the effect of subsequent technical processing. The main methods of image preprocessing are smooth filtering of the image, gray level change, image threshold setting, gray level histogram equalization, geometric transformation, edge detection, normalization, and image enhancement. The main purpose is to highlight the desired image features of the target object, remove other unrelated features and noise as much as possible, and prepare for subsequent work.

(3) Feature extraction

Feature extraction is a primary operation in the binocular stereo vision system. The feature extraction is to highlight the image data of a certain type of feature and extract such features from the image, mainly through image analysis. Generally, such features have digital attributes that are different from other image features, such as pixel values, image edges, specific colors, brightness, and shape features. The process is for the computer to identify the digitized image and look for specific features. The time of feature extraction can be controlled by setting the hierarchy of feature extraction. Many feature extraction algorithms have been developed, and the features they target are also diverse, with varying degrees of complexity and accuracy [36].

(4) Stereo matching

Stereo matching is a key step in the binocular stereo vision system. By establishing a cost function, stereo matching is performed on the target objects in the images taken by the left and right cameras. The cost function is used to minimize the estimated pixel disparity value. It is actually the parallax value that solves the optimization problem. The generation of parallax values is mainly caused by the different positions of the left and right cameras. Second, environmental noise, occlusions, and repeated textures all have a greater impact on the calculation of parallax values. Stereo matching methods are generally divided into three categories: stereo matching based on image features, stereo matching of regions based on image gray values, and stereo matching based on image phase.

The stereo matching based on image features has little dependence on the gray value of the image, so it has strong robustness to the influence of noise such as illumination. Another advantage is that the matching method is based on image features for stereo matching. It is small, and the matching speed is fast. However, stereo matching based on image features also has some disadvantages. For example, feature matching can only match feature points between image pairs, so the parallax field obtained after matching will be sparse. Besides, the method requires a high selection of features and extraction, because this directly affects the accuracy of the matching.

The stereo matching of regions based on image gray values makes full use of the gray value of each pixel of the image, and can obtain a relatively dense parallax field. Moreover, the algorithm can achieve good matching precision in images with rich details, and has the advantages of easy implementation and good real-time performance. It is the most widely used local area stereo matching algorithm. However, the algorithm also has certain disadvantages, such as sensitivity to noise and poor matching accuracy in the case of parallax discontinuity.

Image phase-based matching is a relatively new matching algorithm. The matching primitive of the algorithm is the phase, which gives the algorithm the advantage of suppressing high-frequency noise. Besides, the algorithm can also obtain parallax with subpixel accuracy. The disadvantage of the algorithm is the problem of phase singularity and phase winding.

(5) 3D reconstruction

Three-dimensional reconstruction is to construct a 3D image of the target object from the two-dimensional (2D) image in the binocular by using the obtained depth information of the 2D image. The difficulty lies in how to obtain the distance information of the target object. Usually, the intersection of the two lines is determined by the calculation of the triangle method. The target position information is obtained, and then a 3D model is constructed [37].

6.2.4 Wireless recharging devices

Most of the existing automatic recharging technologies for mobile robots use contact recharging methods. To dock the robot with the recharging device, laser range finder, vision sensor, and infrared detector are usually used. This kind of method has the following problems: first, the robot needs to perform navigation behavior to move from the current position to the recharging device, which is limited by positioning errors and the navigation accuracy is low. Second, the docking of the robot and the recharging device requires high precision, which increases the complexity of the design and the difficulty of control. At the same time, the docking operation process is very complicated and takes a long time. Finally, due to structural design and other reasons, the robot and the recharging device cannot be automatically disconnected once they are connected. Frequent docking can also easily affect the reliability of the system. For example, multiple plugging and unplugging operations can cause mechanical wear and tear, resulting in loose contacts and ineffective transmission of electrical energy.

If the connecting parts are dirty, it will cause poor contact or electrical connection failure; it is also extremely likely to cause a short circuit in a humid environment or an environment full of conductive medium. It is difficult for the robot to achieve the recharging requirements of high navigation accuracy, high-speed positioning, and highly reliable docking. In the rail transit inspection process, wireless recharging is preferred as automatic and efficient recharging methods [38].

There are three types of radio energy transmission methods: microwave, electromagnetic induction, and magnetic coupling resonance. First, the basic principle of microwave radio energy transmission is electromagnetic radiation, which converts the power signal at the transmitting end into an electromagnetic wave signal, and conducts directional propagation through the antenna at the transmitting end [39]. The transmission distance is relatively long, which is more suitable for energy transmission in space. However, electromagnetic waves are easily scattered during transmission, they cannot emit energy in a concentrated manner, and they cannot pass through obstacles. The transmission efficiency is extremely low, and the power can usually only reach the milliwatt level, which is not suitable for daily radio energy research. Electromagnetic induction radio energy transmission is more efficient, but the transmission distance is usually at the centimeter level. This method has strict requirements on the position of the recharging device, and the recharging efficiency will rapidly be degraded with a slight angular deviation. There are already many products using this kind of radio energy transmission, such as electric toothbrushes, electric cars, and sweeping robots.

The basic principle of magnetic coupling resonant radio energy transmission is electromagnetic induction and electromagnetic resonance. In the two coils separated by a certain distance, the magnetic field coupling generates electromagnetic resonance, and establishes an energy transmission channel. The transmitting coil is connected to the power supply to continuously provide the resonant system with energy, and the receiving coil receives energy under strong coupling resonance. In this way, efficient power transmission from the transmitting end to the receiving end is achieved. Compared with microwave energy transmission, this technology has higher transmission efficiency, less impact on living organisms, and is not affected by general obstacles. Compared with the electromagnetic induction transmission, the energy transmission distance of this technology is farther, it can theoretically reach several tens of centimeters to several meters, and the requirements for the position are not so harsh.

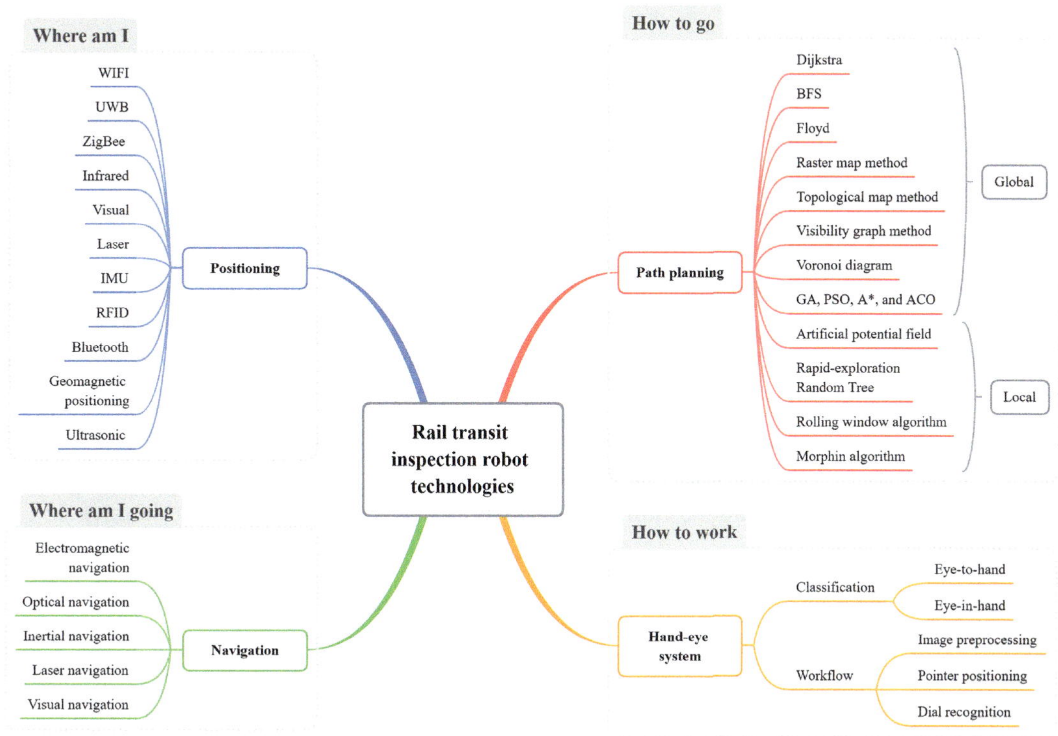

Figure 6.4
Rail transit inspection robot technologies.

6.3 Key technologies in rail transit inspection robots

The main technologies of the rail transit inspection robot are positioning technology, navigation technology, path-planning technology, and hand-eye system, which are shown in Fig. 6.4.

6.3.1 Navigation

The positioning of the mobile robot is to sense the environment and obtain environmental information through some sensors in a 2D environment, and then determine the coordinates and posture of the robot itself after processing the information. At present, positioning methods for robots mainly involve the following technologies: Wireless Fidelity (WiFi) signal, RFID, ultrawideband (UWB), ZigBee, Bluetooth, Infrared, Ultrasonic, Inertial Navigation, Geomagnetic Intensity, visual Simultaneous Localization and Mapping (SLAM), and laser SLAM [40,41]. Navigation methods for robots mainly include the following: electromagnetic navigation, optical navigation, inertial navigation, laser navigation, visual navigation, and GPS navigation [42].

6.3.1.1 Positioning methods

(1) WiFi positioning

Since WiFi devices are widely used indoors, it is relatively convenient to deploy WiFi-based positioning methods without adding more additional hardware devices. The WiFi positioning algorithms can be classified into two categories according to whether it is based on the Received Signal Strength Indication (RSSI) [43].

The RSSI-based positioning method can be further divided into two types: triangular positioning and fingerprint positioning. Assuming that the positions of the Access Points (APs) are already known, the distances between the robot and each AP can be calculated based on the signal attenuation model using the triangular positioning method. Based on the obtained distances, circles can be drawn around each AP, and the intersection of these circles is the position of the mobile robot. This method requires measuring the position of the AP in advance, so it is not suitable for occasions where the environment changes rapidly.

The fingerprint positioning method has two phases: the offline training and online positioning. During the training phase, the room is divided into small area blocks to establish a series of sampling points. During the training phase, the room is divided into small blocks to create a series of sampling points. The WiFi receiving device is used to sample these points one by one and record the corresponding RSSI value and AP address. Using these recorded data, a database can then be established. During the online positioning phase, the mobile robot uses its WiFi signal receiving device to obtain current RSSI values and AP addresses of surrounding APs. After the information is transmitted to the server and matched with the established database, the estimated position can be obtained. Commonly used matching algorithms mainly include K-Nearest Neighbors (KNN), neural networks, and so on. The workload of fingerprint collection is heavy, and the fingerprint needs to be collected again after the location of the robot is changed. The advantage of WiFi positioning is that the arrangement of the WiFi signal generator is easy and the cost is low. The disadvantages of WiFi positioning include high power consumption, vulnerability to interference, and vulnerability to the movement of WiFi devices [44].

(2) ZigBee positioning

Like WiFi positioning, ZigBee positioning is also based on the principle of RSSI. The technology is characterized by a wide signal range, high reliability, low data rate, and good topological capability [45].

(3) Ultrawideband (UWB) positioning

As a wireless communication technology, UWB technology uses nonsinusoidal narrow-band pulses with very short periods to transmit data. The UWB can be used for positioning in industrial warehouses, nursing homes, airports, and other indoor areas. The positioning of the robot can be achieved by using multiple UWB modules. The UWB positioning technology has obvious antiinterference ability, strong penetrating ability, accurate positioning ability, high data transmission rate, strong multipath resolution, and low power consumption. The theoretical positioning accuracy can reach the centimeter level. However, to achieve high-precision positioning, the number of layout points needs to be very large, which leads to high costs [46]. The UWB positioning technology mainly includes two methods: ranging technology and positioning technology [47].

The UWB can be implemented by relying on two methods, the signal strength received by the receiver and the time of arrival. However, after repeated refraction of UWB signals, the signal attenuation phenomenon is serious. The method of time of arrival is more suitable for ranging. First, the delay of the direct path or the optimal path of the UWB signal is detected [48]. The UWB signals are used as low-duty-cycle shock pulses. Various noises in the environment may cause interference with the transmission of the UWB signals. Multiple return paths of the same signal in an indoor obstacle-dense environment can also interfere with the detection of delays. Then, the UWB technology can perform data transmission over a very wide bandwidth, and the signal interception rate is low. Compared with the characteristics of GPS positioning that must be within its visible range, the UWB signals can penetrate walls of a certain thickness to achieve centimeter-level high-precision indoor positioning. With the application of the existing noncoherent detection technology, the signal receiving technology does not need to process the channel, reduces the calculation time, and ensures real-time positioning. However, as the computational difficulty decreases, the effective information used also decreases. To ensure the unity of the two is now a difficult point.

The UWB positioning method is an improvement of the ranging method. The information of all points for UWB ranging is integrated to determine the position of the target point. At present, the UWB positioning methods are mainly divided into two types: noncooperative positioning and cooperative positioning. In the noncooperative positioning method, target nodes do not communicate with each other. The target node only completes the position estimation based on the collected distance information to a reference node with a known position. In the cooperative positioning method, the target node not only measures the distance to the reference node, but measures the distance to other target nodes, and use all the collected ranging information to estimate the position.

Compared with the traditional noncooperative positioning method, the cooperative positioning method improves the positioning accuracy and expands the coverage [48]. As a result, it has received widespread attention. Comparatively speaking, the positioning technology has good real-time performance. In the case of dense multipath, UWB transmission can provide reliable communication, and can resolve the distance resolution with subnanosecond delay, and the positioning accuracy can reach several centimeters. Therefore, the technology has received great attention from the academic and industrial fields, and has become the most competitive technology for achieving reliable indoor communication and precise positioning.

(4) Infrared positioning

Infrared positioning technology can be divided into two types according to whether a signal source is carried. The robot carrying the signal source emits a modulated infrared signal, triggers the receiver installed in the environment, and can derive the position of the robot based on the received information. Another method of carrying a signal source is that the robot emits an infrared signal, the beacon placed in the environment can reflect the infrared signal. The robot receives the reflected infrared information, identifies the beacon ID and direction, and locates itself according to the information.

The positioning technology that does not carry a signal source includes an infrared mesh-based positioning technology, which utilizes a signal detection network composed of a plurality of pairs of infrared emitters and infrared receivers to cover the entire space to be tested, and the objects in the space that needs to be positioned. The high-precision positioning can be achieved without adding any auxiliary equipment. The positioning method is relatively mature at present, and the positioning accuracy is relatively high, but it has poor penetration, and can only be used for line-of-sight measurement and control. Moreover, it is susceptible to environmental influences such as lighting. Therefore, the application of this method is limited [49].

(5) Visual positioning

Visual positioning technology uses a camera to collect information about the surrounding environment and the position can be derived based on the information. At present, the main positioning methods include beacon-based absolute positioning, visual odometry, and visual SLAM method. As computer image processing technology developed rapidly, visual positioning technology has gradually been widely studied in the field of robotics, and it has developed rapidly. The positioning accuracy can reach the centimeter level. However, the stability of this method is not high at present, and it can be seriously affected by light [50].

(6) Laser positioning

LiDAR includes both 2D laser radar and 3D laser radar. For 2D laser radar, only 2D information at the same height can be collected. There are two commonly used positioning methods, one is 2D laser radar SLAM, and the other is based on the positioning method of the reflector. The former has strong adaptability, but the antiinterference ability and stability are poor. For the second positioning method, the accuracy is high and the stability is strong, but the reflectors need to be accurately set in advance. The 3D laser radar is generally used in unmanned vehicles, but there are not many indoor applications [51].

(7) RFID positioning

Radio frequency identification (RFID) and positioning technology can also be used for positioning. There are some advantages of RFID and positioning technology. Multiple RFID transmitters can be read at the same time, and the scanning speed is fast. The body shape is small and diverse. Memory data can be modified based on popular needs. It can penetrate nonmetallic substances such as wood to achieve penetrating communication [52].

Nevertheless, there are also disadvantages of RFID and positioning technology. The cost is too high and the working distance is short. The remote scanning device can scan the identification number of all transmitters, which can cause privacy problems.

(8) Bluetooth positioning

Bluetooth positioning technology is based on signal strength. The longer the distance between the transmitter and the receiver, the lower the signal strength. After receiving the signal strength, the receiving device can estimate the distance between the receiver and the transmitter according to the strength of the received signal. Based on the distance, the coordinates of the measured object can be obtained. The Bluetooth positioning technology mainly uses the broadcast function of Bluetooth beacon to measure the signal strength. The Bluetooth beacons installed at multiple fixed points indoors continuously send broadcast messages that contain the transmit power. After receiving these messages, the receiver can use the information to estimate the distances between the receiver and the beacons. The positioning data can be derived from the calculated distances.

The cost of Bluetooth positioning is low. The positioning accuracy of the Bluetooth positioning can reach up to 2 m, which can meet the needs of ordinary indoor positioning. The deployment of Bluetooth devices is relatively simple. Moreover, the penetration and antiinterference ability of Bluetooth signal are relatively strong. But for the complicated environment, the Bluetooth positioning is susceptible to the interference of noise signals, which will lead to slightly worse stability [53].

(9) Geomagnetic positioning

Earth itself is a huge magnet that forms a geomagnetic field that runs through the north and south poles. The geomagnetic field has different magnetic field strengths in different latitudes and longitudes [54]. The geomagnetic field has parameters such as geomagnetic total field, magnetic dip, magnetic declination, magnetic field gradient, and so on. The geomagnetic field can be interfered with by the metal objects and the parameter information of the magnetic field will be changed, resulting in the formation of a new magnetic field. For buildings that use steel bars, each of these buildings has its own unique indoor magnetic field because these metals have different effects on the geomagnetic field.

The indoor geomagnetic positioning is realized by measuring the regular characteristics of the indoor magnetic field. The geomagnetic sensor is used to collect the parameter of the indoor magnetic field, and identify the differences of the magnetic field parameters at different positions. Through these differences, the relative position of the sensor can be matched. However, the measurement accuracy of the geomagnetic field is not high, and the geomagnetic positioning is usually used in combination with inertial navigation. The combination can also correct the accumulated errors in inertial navigation.

In the implementation of geomagnetic positioning, no additional hardware and equipment are required for construction, and only geomagnetic sensors are required. The errors of geomagnetic positioning do not cumulate over time. As mentioned above, the geomagnetic positioning can be combined with the inertial navigation system to form an integrated navigation system. Through information fusion, more comprehensive external information can be obtained, and the cumulative error generated by the inertial navigation system can be corrected in time.

Nevertheless, geomagnetic positioning still has some shortcomings. The use of the geomagnetic positioning begins with the collection of geomagnetic data on-site, but the geomagnetic field can be changed by interference with metal objects. If the indoor environment changes greatly, the indoor magnetic field will also change. Therefore, the geomagnetic data will also be changed due to the change of the indoor magnetic field, then the geomagnetic data needs to be reacquired and updated. Moreover, the outdoor environment also has effects on the indoor magnetic field.

(10) Ultrasonic positioning

Ultrasound can be used to detect the distance for robot positioning. In the positioning system, the ultrasonic transceivers are set at different positions. The mobile robot also carries the transceivers, and sends out ultrasonic waves. The transceivers at specific locations receive the waves sent by the robot and transmit the corresponding signal. After the signal is accepted by the robot, the robot can be located through unilateral positioning algorithm or reverse ranging algorithm. The positioning accuracy of the ultrasonic positioning method can reach centimeter-level and its penetration ability is strong. However, the ultrasound wave is a mechanical wave, and the intensity of the wave will gradually decrease during the transmission process. After repeated reflections, the attenuation of the wave is severe, resulting in a short detection distance and high cost [55].

6.3.1.2 Navigation methods

(1) Electromagnetic navigation

Electromagnetic navigation is one of the traditional navigation methods. The method ambushes a metal wire on a preset travel route, and generates a magnetic field when the wire is energized. The mobile robot runs along the preset route by detecting the strength of the magnetic field during its running. The method is easy to implement. It is not affected by weather and light conditions and has a low cost. The disadvantages are that the path is not easy to be modified when the path needs to be expanded or changed, and the path is susceptible to interference from other magnetic fields [56].

(2) Optical navigation

In the method, reflective bands are set on the path of the inspection robot. The light source installed on the robot illuminates, and the photoelectric sensor installed on the robot receives the light signal reflected by the bands. The position of the path can be determined by the calculation of the received signals, and is used to adjust the running direction of the inspection robot. The construction cost is lower than that of the electromagnetic navigation method, but it can be easily affected by the illumination of other light sources.

(3) Inertial navigation

Inertial navigation technology is called inertial navigation because it uses a sensor with an inertial frame as a reference to get navigation results. The core component of inertial navigation is the inertial sensor. The inertial sensors are used to measure the acceleration, angular velocity, and angular increase of the object being measured. When the same measured object moves under each different inertial reference frame, its displacement and speed may be different, but the acceleration, angular velocity and angle increase of the measured object are the same. The inertial navigation integrates acceleration over time and transforms the results of the integration over a series of transformations to get position, angle, etc.

The main advantage of inertial navigation is strong antiinterference. Inertial sensors will not lose accuracy due to interference from external environmental signals such as sound waves and infrared rays. The inertial sensors work fast, which makes inertial navigation using inertial sensors more suitable for positioning flexible objects than other positioning techniques.

Disadvantages of inertial navigation are as follows: the positioning error of inertial navigation is cumulative, that is, the previous positioning error will affect the next positioning, which will cause the error to expand, and the positioning data of inertial navigation over time will be very inaccurate. Therefore, other navigation methods are often combined with inertial navigation [57]. This will make the positioning method retain the advantages of strong antiinterference and high speed of inertial navigation, and will also limit the error to a certain range.

(4) Laser navigation

In this method, reflectors are installed next to the operating route of the mobile robot. The mobile robot can detect the distances between itself and the different reflective plates by the carried LiDAR, and calculate the position and direction of itself for navigation according to the triangular geometric operation. The method has high precision and can adapt to various complicated situations and working conditions. The change of the path is also relatively easy. However, the method has certain limitations in resisting light interference [58].

(5) Visual navigation

In the method, the image sensor such as a camera is used to detect the information carried by the ribbon tag or the 2D code on the road surface. The relative position between the mobile robot and the mark is calculated by image processing technologies. Then, the position of the mobile robot can be determined and the running direction for navigation can be obtained and controlled. This method can easily change the guiding path after the marking is completed, which is easy to maintain and has high positioning accuracy. However, it is susceptible to light, and the position accuracy and density of the marks are required to be high, which can lead to troublesome laying [59].

6.3.1.3 Path-planning methods

The path-planning of mobile robots has always been the focus and difficulty of research in the field of robotics. The path-planning is to find a barrier-free, collision-free, relatively optimal path for the robot's operation to ensure the safety of the robot and the surrounding environment [60]. In general, the cost function of path-planning is to target the shortest path or the shortest running time from the start point to the end point. In the past research, the path-planning of mobile robots is divided into two types according to the robot's known information about the surrounding environment: global static path-planning and local dynamic obstacle avoidance. The global path-planning can handle the path-planning under the completely known environment (the position and shape of the obstacle are predetermined). The premise of the global path-planning is to establish a global map model of the environment in which the mobile robot is located. The search for optimal path is performed on the established global map model and the optimal path is obtained by optimization algorithm. Therefore, global path-planning involves two parts: the establishment of an environmental model and the path search strategy. The local path-planning mainly involves how to avoid obstacles safely [61].

6.3.1.3.1 Global path-planning method

(1) Dijkstra algorithm

The Dijkstra algorithm uses the labeling principle to achieve the search for the shortest path. The Dijkstra algorithm takes the approach of constantly naming the vertices. After the algorithm starts, each time a vertex is labeled, the value of the label is the shortest path weight from the given source point to the vertex. First, the vertex closest to the source point among all the vertices connected to the fixed source point with edges is found. Recursively, suppose a part of the current shortest path has been found (consists of n nodes with the shortest distance from the source point and the corresponding n shortest paths), the shortest path weights from n nodes to the source point will become their respective permanent labels. For all vertices u that are unlabeled, m paths are formed from the source point to the unlabeled point. The one with the smallest weight among these paths is selected, and its weight will be the temporary label of the vertex u. Then, in the same manner, the temporary labels for the other unlabeled vertices can be obtained, and the one with the smallest weight among all the temporary labels is finally selected. The vertice with the smallest label (set to x) is the $(k+1)$-*th* vertice to find. The label of x is its permanent label, and the shortest path spanning tree generates to x. Repeat the above process until all vertices have permanent labels [62].

(2) Best-first search (BFS)

The best-first search (BFS) algorithm is a heuristic function [63]. It has some similarities with the form of the Dijkstra algorithm, and the difference is mainly focused on the size of the cost of the algorithm between the node and the target point. It does not pick the position closest to the initial point, but rather biases the position near the target point. The disadvantage of this algorithm is that it may not be able to jump out of the local optimal path. This algorithm is significantly faster than Dijkstra's algorithm.

Heuristic search is different from blind search. The blind search is a comprehensive search without a purpose based on a fixed pattern. In the heuristic search, in the optimal path-planning, some favorable information that can be calculated in advance is used as the cost function, and the program uses the information provided by the cost function to perform a more intelligent search. Algorithms for path optimization through heuristic information are collectively called heuristic search algorithms.

(3) Floyd algorithm

The Floyd algorithm is also called the interpolation method, which is one of the most classic algorithms for path-planning optimization. The Floyd algorithm first needs to establish a point-to-point distance library, and then perform optimization operations through an algorithm containing three loops. The Floyd algorithm is simple and can get the optimal path, which is suitable for map optimization with fewer path points.

However, once the map is huge, the calculation volume of the Floyd algorithm will be very huge [64].

(4) Raster map method

The raster map method is currently the most widely used and most effective path-planning method. The basic idea is to use certain size rasters to represent obstacles and passable areas in the environment of a mobile robot. Obstacle rasters are typically displayed in black in raster maps and passable rasters are displayed in white. The mobile robot uses each raster as a node, and searches for the path starting from the starting node to find the target node. The raster map method is simple to represent the environment, and it is easy to implement and maintain. However, there are also disadvantages that too large rasters lead to low map accuracy, and too small rasters lead to large computational complexity. Therefore, choosing a reasonable raster size is a key issue [65].

(5) Topological map method

The topological graph method represents the motion space of a mobile robot with a series of nodes with topological features. Based on the connectivity between the nodes, a topology network is constructed, and the geometric path of the mobile robot from the starting point to the end point is searched on the constructed relationship network. This method has small requirements and simple planning, and is more suitable for simple environment models and large search space. At the same time, it is not necessary to consider the precise positioning information of the mobile robot, and it is robust to the errors generated. However, when the topology model is used to establish an environment model, the process may be complicated and difficult to implement. In particular, the existence of multiple similar obstacles in the environment cannot effectively determine the node's ownership, which greatly limits the application scope [66].

(6) Visibility graph

The visibility graph method is to connect the starting point, the target point of the mobile robot, and the vertex of obstacle. It finds an optimal path from the starting position to the target position under the premise of ensuring that the line does not pass through obstacles. The visibility graph method is simple to implement and is suitable for path optimization in sparse environments. However, when the starting position and the target position change, the drawing needs to be remapped. When the obstacle increases, the complexity of reconstructing the graph also increases, resulting in a lack of flexibility.

(7) Voronoi diagram

The Voronoi diagrams method can be regarded as the method to improve the visibility graph method. By fully considering the distance between the point on the path and the obstacle, the mobile robot can be guaranteed to be as far away from the obstacle as possible. Therefore, it has better performance than the visibility graph method, but there are also deficiencies in finding the optimal path frequently. Based on the built environment model, a reasonable and effective search algorithm is used for path search [67].

(8) Genetic Algorithm (GA)

The genetic algorithm is a bionic intelligent path-planning algorithm. The basic idea is to use "chromosomes" to encode every possible solution to the problem. In a population composed of multiple chromosomes, the optimal solution to the problem is gradually obtained through operations such as selection, crossover, and mutation in accordance with the principles of biological evolution. The method is easy to use and has good global search ability. Its main advantage is that it is easy to jump out of the local optimal solution and is suitable for global path-planning. However, there are also defects such as long computation time and slow convergence speed [68].

The genetic algorithm equates the problem of path-planning optimization to the evolution process of species. In the early stages of species formation, there will be a fixed population called the initial population. In path-planning, it is equivalent to using some methods, such as random sampling, to generate some initial paths. The ability of the newly produced species to adapt to the environment varies, and highly adaptive individuals will continue to thrive. The algorithm measures the strength of the individual's adaptability to the environment by setting a fitness function. This function can calculate the fitness value of each individual in the initial population. The path with a high fitness value will be retained by the probability method or the "roulette" method. This process corresponds to the aforementioned selection operation.

The selected generation of species will produce crossovers and mutations in genes, which is a manifestation of species diversity in biological evolution, allowing species to better adapt to the environment and survive. In path-planning, crossover and mutation are also required to achieve diversity of path, and to prevent local optimal solutions and premature phenomena. Although crossovers and mutations are random and subtle changes, they are also subject to artificially set probabilities. This creates a new generation of individuals that are gradually optimized. Through continuous iteration, the path with low fitness is gradually eliminated, and it is possible to generate individuals with high fitness after the evolution of the generations, that is, the path with the shortest distance [69].

The advantages of the genetic algorithm are presented as follows:

(1) Parallelism: Many individuals can be evaluated at the same time, making the global search ability of the genetic algorithm stronger.
(2) Easy operation: The genetic algorithm does not optimize the problem itself, but the structural objects obtained by encoding the problem, which makes the computer easier to process and calculate.

(3) Heuristic random search ability: The genetic algorithm uses the probabilistic criterion to control search objects instead of using the deterministic criterion, which makes the search faster and more accurate.
(4) Convergence: The genetic algorithm evaluates the individual through the fitness function. In other aspects, the genetic algorithm is easy to integrate with other algorithms, thereby improving the overall effect.

(9) Particle Swarm Optimization (PSO) algorithm

The particle swarm optimization (PSO) is an intelligent global optimization algorithm that mimics the foraging behavior of birds. The parameters are modeled as "bird swarm" individual behavior, and the optimal target is "food." Like the genetic algorithm, each individual is evaluated by setting a fitness function. The PSO algorithm has the characteristics of fast operation speed and strong local search ability. But it may not be possible to jump out of the local optimal solution. The idea of security and smoothness is introduced into the path length, and the optimal path is obtained by establishing a fitness function with dynamic adjustment function [70].

(10) A* algorithm

The A* algorithm belongs to the heuristic search algorithm in the field of artificial intelligence, and is the most widely used algorithm in the field of path-planning. It calculates the initial node-to-node cost and the node-to-target node cost to analyze the nodes. The evaluation function of the A* algorithm is defined as [71]:

$$F(n) = G(n) + H(n) \tag{6.9}$$

where $F(n)$ is the evaluation function of all nodes. $G(n)$ is the true cost between the initial node and a certain node in the state environment. $H(n)$ is the budgeted cost of the process from the current node n to the target node. If $H(n) = 0$, only $G(n)$ has an effect, and then the A* algorithm becomes the Dijkstra algorithm, so that the shortest path can be found. If $H(n) < G(n)$, the A* algorithm can still search for the shortest path. If $H(n)$ is smaller, the number of the node that is extended by the A* algorithm will increase, and the operating rate will decrease at this time. If $H(n) = G(n)$, then the A* algorithm can find the shortest path faster and it will not perform additional expansion. The rate will be the fastest. If the price paid by $H(n)$ is higher than the cost of a node and the target node, then the best path may not be found at this time, but the rate is increased.

The A* algorithm needs an open list and a closed list in the process of finding the optimal path. The open list stores the nodes to be checked and the estimated value of these nodes. The closed list stores the nodes that have been checked and the path from the starting point to the node. The specific steps of the A* algorithm are given as follows:

(1) Create an open list and an empty closed list. The starting point is stored in the open list as the detection node.

(2) Find the nodes around the detection node that minimize the total cost in the open list, and use this node as the detection node.
(3) Detection of nodes around the detection node. If these nodes are feasible and not in the closed list, the total costs of the nodes are calculated. If the value of this node is less than the current minimum, then it will be updated.
(4) Add detection points to the closed list and delete detection points from the open list.
(5) If the end point is added to the open list, the optimal path is found and the algorithm ends. If the end conditions are not met, repeat step (3).

(11) Ant colony optimization (ACO)

Ant colony algorithm is a heuristic path-planning algorithm that simulates ant colony foraging. The positive feedback mechanism is used to release the pheromone from the individual to change the surrounding environment. It has the ability to jump out of the local optimum and excels in global optimization. In recent years, there have been a lot of studies on the ant colony algorithm, which have obtained good application results in the fields of transportation, chemical industry, and electric power [72].

The principle of ant colony algorithm can be described as: In nature, the behavior of ants looking for food is not an individual behavior, but a group behavior. When the ants search for food, each ant is responsible for transmitting information to others. First, without telling the ants specific food locations, the ants searched aimlessly. When one of the ants finds food, it releases pheromones. The pheromones will attract other ants. The pheromones can evaporate over time, therefore, the size of the pheromone concentration indicates the distance of the path. Then, more and more ants will gather along a path along with the pheromone, and eventually find the food along this path. However, some ants do not search for food by the pheromone, they will find food through other paths and secrete pheromone. If the path is shorter than the previous path, more ants will be attracted to the shorter path. Eventually, after multiple iterations, most ants will find food along the shortest path. This is a positive feedback process [73].

The rules for the ant colony algorithm are presented as follows:
(1) Scope

The ant cannot see and smell the food in the distance, and it is not sensitive to the smell. In fact, the range an ant can sense is very small.
(2) Environment

The environment in which ants are located is random and variable. For example, there are obstacles, ants, pheromones in the environment. The ants can only feel the environmental information within its scope. In the process of searching for food, the ants mainly advance according to the secretion of pheromone, but the pheromone disappears with a certain probability.

(3) Foraging rules

Ants can feel the environment within their range during the foraging process. If the food is in the range, the ants will find the food directly. Otherwise, the ants will advance according to the pheromone. In the process of searching for food, the ants always go in the direction with more pheromones. However, some ants do not follow the pheromone. If these ants search for a shorter path, they will attract other ants and eventually form the shortest path. In short, the foraging process is a positive feedback process.

(4) Movement rules

When the ant moves, it moves in the direction of more pheromone. If there is no pheromone secretion in its range, it will continue to advance according to the original direction. Once the direction needs to be changed, the direction with more pheromone is selected.

(5) Obstacle avoidance rules

The ants will advance according to the foraging rules. Once there are obstacles in front, the direction is changed according to the size of the pheromone and the obstacles can be avoided.

(6) Pheromone rules

According to these rules, ants are not directly related to each other before they search for food, but each ant interacts with its environment, secreting pheromones in the process of searching for food, and the ants are related based on pheromones.

6.3.1.3.2 Local path-planning method

The local path-planning method is applicable when the environment information of a mobile robot is unknown. It is a planning method that uses the sensor carried by the robot to detect the obstacle information in the current environment and combines local algorithms (such as Artificial Potential Field [APF] method, rapid-exploration random tree algorithm, rolling window method, and morphin algorithm) to carry out obstacle avoidance.

(1) Artificial Potential Field (APF) algorithm

The APF algorithm is a path-planning method of the virtual force method [74]. The algorithm assumes the environment of a mobile robot as the environment of a virtual force field. A gravitational force is generated between the mobile robot and the target node, and a repulsive force is generated between the mobile robot and the obstacle node. The resulting geometric curve is the moving path of the mobile robot. This method is mainly used for robots to avoid obstacles. It has the following advantages and disadvantages [75].

The advantages are presented as follows:

The APF method is simple and can directly map the environment into the model without changing the working environment of the robot. When using the APF method, the parameters of the model are mainly divided into the following types: the area of the obstacle, the direction of movement of the robot, and the distance between the robot and the obstacle. Thus, a path is obtained according to a simple mechanical equation.

The APF method has fast calculation speed because it has few calculation equations, small calculation amount and simple calculation. The computer can build a model and get an obstacle avoidance path in a very short time. This ensures that the robot can perform safe obstacle avoidance processing in real-time when it encounters obstacles.

According to the concepts of gravity and repulsion introduced by the APF method, the path obtained by the APF method is third-order derivable. The direction and derivative of the resultant force is the fastest gradient descent paths, so the paths are guaranteed to be smooth [76].

The disadvantages are presented as follows:

The APF method is also accompanied by some problems due to its simple operation and small amount of calculation. The APF method can only build maps in small areas when modeling the environment. The model established by the APF method does not take the global information as a reference, so the obtained path is easy to fall into a local optimum. Because the APF algorithm is simple, it has no ability to jump out of local traps. Especially when the distance between the target point and the obstacle is very close. Since the repulsive force between the robot and the obstacle increases as the distance decreases, it may appear that the repulsive force is always greater than the attraction of the target point to the robot, and the robot cannot reach the target.

A typical problem of APF method is that it is prone to local minimum points. Local optimization refers to the stagnation of the robot in areas that are not the target point. The main reason for the stagnation phenomenon is that in the concept of gravity and repulsion by the APF method, there may be points where the gravity and repulsion are equal. The emergence of such a singularity will cause the acceleration to drop to zero when the robot reaches this position. When such singularities appear in large numbers in a small range, the robot will stagnate. This area usually occurs in a narrow area with a lot of obstacles. Around the local optimal point, the direction of the resultant force usually points to the local optimal point, so that the robot cannot get out of the local optimal point.

If there is a concave obstacle or a large obstacle between the robot and the target, the local minimum point is also prone to be generated in the cavity of the concave obstacle [77]. The robot enters the concave shape under the gravity and can easily fall into local minima.

When there are large planar obstacles between the robot and the target, local minimum points are also prone to occur on the vertical line between the target and the obstacle. The local minima defects formed by discrete obstacles can be solved by modifying the gravitational field and repulsion field functions. The local minima defects formed by continuous obstacles need to be processed by other methods.

(2) Rapid-exploration Random Tree (RRT) algorithm

The rapid-exploration random tree (RRT) algorithm occupies an important position in the field of path-planning, especially in high-dimensional environments. It takes points from the surrounding environment at random intervals and then connects them. The RRT algorithm does not require the modeling of space and has an efficient search capability. However, the algorithm also has the disadvantages that the real-time performance is poor when it is used in a known environment, and the obtained path is often not the optimal path [78].

(3) Rolling window algorithm

The rolling window algorithm is an effective method to solve the path-planning problem in an unknown environment. It relies on the sensors of the mobile robot to detect local information in real-time and adopts the rolling window to perform path-planning. The method has the advantages of small calculation amount, fast response and strong operability, and is suitable for a dynamic unknown environment [79].

(4) Morphin algorithm

The morphin algorithm is a local obstacle avoidance algorithm based on raster images. It analyzes the ground information collected by sensors to obtain several arc paths, and then selects the optimal path based on the evaluation function. The algorithm has fast calculation speed and is suitable for mobile robots that need to find obstacle avoidance paths in a short time [80].

6.3.2 Hand-eye systems

6.3.2.1 The role of the hand-eye system

Intelligent rail transit requires the daily inspection of rail transit to have technical features such as information digitization, functional integration, compact structure, and state visualization. To realize automatic control of automatic inspection and unmanned operation management, regular reading of the monitoring instruments in rail transit (such as voltmeters, ammeters, barometers, thermometers, and arresters) is required.

The rail transit monitoring instruments are mostly pointer type instruments, because the pointer type instruments have the advantages of simple structure, convenient maintenance, convenient use, high reliability, low price, waterproof, antifreeze, and dustproof. This type of instrument does not have a data interface. Therefore, automatic acquisition and transmission of measurement parameters cannot be achieved. At present, the reading and recording of such data are mainly carried out manually. The monitoring instruments and meters for rail transit equipment are read by human inspectors, and then the operating state of the rail transit equipment is judged. This mode is labor-intensive and slow in reading, and is subject to human beings. The influence of subjective factors and objective factors are as follows: (1) The observation distance and observation angle between the human inspector and the pointer meter. (2) When the pointer points between the minimum scale lines, the number of instrument representation can only be estimated by experience, and the reading results read by different inspectors are different. As a result, the error will is large. (3) After a lot of repeated work, the fatigue of the human inspectors will cause certain errors. (4) When the weather conditions are bad, the manual inspections may have greater safety hazard, while the rail transit inspection robots can carry the monitoring devices to monitor the rail transit equipment [81].

The workflow of the rail transit inspection robot to inspect equipment and instruments is shown as follows: first, based on the inspection task, the predesigned path-planning algorithm is used to plan the inspection path for the inspection robot. Then, the predesigned navigation algorithm is used to move the inspection robot to the inspection point. The navigation algorithm moves the inspection robot to the inspection point; second, adjusts the pan/tilt that controls the angle of the sensor to make the sensor align with the monitoring instrument of the rail transit device to be detected; finally, transmits the collected instrument image to the background through the wireless network. The number of monitor representations is automatically obtained using a predesigned meter recognition algorithm. During the entire inspection process, the instrument identification algorithm plays a decisive role, and its performance directly affects the quality of the inspection robot to complete the inspection task.

6.3.2.2 Hand-eye system

A mechanical structure with a camera installed can be seen as a hand-eye system. Based on the installation of the camera, the commonly used hand-eye system can be divided into two types: eye-to-hand and eye-in-hand [82].

6.3.2.2.1 Eye-to-hand

In this type of hand-eye system, the camera is separated from the robot. The camera is fixedly mounted at a certain position, monitoring the movement and current position of the robot from time to time. By feeding back the deviation between the target object and the end of the robot actuator, the robot can perform the corresponding operation to realize the interaction between the end effector and the target object. The biggest advantage of this type of hand-eye system is that it can observe the global environment, so it has been widely used in the field of mobile robots. At the same time, the shortcomings are also obvious. When the robot moves between the target object and the camera, the image of the target object will be hidden. Under this situation, the camera cannot obtain the image and cannot feedback useful feature information. Therefore, robot systems that rely solely on the feedback of images for control are not suitable for using this type of hand-eye system.

6.3.2.2.2 Eye-in-hand

In this type of hand-eye system, the camera is integrated with the robot. The camera is fixedly mounted on the robot. The image in the current field of view of the camera is observed in real-time and is provided to the robot for the control of the next movement. The biggest advantage of this type of hand-eye system compared to the eye-to-hand type is that it does not produce image occlusion, so it is more suitable for control systems based on image feedback. Of course, this type also has shortcomings. The serious one is that the camera will move together with the robot during the movement of the robot. This combined movement is likely to cause the target object to exceed the camera's field of view, and the robot will not be able to get the correct information for its control.

The automatic recognition tasks conducted by rail transit inspection robots for the pointer instrument mainly include image preprocessing, pointer positioning, and dial recognition. The image preprocessing is to process the noise generated by external factors during the acquisition and transmission of the target image [83]. It provides main information for the subsequent pointer positioning and dial recognition by filtering unnecessary information and highlighting useful information. The main techniques of the pointer positioning are template matching technology and edge detection technology in image processing. The parameters of the pointer are extracted by machine vision algorithms, and the position of the pointer is determined from the image. The dial recognition is also processed by techniques in image processing. The main characteristics of the dial are the scale, the letters of the scale, and the pointer.

(1) Image preprocessing

It is necessary to reduce the noise in the instrument image introduced by the acquisition and transmission process of the image and the effect of changes in light under outdoor conditions. For removing image noise, some general image processing techniques, such as graying and geometric transformation can be employed.

(2) Pointer positioning

The main methods for pointer positioning are as follows: subtraction method, improved Hough transform, active shape model (ASM), and saliency detection. The core idea of the subtraction method is to use the difference between the current instrument image and the instrument image without a pointer, and subtract the two images to obtain the corresponding pointer area. This method is easily affected by external environmental factors, such as changes in light or weather, because the corresponding background area of the pointer instrument image acquired will also change accordingly, making the pointer area obtained by the subtraction method inaccurate. The core idea of the Hough transform is to transform the contour position of the instrument dial in the Cartesian coordinate system to the corresponding polar coordinate system, and then analyze the linearity of the contour on the topology to obtain the corresponding line segment characteristics. Using the prior knowledge of the dial pointer distribution, the corresponding parameter space of the Hough transform is reduced to improve the real-time performance and accuracy of the algorithm. Multi-scale homomorphic filtering is used to restore the color of different brightness levels. The perspective view is used to obtain an arbitrary angle instrument view. The Hough transform is used to obtain the pointer position. The morphological segmentation is used to extract the dial from the background, and then the improved Hough transform is used to locate the pointer [84].

(c) Dial recognition

The existing ideas of dial recognition for pointer instruments are mainly divided into template matching and the use of classifiers. The idea of template matching is to store an image template in advance for each dial to be recognized. When the dial recognition is required, the corresponding template of the current instrument is matched with the acquired image to obtain the dial position. The template matching method for locating of the dial position is usually divided into two types: feature point matching and direct template matching. The idea of designing a classifier is to treat the dial recognition problem as a classification problem. that is, to generate a series of candidate areas from the current instrument image, and then classify these candidate areas to obtain the corresponding position of the dial.

In summary, the research on automatic identification of pointer instruments is mainly focused on image preprocessing, pointer positioning, and dial recognition to extract the corresponding line segment characteristics of pointers. The dial scale recognition depends on manual calibration parameters on the instruments. When the inspection robot travels to the position of the observation point and the observation angle deviates, the recognition algorithm may produce a large error. Therefore, it is of high practical value and research significance to study more intelligent methods to identify each scale and its direction.

6.4 Conclusion and outlook

With the continuous development of cities in some countries, rail transit has developed rapidly and is currently the most important mode of transportation within and between cities. Nowadays, the development speed of mobile robot technology and communication system is extremely fast. To realize the unmanned rail transit inspection work, the inspection robot technology came into being.

This chapter mainly introduces the development overview, hardware structure, and main technologies of the rail transit inspection robot. First, the situation and importance of rail transit development are described, and the development of the robot field and several major inspection robots are introduced. Then the main hardware structure and sensors of the rail transit robots adopted to carry the inspection work in the rail transit environment are given. The inspection robots can replace the human staff to inspect the rails, trains, and the supporting equipment in the rail transit environment. Finally, the key technologies that support the inspection robots to complete the task are introduced. These technologies are positioning, navigation, path-planning, and hand-eye system. The details of this chapter are summarized as follows:

(1) The development status of rail transit is explained. The urbanization process has brought about the rapid development of urban traffic. The mileage and scale of rail transit operations have gradually increased in recent years. The role of rail transit in daily life and work is becoming more and more important. Once an accident occurs in rail transit, it may have a huge impact on people, economy, and environment. Moreover, the drawbacks of workforce inspection are obvious. The inspection results depend on the business proficiency and subjective initiative of the inspectors, and the long-term labor under mechanical repetition. It is easy to cause negligence, so the appearance of the inspection robots conforms to the actual demand for rail transit.
(2) The development history and current situation of inspection robots are reviewed.
(3) The main hardware devices of the rail transit inspection robots are as follows: driving wheel, laser sensor, pan/tilt, ultrasonic sensor, RFID sensor, sound collection device, wireless recharging device, etc. The working principles of these devices are introduced. The devices are necessary for the inspection robots to perform inspection work.
(4) The positioning technology, navigation technology, path-planning technology, and hand-eye system required for the inspection robot for the inspection work are explained. At present, the positioning methods for robots mainly include the following: Wireless Fidelity (WiFi) signal positioning, RFID positioning, UWB positioning, ZigBee positioning, Bluetooth positioning, infrared positioning, ultrasonic positioning, inertial positioning, geomagnetic positioning, visual Simultaneous Localization and Mapping (SLAM) positioning, and laser SLAM positioning. The navigation methods for robots include the following: electromagnetic navigation, optical navigation, inertial navigation, laser navigation, visual navigation, and GPS navigation.

The main path-planning algorithms are divided into global path-planning and local path-planning. The global path-planning algorithms mainly include the following: Dijkstra,

BFS, Floyd, raster map method, visibility graph, Voronoi diagram, GA, A*, and ACO algorithm. The local path-planning algorithms are as follows: APF, RRT, rolling window algorithm, and morphin algorithm.

Combining the literature referenced in this chapter and the research on inspection robots, it should be noted that although this chapter has introduced the existing inspection robot technologies, there are still many possible future development directions to understand. The future development directions of the inspection robots are as follows:

(1) The mechanical structure of the inspection robots can be further optimized and lightened. In the driving devices, it is possible to use a crawler with stronger off-road capability for the inspection robots. With the development of embedded technology, the sensors installed on the inspection robots can be more simplified and lightweight.
(2) The obstacles encountered in the path-planning are also the problems for the inspection robots. For the detected dynamic obstacles, the ordinary path-planning methods are not applicable and more advanced algorithms are required.
(3) Due to the complexity of the outdoor environment, the positioning and navigation of the inspection robots are susceptible to interference. In the case of loss of signal, how to continue safe navigation based on historical data is also a research direction worth exploring.
(4) The important part of the hand-eye system is image processing, especially the recognition for the scale of the instruments and meters. Its technical requirements are very high. The image processing mainly includes the following: image color space model, image filtering technology, edge detection technology, and dial pointer recognition.

References

[1] M. Tian, Acceptance check scheme of integration test and commissioning for urban rail transit, in: International Conference on Electrical and Information Technologies for Rail Transportation, 2017, pp. 747–758.
[2] C. Bernard, The King's Cross Underground fire and the setting up of the investigation, Fire Saf. J. 18 (1992) 3–11.
[3] P.W.B. Semmens, Railway Disasters of the World: Principal Passenger Train Accidents of the 20th Century, 1994. Patrick Stephens.
[4] J.-G. Zhang, L. Cong-Fei, Substation patrol robot based on wireless technologies, Electr. Autom. 36 (2014) 36–38.
[5] F. Gao, W. Guo, Thinking of the development strategy of robots in China, J. Mech. Eng. 52 (2016) 1–5.
[6] Q.-M. Hu, Run-Zi Hu, P. Zhou, Application of inspection robot technology in substation, Cent. China Electr. Power 5 (2011) 36–40.
[7] S. Lu, Q. Qian, B. Zhang, et al., Development of a mobile robot for substation equipment inspection, Autom. Electr. Power Syst. 30 (2006) 94–98.
[8] R. Guo, B. Li, Y. Sun, et al., A patrol robot for electric power substation, in: 2009 International Conference on Mechatronics and Automation, vols. 55–59, 2009.

[9] N.J. Nilsson, A Mobile Automaton: an Application of Artificial Intelligence Techniques, Sri International Menlo Park Ca Artificial Intelligence Center, 1969.
[10] H. Takahashi, Japan IERE Council Special Document R-8903, Development of Patrolling Robot for Substation, vol. 10, 1989, p. 19.
[11] N. Pouliot, S. Montambault, Geometric design of the LineScout, a teleoperated robot for power line inspection and maintenance, in: 2008 IEEE International Conference on Robotics and Automation, 2008, pp. 3970–3977.
[12] J. Beaudry, J.F. Allan, Electrical substation inspection and intervention robot, field experiments, In: Proceedings of the 2014 3rd International Conference on Applied Robotics for the Power Industry.
[13] L. Zheng, Y. Ruan, Autonomous inspection robot for power transmission lines maintenance while operating on the overhead ground wires, Int. J. Adv. Rob. Syst. 7 (2010) 107–112.
[14] Y. Jia, Y. Zou, P. Jinsong, et al., The scientific objectives and payloads of Chang'E– 4 mission, Planet. Space Sci. 162 (2018) 207–215.
[15] A.V.S. Bhadoriya, V.K. Gupta, S. Mukherjee, Development of in-pipe inspection robot, Mater. Today Proc. 5 (2018) 20769–20776.
[16] C.-C. Chang, C.-J. Lin, LIBSVM: a library for support vector machines, ACM Trans. Intell. Syst. & Technol. 2 (27) (2011).
[17] G. Mourioux, C. Novales, G. Poisson, et al., Omni-directional robot with spherical orthogonal wheels: concepts and analyses, in: Proceedings 2006 IEEE International Conference on Robotics and Automation, 2006, pp. 3374–3379, 2006 ICRA 2006.
[18] W.I. Liu, Y. Li, Error modeling and extrinsic–intrinsic calibration for LiDAR-IMU system based on cone-cylinder features, Robot. Autonom. Syst. 114 (2019) 124–133.
[19] J. Sankar, S. Adarsh, K.I. Ramachandran, Performance evaluation of ultrasonic and infrared waves on human body and metal surfaces for mobile robot navigation, Mater. Today Proc. 5 (2018) 16516–16525.
[20] A. Rocchi, E. Santecchia, G. Barucca, et al., Optimization of distances measurement by an ultrasonic sensor, Mater. Today Proc. 19 (2019) 33–39.
[21] F. Tong, A processing method with high precision for ultrasonic distance measurement, J. Xiamen Univ. Nat. Sci. 37 (1998) 507–512.
[22] G. Feng, Z.-R. Jing, C.-X. Chen, et al., Study on the choice of the triangulation parameter, Electr. Meas. Technol. 30 (2007) 29–31.
[23] J.-H. Li, Z.-Y. Sun, Y.-L. Wang, et al., RFID positioning system construction method based on virtual reference tags, Comput. Sci. 38 (2011) 107–110.
[24] Y. Zhang, Y. Fan, Q. Wang, et al., An anti-collision algorithm for RFID-based robots based on dynamic grouping binary trees, Comput. Electr. Eng. 63 (2017) 91–98.
[25] S. Lu, C. Xu, R.Y. Zhong, et al., A RFID-enabled positioning system in automated guided vehicle for smart factories, J. Manuf. Syst. 44 (2017) 179–190.
[26] F. Chetouane, An overview on RFID technology instruction and application, IFAC-PapersOnLine. 48 (2015) 382–387.
[27] W.-T. Lan, W.-Q. Zhang, J.-Y. Luo, et al., Design of RFID indoor positioning system based on phase difference ranging, Transducer & Microsyst. Technol. 36 (2017) 85–88+99.
[28] C. Wu, X. Wang, M. Chen, et al., Differential received signal strength based RFID positioning for construction equipment tracking, Adv. Eng. Inf. 42 (2019) 100960.
[29] X.-S. Que, H. Yang, R. Tang, et al., Application of electret microphone in the heart sound acquisition, J. Chongqing Univ. Nat. Sci. Ed. 10 (2007) 33–35+41.
[30] G. Benet, F. Blanes, J.E. Simó, et al., Using infrared sensors for distance measurement in mobile robots, Robot. Autonom. Syst. 40 (2002) 255–266.
[31] Z. Hu, H. Hei, X. Li, et al., Modeling background response and applications for mid-infrared remote sensing camera, Infrared Phys. Technol. 103 (2019) 103082.
[32] M. Vollmer, K.-P. Möllmann, Infrared Thermal Imaging: Fundamentals, Research and Applications, John Wiley & Sons, 2017.

[33] Y. Yu, H. Zhang, M.-S. Lin, Research and design of the 3D reconstruction system based on binocular stereo vision, J. Comput. Technol. & Dev. 6 (2009) 127–130.

[34] W. Zhang, H. Cheng, W. Sui, Design and implementation of camera calibration system, J. Comput. Eng. 2 (2007) 255–257+276.

[35] Z. Zhang, A flexible new technique for camera calibration, IEEE Trans. Pattern Anal. & Mach. Intell. 22 (2000) 1330–1334.

[36] L. Wan, S. Huang, T. Zhang, Object recognition system for an autonomous underwater vehicle based on the wavelet invariant moment, J. Harbin Eng. Univ. 35 (2014) 148–153.

[37] A. Anand, H.S. Koppula, T. Joachims, et al., Contextually guided semantic labeling and search for three-dimensional point clouds, Int. J. Robot Res. 32 (2013) 19–34.

[38] Y.-J. Niu, J.-W. Leng, Research on wireless charging technology for intelligent robots, Comput. Simul. 34 (2017) 348–352+403.

[39] A. Kurs, A. Karalis, R. Moffatt, et al., Wireless power transfer via strongly coupled magnetic resonances, Science 317 (2007) 83–86.

[40] M. Tao, X. Chen, Z.-P. Sun, Positioning techniques of mobile robot [J], Fire Control & Command Control 7 (2010) 169–172.

[41] D. Liu, B. Sheng, F. Hou, et al., From wireless positioning to mobile positioning: an overview of recent advances, IEEE Syst. J. 8 (2014) 1249–1259.

[42] S.G. Tzafestas, Mobile robot control and navigation: a global overview, J. Intell. Rob. Syst. 91 (2018) 35–58.

[43] P. Prasithsangaree, P. Krishnamurthy, P. Chrysanthis, On indoor position location with wireless LANs, in: The 13th IEEE International Symposium on Personal, Indoor and Mobile Radio Communications, vol. 2, 2002, pp. 720–724.

[44] N. Le Dortz, F. Gain, P. Zetterberg, WiFi fingerprint indoor positioning system using probability distribution comparison, in: 2012 IEEE International Conference on Acoustics, Speech and Signal Processing, ICASSP), 2012, pp. 2301–2304.

[45] M. Ye, T. Chen, C. Yu, ZigBee-based positioning and navigation system for robot, J. Converg. Inf. Technol. 6 (2011) 135–146.

[46] S. Huang, Y. Guo, S. Zha, et al., A real-time location system based on RFID and UWB for digital manufacturing workshop, Procedia Cirp 63 (2017) 132–137.

[47] B. Firoozbakhsh, T.G. Pratt, N. Jayant, Analysis of IEEE 802.11 a interference on UWB systems, in: IEEE Conference on Ultra Wideband Systems and Technologies, 2003, 2003, pp. 473–477.

[48] P. Neal, J.N. Ash, S. Kyperountas, et al., Locating the nodes: cooperative localization in wireless sensor networks, IEEE Signal Process. Mag. 22 (2005) 54–69.

[49] E.M. Gorostiza, L. Galilea, J. Luis, et al., Infrared sensor system for mobile-robot positioning in intelligent spaces, Sensors 11 (2011) 5416–5438.

[50] K. Yousif, A. Bab-Hadiashar, R. Hoseinnezhad, An overview to visual odometry and visual SLAM: applications to mobile robotics, Intell. Ind. Syst. 1 (2015) 289–311.

[51] U. Weiss, P. Biber, Plant detection and mapping for agricultural robots using a 3D LIDAR sensor, Robot. Autonom. Syst. 59 (2011) 265–273.

[52] J. Guo, H. Zhou, Q. Liu, Study of Mobile-Robot Rfid Position System Based on Extended Kalman Filter, vol. 31, Wuhan University of Technology (Information and Management Engineering), 2009, pp. 917–921.

[53] J. Li, J. Zhang, B. Zhang, et al., Fuzzy fingerprint location for bluetooth specification version 4.0, J. Shanghai Univ. Nat. Sci. 19 (2013) 126–131.

[54] F.C. Teixeira, J. Quintas, António Pascoal, Robust methods of magnetic navigation of marine robotic vehicles, IFAC-Papers OnLine 50 (2017) 3470–3475.

[55] Y. Zhao, W. Shi, X. Ai, Fusion algorithm in indoor integrated position system for pedestrian dead reckoning/ultrasonic, J. Cent. S. Univ. 47 (2016) 1587–1598.

[56] W. Wang, Y.-C. Bai, G.-P. Wu, et al., An electromagnetic navigation method based on information fusion for inspection robot, Autom. Electr. Power Syst. 37 (2013) 73–79.
[57] C. Wang, W. Sun, D. Bu, et al., Simultaneous localization and mapping research for air-duct cleaning robot based on inertial navigation and stereo vision, J. Mech. Eng. 23 (2013) 59–67.
[58] P. Xiao, Y.-Q. Luan, R. Guo, et al., Research of the laser navigation system for the intelligent patrol robot, Zidonghuayu Yibiao/Autom. & Instrum. 27 (2012) 5–9.
[59] F. Bonin-Font, A. Ortiz, G. Oliver, Visual navigation for mobile robots: a survey, J. Intell. Rob. Syst. 53 (2008) 263–296.
[60] G. Zhang, X. Hu, J. Chai, et al., Summary of path planning algorithm and its application, Mod. Mach. 5 (2011) 85–90.
[61] B.K. Patle, L. Ganesh Babu, A. Pandey, et al., A review: on path planning strategies for navigation of mobile robot, Def. Technol. 15 (2019) 582–606.
[62] Y. Yang, J. Gong, An efficient implementation of shortest path algorithm based on Dijkstra algorithm, J. Wuhan Tech. Univ. Surv. & Mapp. 24 (1999) 209–212.
[63] H. Chen, W.-Q. Huang, Best-first search algorithm for solving HP Lattice model, J. Comput. Eng. & Appl. 36 (2006) 49–50+72.
[64] D. Wei, An optimized floyd algorithm for the shortest path problem, J. Network. 5 (2010) 1496.
[65] H. Moravec, A. Elfes, High resolution maps from wide angle sonar, in: Proceedings 1985 IEEE International Conference on/Robotics and Automation, vol. 2, 1985, pp. 116–121.
[66] A.A. Maciejewski, J.J. Fox, Path planning and the topology of configuration space, IEEE Trans. Robot. Autom. 9 (1993) 444–456.
[67] O. Takahashi, R.J. Schilling, Motion planning in a plane using generalized Voronoi diagrams, IEEE Trans. Robot. Autom. 5 (1989) 143–150.
[68] C. Liu, X. Yan, C. Liu, et al., Dynamic path planning for mobile robot based on improved genetic algorithm, Chin. J. Electron. 19 (2010) 2010–2014.
[69] G. Yuan, Y. Zhai, C. Ding, et al., AGV path planning based on improved genetic algorithm, J. Beijing Union Univ. 32 (2018) 65–69.
[70] H.I. Kang, B. Lee, K. Kim, Path planning algorithm using the particle swarm optimization and the improved Dijkstra algorithm, in: 2008 IEEE Pacific-Asia Workshop on Computational Intelligence and Industrial Application, vol. 2, 2008, pp. 1002–1004.
[71] F. Duchoň, A. Babinec, K. Martin, et al., Path planning with modified a star algorithm for a mobile robot, Procedia Eng. 96 (2014) 59–69.
[72] E. Shi, M. Chen, J. Li, et al., Research on method of global path-planning for mobile robot based on ant-colony algorithm, Trans. Chin. Soc. Agric. Mach. 45 (2014) 53–57.
[73] H. Duan, D. Wang, J. Zhu, et al., Development on ant colony algorithm theory and its application, Control Decis. 19 (2004) 1321–1326.
[74] Y. Huang, H. Hu, X. Liu, Obstacles avoidance of artificial potential field method with memory function in complex environment, in: 2010 8th World Congress on Intelligent Control and Automation, 2010, pp. 6414–6418.
[75] M. Phillips, M. Likhachev, Sipp: safe interval path planning for dynamic environments, in: 2011 IEEE International Conference on Robotics and Automation, 2011, pp. 5628–5635.
[76] M.C. Lee, M.G. Park, Artificial potential field based path planning for mobile robots using a virtual obstacle concept, in: Proceedings 2003 IEEE/ASME International Conference on Advanced Intelligent Mechatronics (AIM 2003), vol. 2, 2003, pp. 735–740.
[77] M.G. Park, J.H. Jeon, M.C. Lee, Obstacle avoidance for mobile robots using artificial potential field approach with simulated annealing, in: ISIE 2001 2001 IEEE International Symposium on Industrial Electronics Proceedings (Cat No 01TH8570), vol. 3, 2001, pp. 1530–1535.
[78] J. Song, B. Dai, E. Shan, et al., An improved RRT path planning algorithm, Acta Electron. Sin. 38 (2010) 225–228.

[79] M. Chen, W. Lin, B. Zeng, Path planning for robot autonomous map building based on rolling window, Comput. Eng. 43 (2017) 286–292.
[80] Z. Zhang, W. Zhonghui, 3-D real-time deviation method for avoiding hazardous weather in terminal Airspace based on morphin planning algorithm, J. Nanjing Univ. Aeronaut. Astronaut. 47 (2015) 467–473.
[81] G.-F. Guo, R.-H. Wang, H. Wang, A pretreatment of image based on reformative homomorphic filtering, J. Guangdong Univ. Technol. 26 (2009) 57–59.
[82] S. Hutchinson, G.D. Hager, P.I. Corke, A tutorial on visual servo control, IEEE Trans. Robot. Autom. 12 (1996) 651–670.
[83] M. Hanmandlu, D. Jha, An optimal fuzzy system for color image enhancement, IEEE Trans. Image Process. 15 (2006) 2956–2966.
[84] H. Xing, Z. Du, B. Su, Detection and recognition method for pointer-type meter in transformer substation, Chin. J. Sci. Instrum. 38 (2017) 2813–2821.

CHAPTER 7

Rail transit channel robot systems

7.1 Overview of rail transit channel robots

7.1.1 Development progress of rail transit equipment

7.1.1.1 Background

The manufacturing industry is the main part of the national economy and the main driving force of national development. To enhance the comprehensive strength of China, it is necessary to build a number of internationally competitive manufacturing enterprises. With rapid progress in science and technology, the rail transit equipment manufacturing industry has gradually become an excellent example of the field of advanced equipment manufacturing in China. The rail transit industry is composed of urban rail transit and high-speed rail, both of which are developing rapidly [1].

Since 1995, Chinese rail transit equipment enterprises have realized the importance of their products going abroad and accelerated the pace of the "Going Global" strategy; they successfully won the Tehran Metro Project in Iran [2], the first rail transit export project in China. Chinese rail transit products have been exported to more than 20 countries in Asia, Africa, Europe, and the United States. The rapid development of high-speed rail technology has since risen to the level of a national strategy with the strong support of the state and government. It is the general trend for high-speed rail products to go abroad more thoroughly.

According to statistics, Chinese rail transit equipment exports accounted for only a small portion of world rail transit equipment products export in 2002. In Oct. 2014, the China Northern Locomotive Rolling Stock Industry Group (CNR) and Massachusetts Bay Transit Authority signed a contract for the US Boston Metro System project. This was the first transaction for Chinese rail transit equipment to be exported to the United States.

With the support of policy and relevant departments, the China Southern Locomotive Rolling Stock Industry Group and CNR formally merged into the CRRC in 2015 after many consultations; it is the leading enterprise of rail transit equipment production in China [3]. The CRRC has the full strength to compete with leading industries such as the Bombardier Company of Canada, the Alstom Company of France, and the Siemens Company of Germany in the international rail transit equipment market [4].

Robot Systems for Rail Transit Applications. https://doi.org/10.1016/B978-0-12-822968-2.00007-3

High-speed rail has become a main diplomatic content of the Chinese government in its foreign policy [5]. In addition, a series of strategies, such as the "13th Five-Year Plan," the "Belt and Road Initiatives," and "A Country With Strong Transportation Network," involve the Chinese rail equipment manufacturing industry. Therefore, rapid development of the rail transit equipment manufacturing industry has a stable political foundation and strong national support, which can enable the development of the industry in China [6].

7.1.1.2 Development of the rail transit industry

In the 19th century, the first subway in the world was built in Britain, which greatly relieved the traffic pressure of large cities [7]. As a modern means of transportation, urban rail transit has the advantages of safety, reliability, a large volume, savings in energy, and environmental protection, and it has great significance for the populous countries in the world. In the 1960s, China began to construct subways. By the end of 2017, many cities had opened subway operations. With the vigorous development of the metro industry, well-known rail transit equipment manufacturing enterprises have emerged in Japan, Canada, Germany, and other countries. The rail transit products manufactured by Chinese enterprises such as CRRC are gradually being exported to Singapore, Malaysia, and other countries.

In 1825, the first modern railway in the world was built in Britain, between Stockton and Darlington [8]. Since then, railways have gradually entered people's lives and become the main means of transportation for travel. Construction of Chinese railways began in the middle and late 19th century. After years of vigorous development, the passenger turnover on railways has become much higher than passenger turnover on roads, waterways, and airlines. In the middle of the 20th century, the Japanese economy began to recover and develop rapidly after the depression.

At the same time, the backwardness of rail transit hindered the development of the Japanese economy. To alleviate the pressure of transportation and improve passenger transport capacity, the Shinkansen Line between Tokyo and Osaka in Japan was opened as the world's first high-speed railway in 1964. The high-speed railway in Japan has greatly reduced transport pressure and laid the foundation for the development of its economy; it also promoted the development of the global high-speed railway industry [9].

Germany and Japan have always been traditional powers in the field of rail transit equipment manufacturing. Although China started late in the field of rail transit, the momentum of development is rapid and China has become a powerful country in the field of rail transit equipment manufacturing. As a major rail transit equipment manufacturing country, Germany has Siemens and other international well-known brands. With the proposal of the "Belt and Road Initiatives" and the "A Country With Strong Transportation Network" plan, China has adjusted its position in the development of foreign trade and has taken the initiative to carry out multidirectional trade cooperation with the "Belt and Road Initiatives" country, thus greatly promoting the trade development of rail transit equipment.

7.1.1.3 Significance of vigorously developing rail transit equipment

Global rail transit is developing vigorously. Taking China as an example, the Chinese rail transit equipment industry is facing important development opportunities and challenges. Within the background of the national policy of the "Belt and Road Initiatives" and the "13th 5-Year Plan," Chinese rail transit equipment industry needs to seize the opportunity and actively promote internationalization, which has an important role in the process of Chinese rail transit equipment industry going out. Its significance is mainly reflected in the following aspects [10].

7.1.1.3.1 Helping the Chinese economy grow steadily

Vigorously developing Chinese rail transit equipment inevitably leads to an increase in the export of the rail transit equipment industry. Therefore, the upgrade of technology can be promoted, the level of development of the Chinese high-end manufacturing industry and labor productivity can be improved, and the goal of reducing operating and management costs can ultimately be achieved. On the other hand, with the increase in the export volume of the rail transit equipment industry, an economic series will be created. Economic changes such as stimulating domestic demand, stimulating investment, and increasing consumption will promote the steady growth of the Chinese economy.

7.1.1.3.2 Improving the internationalization of the Chinese rail transit equipment industry

The "13th 5-Year Plan" put forward by China clearly brings the rail transit industry into key development industries. Relying on leading technology, an excellent brand and reliable quality assurance, and other industrial advantages, the Chinese rail transit equipment industry should aim to increase the international market share, give full play to its advantages, achieve specialization and standardization in the international market, and constantly enhance international influence and competitiveness.

7.1.1.3.3 Promoting the development of the industrial chain of rail transit equipment industry

The rail transit equipment industry is a strategic emerging industry supported by the state. Its industrial chain is diversified and complex, covering a series of industrial chain enterprises such as raw materials and accessories, electronic components, mechanical processing, logistics and transportation, and information software. The development of enterprises in the industry chain will have a positive or negative impact on the rail transit equipment industry. Therefore, with the rapid development of the Chinese rail transit equipment industry in the international market, enterprises in the entire industrial chain will surely be driven to compete or cooperate with other relevant enterprises worldwide. This has a positive effect on promoting the development of enterprises in the industrial chain.

7.1.1.3.4 Promoting the optimization of export trade structure

The rail transit equipment industry is different from the traditional manufacturing industry and belongs to the modern high-tech industry. An increase in its exports will optimize the Chinese trade structure. If the export of rail transit equipment industry increases, it will inevitably promote the transformation of Chinese export products to technology-intensive ones, so as to optimize the export trade structure.

7.1.2 Development progress of rail transit channel robots

The robot is the most typical mechatronics digital equipment; it integrates machinery, automation, computers, sensors, artificial intelligence, and other disciplines. It has high application prospects and value. In 1959, the first robot, Unimate, was developed [11]. The robot weighs nearly 2 tons and is primarily used for handling. Early robots resembled human hands and arms, but they were similar to computerized numerical control machine tools in control mode. Prominent technical features of robots are that they are programmable and have teaching and reappearance qualities.

With the extensive application of sensors, a robot system that could automatically identify and locate appeared. In the 21st century, manufacturing enterprises increasingly established automatic production lines of robots, such as automobile manufacturing, electronic consumer goods manufacturing, machine manufacturing, and other industries. According to statistics from the United Nations Economic Commission for Europe and the International Federation of Robots, the world robotics industry has been growing annually.

In early days, the research direction of robots mainly focused on closed space, most of which belonged to a simple structured environment. From 1966 to 1972, a researcher and his team at Stanford International Research set up a project to develop a mobile intelligent automatic robot, Shakey, for indoor environments [12]. Subsequently, research institutions and organizations around the world, especially in developed countries, joined the research team of outdoor mobile robots.

Britain, Italy, and other European countries put forward the Eureka plan to jointly develop underwater vehicles [13]. In the early 1990s, the focus of robotics research changed to a more practical field with the help of advanced technologies such as computers and sensors. In 1994, the National Aeronautics and Space Administration (NASA) provided a large number of funds to develop a walking robot, Danty II, with mobile flexibility and long-distance inspection capability, and successfully demonstrated it in Spur crater [14].

Three years later, the planetary exploration mobile robot Sojourner, developed by NASA, successfully landed on Mars to complete a scientific expedition mission; later, Rocky 7 traveled farther [15]. The wheelchair robot developed in Germany has field exercises in public gathering places with large traffic.

In the 21st century, the development of mobile robots has opened a new chapter. The mobile robot developed by Patol Bot Robot launched in 2005 in the United States has the powerful functions of accurate positioning, path planning, and autonomous navigation in complex environments, which is widely used in many locations such as shopping malls, office areas, and factories [16]. The robot is also used as the basic platform for the robot, which can be employed for secondary development research. For the early research of robots, most of it is led by relevant departments of states and the research structure of universities. With the increasing investment of enterprises in robotics research and development, the accumulation of technology has gradually caught up with national robotics research institutions, and robotics technology companies have emerged in many countries.

The production of robots for half a century has proved that to achieve large-scale automated production, improve the level of manufacturing, and improve social industrial production efficiency and social productivity, robots must be vigorously developed [17]. After more than 50 years of development, robots have had an important role in many fields and even brought changes to many industries. In the manufacturing industry, especially in the automobile industry, robots have been the most widely used for mechanical processing, welding, heat treatment, painting, loading and unloading, assembly, and other operations.

Undoubtedly, with the development of science and technology, the intelligent degree of robots continues to improve, and the application of robots will continue to expand from the automobile manufacturing industry to others, and eventually to nonmanufacturing industries such as underwater robots, mining robots, construction robots, and inspection robots. Robots have been integrated into all aspects of human life and are continuously improving the quality of humanity.

German robotics manufacturing has been at the forefront of the world. Its global market share of robots is second only to that of Japan and the United States. Germany formulated the strategy of Industry 4.0 at the Hanover Industrial Expo in 2013, which attracted worldwide attention. The most famous robot in Germany is the KUKA robot, which is popular; it mainly involves logistics, consumer goods, and medical and other fields.
In addition, Japan belongs to one of the robotics manufacturing powers in the world. For a long time, Japan has accounted for the largest sales of robots in the global robotics industry. The robots produced by FANUC are used in various industries and have an

important role in the world robot market. In addition, Japan's OTC, Panasonic, and other brands are also world famous.

In terms of technology, there is no gap between Chinese robotics technology and developed countries. The mobile robot developed by the Institute of Automation, Chinese Academy of Sciences in 2003 uses multisensor fusion technology to sense the environment, thus realizing the functions of autonomous movement and trajectory tracking [18]. Among the Chinese robot brands, SIASUN robots are the most famous and powerful in China. The company is affiliated with the Chinese Academy of Sciences and has independent property rights and independent research and development capabilities.

With the rapid development of robotics technology, dual-arm robots were first developed and manufactured in the 1990s and have been gradually applied to the rail transit equipment manufacturing industry [19]. Many manufacturers of industrial robots have developed a number of humanoid dual-arm robots. In 2012, Rethink Robotics of the United States created Baxter robots, equipped with two flexible seven degrees of freedom robotic arms [20]. The head and chest constitute the vision-guided motion detection system of the robot, which ensures the safety of human—computer cooperation and can be used to assemble parts on the production line.

In 2015, the ABB Group exhibited a Yumi dual-arm robot equipped with a flexible manipulator and vision system at the Hanover Industrial Exposition [21]. It can cooperate with people to complete tasks in the production environment of rail transit equipment. The dual-arm robot system is composed of two serial manipulators. The structure of each single manipulator is basically the same as that of a single-arm robot. The previous chapters of this book give a detailed description of the single-arm robot, and they will not be repeated here. However, unlike the one-armed robot, only the end structure and the pose of the target object need to be considered. The dual-arm robot needs to ensure the relative relationship between the two arms based on the relative motion trajectory of the two arms in accordance with the constraints of the task, ultimately forming a system in which the two arms are coupled and coordinated.

Considering the characteristics of coordinated operation tasks of dual-arm robots, it is difficult for single-arm robots to perform complex tasks such as axle hole assembly, smooth handling, broaching, and so on. In view of the characteristics of these tasks, considering the use of the dual-arm system to complete the operation, the cooperative operation of dual-arm robots has the following characteristics [22]: (1) In the process of planning the movement of a dual-arm, it is necessary to consider interference between the two arms or collision with surrounding objects; (2) the two-arm robot has high flexibility and easily avoids singularities in the process of kinematic coupling; and (3) in some specific cases, dual-arm robots can also control the two manipulators separately according to the task requirements, so that they can complete the task independently. The advantages of dual-arm robots are shown in Fig. 7.1.

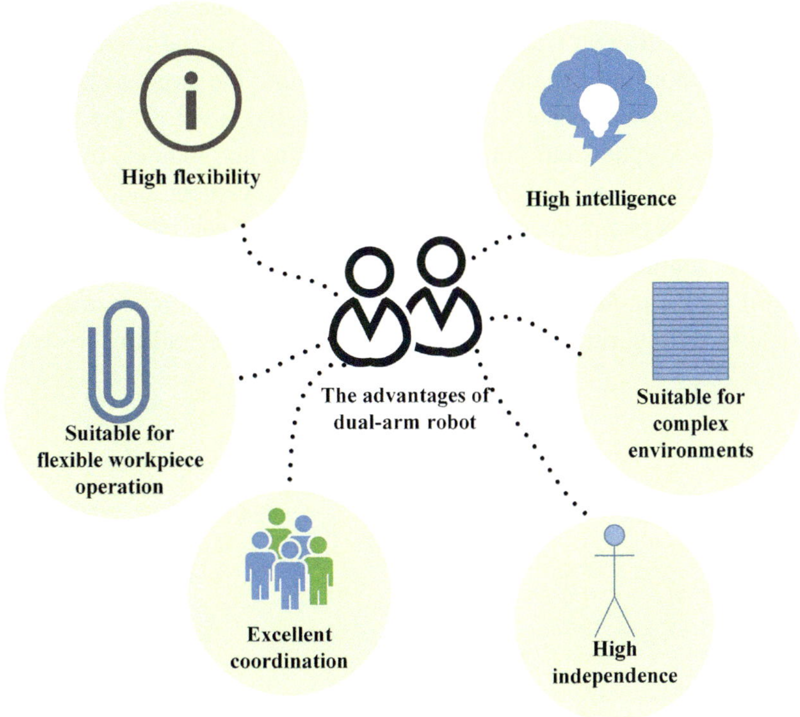

Figure 7.1
Advantages of dual-arm robots.

7.1.3 Definition of rail transit channel robots

With the rapid development of rail transit, an increasing number of trains are serving on the line, and functional robots based on rail transit maintenance are becoming necessary. As a result, rail transit channel robots have emerged. Compared with industrial, service, and power inspection robots, rail transit channel robots have appeared for a relatively short time, but they have basically formed a structure that takes intelligent technology as the core and network platform as the carrier, and integrates a variety of sensor equipment. Based on information networks, remote control of robots is realized, so as to replace field people. The main goal of the channel robot is to integrate robot technology, infrared temperature measurement, image recognition, ground orbit technology, large data analysis, and navigation technology scientifically to achieve high coverage and accuracy of the detection target, and ultimately to achieve the goal of reducing the manual workload. Based on the principle that rail transit channel robots can replace manual work on-site independently, remote control can be carried out through network technology, and real-time reliable monitoring of the site can objectively monitor equipment while meeting the requirements of flexibility, high accuracy, and antiinterference.

Generally speaking, rail transit channel robots need to meet the following requirements: (1) channel robots are different from fixed robots in that they need to move according to their tasks; (2) the robots need to obtain information about the target to be detected through a variety of sensors to achieve intelligent fault diagnosis; (3) the function of the navigation system is designed and implemented according to modules. Each module has a low coupling degree, which is convenient for module replacement and system maintenance, as well as communication and interaction between modules; (4) robots need to upload the collected information and corresponding fault types and fault locations to the historical fault database to facilitate megadata analysis later. Combined with hardware structure and software algorithms, the system framework of the rail transit channel robots is shown in Fig. 7.2.

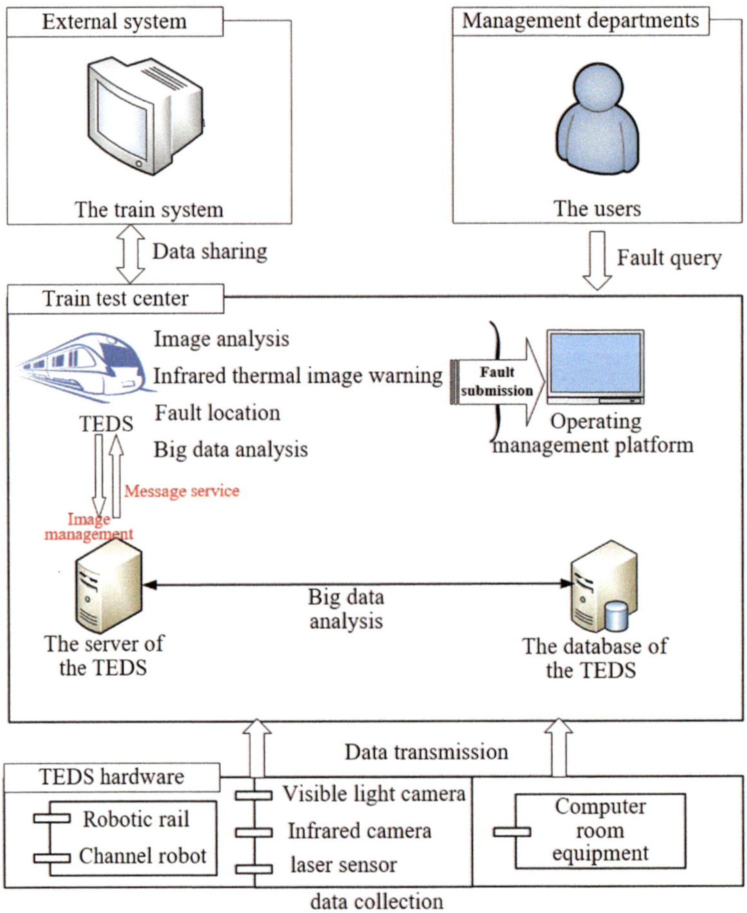

Figure 7.2

Channel robots Trouble of Moving Electric Multiple Units Detection System diagram.

7.1.4 Main components of rail transit channel robots

The rail transit channel robot described in this section is mainly used for the fault diagnosis of high-speed trains. Generally speaking, the workflow is as follows: (1) the maintenance task is acquired and traveled along the track to the checkpoint; (2) information about train maintenance is collected by onboard sensors including a visible light camera, laser sensor, and infrared thermal imager, some of which are installed on the arm of the robot; (c) the collected images are uploaded for fault diagnosis and analysis; (d) fault diagnosis results are reported and stored in the historical fault database. With the growth of the database, the big data analysis module will carry out data mining on the fault information of the train bogie equipment and predict the potential fault of the equipment. In addition, the robot is constantly detecting its power. Once it reaches the minimum power threshold, it will recharge. The hardware structure of the rail transit passage robot mainly includes a ground track, dual-arm robot, laser sensor, visible light camera, infrared thermal imager, and server.

7.1.5 Ground track

Robot ground track technology has recently arisen and covers many disciplines such as computer science, control theory, mechanism, sensing technology, artificial intelligence, and bionics. Application of the robotic ground track is important to measure the level of industrial automation in a country. Most of the ground track can be used for loading and unloading, welding, assembly, spraying, maintenance, handling, palletizing, and other work of machine tools, and can meet the actual needs of factory automation production lines [23]. The robot track is also known as a robot walking axle and robot outer axle. Its most important function is to drive industrial robots so that they can move on preplanned routes, expand the operating radius of industrial robots, and ultimately improve the efficiency of robots. Generally speaking, the walking axle of a robot is a walking system device that installs an industrial robot on a movable base and realizes linear motion by means of a sliding rail device. Robots using this device can increase the scope of handling, spot welding, maintenance, and other operations. In addition, there is an obvious advantage of the ground rail device: the use of a ground track control system to control industrial robots can avoid adding additional control systems.

The main part of the robot ground track is composed of high-strength structural parts welded and is processed precisely. The ground track consists of five parts: robot, ground track base, driving device, power sliding platform, and a long-distance tow chain. The structure of the ground track is shown in Fig. 7.3. The ground track base is the foundation of the ground track, including the supporting floor, origin correction block, linear sliding track, and rack. The power sliding platform is mainly used to support the robots.

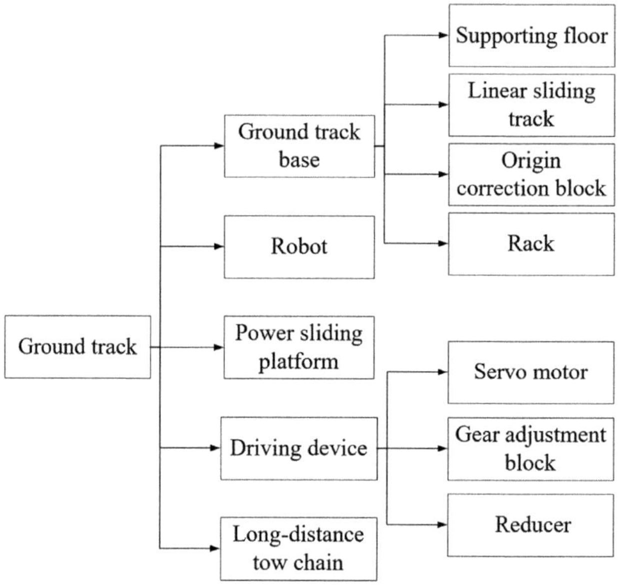

Figure 7.3
Structure of ground track.

The robot is mounted on the base plane of the power slide. The power slide relies on the driving device. The rack mounted on the base of the slide track moves along the direction of the linear guide rail. The range of motion is controlled by a programmable software limit switch. The driving device mainly provides power, which is composed of a servomotor, reducer, and gear adjustment block.

7.1.6 Sensors

7.1.6.1 Laser sensor

The laser sensor obtains information in the environment by means of a laser and realizes the detection of distance, speed, angle and other information. Its structure generally includes a laser emitter, receiver, and measurement circuit. According to its working substance, laser sensors can be divided into four types: solid, gas, liquid, and semiconductor. According to the dimensions of information that are obtained, laser sensors can be divided into two-dimensional and three-dimensional laser sensors [24]. Compared with a three-dimensional laser sensor, a two-dimensional laser sensor has obvious advantages in price, volume and detection speed. Therefore, two-dimensional laser sensors have an important role in indoor scenes, such as target location, map building, and robot navigation.

7.1.6.1.1 Operating principle

The positioning function of the two-dimensional laser sensor is realized by distance measurement. Time of flight (TOF) is commonly used for positioning. TOF-based positioning is mainly achieved by the time difference between laser emission and reception to complete distance measurement and positioning. The laser-based TOF method mainly includes pulse laser and continuous wave (CW) laser. The process of measuring distance by the pulse is thus: (1) when the laser beam is emitted, the laser sensor records the time of emission; (2) when laser beam touches the object, it reflects back to the receiving end of the laser sensor immediately and records the receiving time; (3) the distance between the object in the environment and the laser sensor is calculated according to the speed of light.

Laser sensors can also measure distance by the CW laser, including phase ranging, interferometry ranging, and frequency modulation ranging. For flight time and nonflight time methods, the former has obvious advantages in measuring speed and measuring range and is not easily affected by environmental temperature changes. Therefore, it can be more suitable for vehicles, mobile robots, and obstacle detection applications. In addition, the two-dimensional laser can construct a map by the simultaneous localization and mapping (SLAM) algorithm, thus realizing the function of navigation [25]. In this section, two commonly used laser sensors are briefly introduced.

The S30B-2011 BA laser sensor of SICK Company in Germany has the advantages of a wide scanning range, high angular resolution, and high safety performance [26]. The SICK S30B-2011 BA can carry out active scanning. A mirror rotating at a constant speed in the scanner can deflect the laser beam, enabling the sensor to scan around 270 degrees, and its maximum scanning distance can reach 29.9 m. The composition of laser sensor is shown in Table 7.1.

RPLIDAR A2 laser sensor is a new generation of low-cost laser sensor developed by SLAMTEC Company [27]. It can achieve high-speed laser ranging and sampling capacity of 8000 times per second, and can operate reliably and stably for a long time. The laser sensor is mainly composed of a laser ranging sensor, a high-speed rotating motor, and a universal serial bus adapter. The principle of this laser sensor is to use laser triangulation ranging technology. In each ranging process, the RPLIDAR transmits a modulated infrared

Table 7.1: Composition of laser sensor.

Sensor type	Composition
Laser sensor	Laser transmitter Laser detector Measurement circuit

Table 7.2: Functions and characteristics of two-dimensional laser sensor.

Sensor type	Functions	Characteristics
Two-dimensional laser sensor	Building the map Positioning Navigation Precision ranging	Short sampling period Fast positioning speed Strong antiinterference ability High precision

laser signal to the object, and then the reflection of the signal is received by the radar's visual acquisition system. Because of the existence of the rotating motor, the laser radar can be sampled at 360-degree rotation, and the scanning frequency can from 5.5 to 15 Hz. The functions and characteristics of the two-dimensional laser sensor are shown in Table 7.2.

7.1.6.1.2 Advantages and disadvantages of laser sensor

Advantages of two-dimensional laser sensor are high directivity, brightness, and strong antilight interference ability, which can enable noncontact and accurate measurement in the indoor environment; the monochromaticity of the laser emitted by the laser sensor gives it high resolution; the data transmission speed is fast because the laser travels fast according to the speed of light; the two-dimensional laser sensor can detect various objects of the surfaces. It has good reliability when tested. At the same time, the hardware has many advantages, such as small size, light weight, and low power consumption. Combining these advantages, the two-dimensional laser sensor has the characteristics of easy operation and a simple principle.

A disadvantage of the two-dimensional laser sensor is that accuracy of measurement is limited by hardware equipment. Because of the low frequency of the laser, spatial resolution is poor and it can detect only the surface of objects in the environment. If there are obstacles between the target and the laser sensor, it is difficult to obtain location information about the target that needs to be located. Two-dimensional laser sensing has an important role in many scenarios. Among them, the two-dimensional laser sensor is used to collect environmental information and the SLAM algorithm is used to build a map. It can be applied in autonomous navigation, target tracking, obstacle avoidance, position calibration, and other fields.

In addition, the two-dimensional laser sensor is common in the industrial field, mainly in the production of high-precision products. The two-dimensional laser sensor can detect surface contours, edges, dimensions, and welds of various devices.

7.1.6.2 Infrared thermal imager

7.1.6.2.1 Basic composition and function

The infrared thermal imager is an imaging temperature measuring device. Its basic functions include temperature measurement and imaging [28].

(a) Temperature measurement: Each unit senses infrared radiation and converts the received infrared radiation into corresponding electrical signals, which are then expressed in the form of a gray level according to the size of the electrical signals.
(b) Imaging: The gray level generated is converted into the image data format in the process of temperature measurement. The image data are displayed on the display as a thermal image. In the infrared thermal imager, the conversion function of the infrared photoelectric signal and electric signal to visible light is completed by each component of the infrared thermal imager. The infrared thermal imager consists of a thermal image detection device, a control device, and an image processing system. The thermal image detection device is an infrared camera, including an optical system, optical machine scanning mechanism, detector, and scanning synchronization mechanism. The image processing system is composed of an image analysis system, storage system, and image output device.

7.1.6.2.2 Working principle

Any object radiates energy outward in the form of electromagnetic waves. The infrared radiation intensity of different objects or different positions of the same object is different. The infrared thermal imager is an imaging temperature measuring device. The principle of infrared thermal imaging is to display the distribution of infrared radiation energy density according to the differing thermal contrast produced by the difference of temperature and emissivity between the target and its surrounding environment [29]. In fact, human vision is completely insensitive to infrared light, and the thermal image that can be seen by the human eye is converted from the infrared radiation emitted by the thermal imager according to the object. Different colors in infrared images represent different temperatures of objects, and the depth of color represents the surface temperature of objects. An infrared thermal imager converts infrared images into visible images in two steps.

Specifically, each unit senses the infrared radiation of the object and converts the received infrared radiation into the corresponding electrical signal. Then, according to the size of the electrical signal, the gray level generated in the temperature measurement process is converted into an image data format and displayed. The size of infrared radiation determines the temperature. Temperature and gray scale have a corresponding mapping relationship. The infrared thermal imager can not only generate a thermal image, it can calculate the temperature corresponding to each pixel, which is convenient for postprocessing. The hot image displayed on the display can qualitatively show the temperature. The process of infrared image generation in an infrared thermal imager is shown in Fig. 7.4.

7.1.6.2.3 Selection of infrared thermal imager

Infrared thermal imagers are made by many manufacturers and come in models; different models are suitable for various occasions. In actual use, it is necessary to select the appropriate infrared thermal imager according to the specific application and the target to be measured. The main principles are [30]:

(a) The temperature measurement range is the most important parameter of the infrared thermal imager. Different types of infrared thermometers have different ranges of temperature measurement. Therefore, the range of target temperature must be accurately estimated before the actual working process. Generally speaking, for power transformers equipped with rail transit equipment, the temperature measurement range of infrared thermal imager meets the temperature range of normal operation and failure.
(b) The monitoring distance of actual use is another important parameter. An infrared thermal imager should keep an appropriate distance with the object under test, so that the area size of the object under test exceeds 50% of the size of the field of view when measuring temperature. It is easy to cause errors if the distance is too far or too close.
(c) Emissivity and surface characteristics of the material to be measured affect the spectral response or wavelength of the thermal imager. In the high-temperature region, the best wavelength for measuring metal materials is near-infrared.
(d) The infrared radiation thermometer needs calibration. Accurate measurement of any measuring instrument is based on good calibration, and infrared thermal imager is no exception. In the course of use, the thermometer used must be recalibrated if there is an excessive difference in temperature measurement.

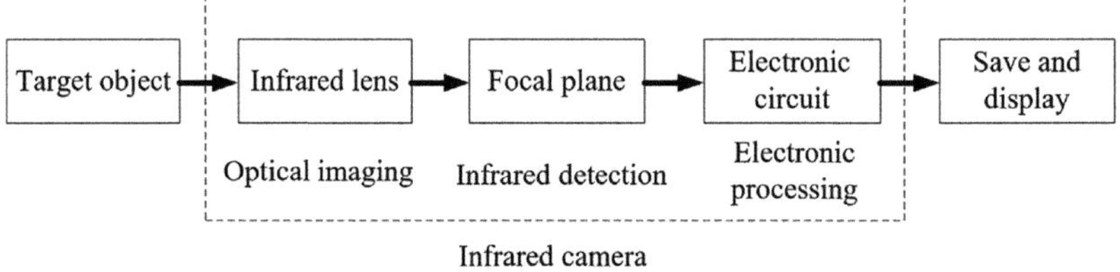

Figure 7.4
Forming diagram of infrared image formation.

7.1.7 Autonomous recharging

After the channel robot inspection task is completed, it is driven into the designated recharging area by the ground rail to charge, or in case the battery is low and the alarm is issued. The robot recharging connector automatically connects with the ground recharging device and starts recharging. After recharging is completed, the robot automatically disconnects from the recharging device and enters the standby state. This is the traditional recharging device; the most common one is contact recharging: that is, it is directly connected to the recharging post through the power cord and plug for recharging [31].

There are many drawbacks to the contact recharging method, because the working environment of the train maintenance station is complicated, and the channel robot may work in an open environment. Repeated insertion and removal of the recharging interface in a harsh environment will accelerate aging deformation, which may cause the recharging interface to fail to connect. In a humid environment, problems such as poor contact and broken lines may cause short circuit and electric shock accidents, which seriously affect the safety of recharging and reliability. At the same time, contact recharging requires accurate alignment between the channel robot and the recharging station to charge successfully, which increases the complexity and cost. In addition, the contact recharging device has a large footprint, and it is generally necessary to construct a separate robot recharging chamber, which greatly increases the recharging cost.

Compared with the contact recharging method, gradually mature wireless recharging technology has a great advantage [32]. There is no electrical connection or leakage interface between the recharging device and the device, contact sparks and wear aging are avoided. The transmitting coil is buried in the ground. Compared with the contact recharging station, the occupied space is significantly reduced, and it can also operate safely and stably in harsh environments. In addition, wireless recharging technology is effective in a certain area, and accuracy of alignment of the channel robot is not high, which also reduces the technical cost of the inspection robot and increases the reliability of recharging. Main advantages of wireless recharging of substation inspection robots relative to the contact recharging method are shown in Table 7.3.

Table 7.3: Comparison of channel robot wireless recharging methods.

Recharging way	Contact recharging	Wireless recharging
Reliability	Lower and lower	High
Versatility	Not universal	Not universal
Manpower requirement	All-day guard	Unattended
Precision requirement	High requirements	Low requirements
Cost	High cost	Low cost

The wireless recharging system of the channel robot is composed of a transmitting end power conversion device, an energy coupling mechanism, and a robot terminal power conversion device. The transmitting end power conversion device mainly adopts a high-frequency power supply technology, including a rectifying circuit, an adjustable frequency high-frequency inverter circuit, a transmitting end control circuit, a liquid crystal display module, and a voltage and current detecting circuit. The energy coupling mechanism is magnetic. The coupled resonant radio energy transmission technology includes a transmitting coil, a receiving coil, an impedance matching circuit, and the like. The robot terminal power converting device is composed of a high-frequency rectifying circuit, a receiving end control circuit, and a voltage and current detecting circuit.

7.2 Channel robot TEDS

The Trouble of Moving Electric Multiple Units (EMU) Detection System (TEDS) is used for the detection and abnormal warning of the status of EMU components, which can improve the quality of EMU maintenance work. Various sensors are installed inside each train, which can enable fault monitoring and fault diagnosis of its operating state. The scope of monitoring contains a power supply, air conditioning, heating, lighting, braking, a door, shaft temperature, a fire alarm, and so on.

However, it is difficult for this monitoring system of the EMU to cover all the positions of the train body, especially the structural parts at the bottom and side. Under the high-speed operation of the EMU, a small stone will also generate a huge impact kinetic energy, which can damage the key position of the train body. At the same time, for the running part of the EMU, the high-speed operation of the EMU may cause a bolt to loosen owing to the large transmission force and braking force of each component. Based on this, if the potential hazard cannot be discovered and eliminated in time, it will lead to an accident.

To enable an intelligent alarm regarding the abnormal structure of the vehicle body, it is necessary to design a set of detection systems to complete the fault diagnosis of the state of each structural member when the train is running or in the state to be inspected. The channel robot is equipped with various sensors, and the structural anomaly analysis and alarm using machine vision and deep learning are a relatively advanced and mature technology mode. Owing to the feature of turning back in operating the EMU, the following information about the train can be obtained by comparing image registration and eigenvalue between the detection image and the recent historical image: change in structural parts, change trend, and change type, to design a hierarchical alarm mode. Moreover, the system obtains an infrared thermal image of some key components and uses the model learning method to identify and alert about the faults, and to enable early about key faults to ensure the safe operation of the EMU.

7.2.1 Visible light image processing

Image recognition technology has had an important role in train fault diagnosis. It aims to realize the judgment and early warning of train failure by analyzing and identifying an image. A visible light camera is used for image acquisition, and image recognition technology is used to find accurate image features. The extracted key information is compared with the standard, thereby completing train fault identification. The specific process is shown in Fig. 7.5. The real-time monitoring and transmission of train status are enabled through the Internet of Things. After the fault information is found, maintenance staff are quickly coordinated to perform vehicle scheduling and fault information analysis and processing. However, considering the actual situation of the vehicle, it is not difficult to find that image recognition technology based on train fault diagnosis has several problems: (1) there are numerous and various in the structure structural equipment parts that need to be monitored and diagnosed; (2) the surface of train bottom parts is in long-term operation so that the state of the middle surface is not clear; (3) there are significant differences in the color of the vehicle body, weather, and brightness between pictures. Therefore, to identify the fault phenomenon accurately, it is necessary to preprocess the image to depict the fault feature correctly.

7.2.1.1 Image registration

In the field of train fault diagnosis, the robot is usually equipped with necessary sensors, and the intelligent image recognition algorithm is used to compare and analyze the same part of the vehicle image to enable abnormal image recognition and early warning. In theory, this image comparison strategy can solve the problem of motor vehicle fault diagnosis better, but in practical applications, there are many problems such as difficulty in

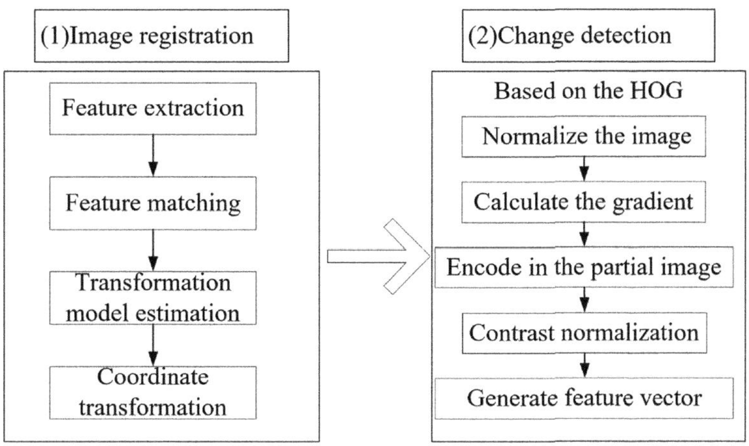

Figure 7.5
Analysis process of visible image. *HOG*, histogram of orientation gradient.

image registration, difficulty in detecting small changes, and the complicated structure of parts [33]. The process of the algorithm is: (1) the feature extraction algorithm is used to find the same feature points of the two images; and (2) least-squares regression is used to establish the corresponding mapping relationship to remove the influence of different external factors of the image, to achieve accurate registration.

The method can enable the positioning of important parts of the train, obtain the specific position of the area by target detection, and realize the abnormal identification and alarm of parts based on accurate registration. In the realization of abnormal alarms of components with complex structures, machine learning is used to deal with classification problems. Among them, the comparison method is used to realize the abnormal alarm of image failure. The main purpose of this registration is to eliminate external factors such as weather, light, and jitter during image acquisition. The resulting image difference is normalized to lay the foundation for the next real image detection.

The image matching algorithm, based on scale-invariant feature transform (SIFT) feature points, is considered one of the best matching methods. It has good stability for the rigid body transformation, illumination fitness, noise intensity, and scale change of objects in the image. Lowe et al. first proposed the SIFT in 1999 [34]. Because of its good feature extraction and description, it has attracted the attention of many scholars. Ratna et al. first sparsely encoded the SIFT feature point method and then used spatial pyramid matching for data retrieval. Finally, the feasibility of the improved algorithm was proved by experiments [35]. Ramu et al. combined the SIFT algorithm with random sample consensus technology to verify the authenticity of the image, and proved its superiority [36].

7.2.1.2 Change detection

After registration of the train component image and the standard image is completed, change detection is performed using the image that was registered. Change detection is based on a comparison of the histogram of orientation gradient (HOG) of the image to obtain the position of the change: that is, the position at which failure may occur [37]. The core idea of the HOG is to describe the edges and shapes of objects in an image in detail using the gradient or the density distribution of the edges. The HOG implementation method is: (1) the single image is divided into a square area of a certain size, and the area is named a unit connected area; (2) the histogram of the edge direction or the gradient direction of each pixel of the connected area of the unit is calculated, and then the obtained histograms are combined to form a feature descriptor; and (3) a larger interval can be divided from the image, and the local histogram can be normalized by contrast.

By normalization, better stability can be obtained for shadow and illumination changes. This can improve the accuracy of the test. Compared with other characterization methods, directional gradient histogram descriptors have many advantages, such as excellent dimensionality for image geometry and optical distortion. This is because the operator's operand is the local square cell of the image, and the geometric and optical deformations need to be displayed on a large area of the imaging unit. The HOG realizes the characterization of local objects in the shape and appearance of the image. It does not necessarily require precise information such as the edge position and correlation gradient, and the image brightness gradient change does not affect the characterization result.

The HOG steps are [38]:

(a) The image is normalized to reduce the effects of lighting factors. In the texture intensity of the image, the surface exposure can locally produce a large contribution weight. Therefore, the surface compression processing can effectively reduce the illumination variation and the local shadow of the image.
(b) The gradient of the image is calculated. The purposes of the derivative are to find the image contour, obtain the edge information and texture information, and further weaken the influence of light. The equation for calculating the gradient and direction is [39]:

$$m(x,y)=\sqrt{(L(x+1,y)-L(x-1,y))^2+(L(x,y+1)-L(x,y-1))^2} \tag{7.1}$$

$$\theta(x,y)=\arctan\left(\frac{L(x,y+1)-L(x,y-1)}{L(x+1,y)-L(x-1,y)}\right) \tag{7.2}$$

(c) The purposes of this step are to generate an encoding in the partial image area and to maintain weak sensitivity to rotation and scaling of the foreground image. The image is divided into a number of small areas, called unit connected areas. Then, pixels in each cell are subjected to one-dimensional gradient histogram or edge direction accumulation, and the basic direction histograms are mapped according to a fixed angle to form a final feature vector. The gradient direction ranges from 0 to 360 degrees. The histogram can be divided into 9 or 10 column levels, and the range is changed according to the direction change range.
(d) Contrast normalization is carried out in this step. Normalization can further compress the light, shadows, and edges to obtain the HOG descriptor.
(e) The final step is to complete the generation and collection of features of the overlapping image blocks in the detection window and combine them into a final feature vector according to certain rules and provide them to the classifier for use.

7.2.2 Infrared thermal images processing

Infrared thermal imaging detection is widely used in the field of train equipment fault diagnosis. Because train equipment is generally accompanied by abnormal high or low temperatures in the event of a fault, infrared spectrum radiation is closely related to the state and temperature of the equipment. The infrared thermal imager inspection equipment can accurately obtain the distribution of the surface temperature field of the equipment. If the parts of the equipment itself are defective or there are faults such as poor contact, corrosion, wear, deformation, fatigue, abnormal vibration, and pollution, there will be obvious cold spot low-temperature zones and hot spot high-temperature zones in the distribution map of the surface temperature field of the equipment.

In this case, fault diagnosis can be performed based on the obtained related data. In many cases, to monitor the fault point better, the infrared camera image can be synchronized and parallel analysis of the visible light camera image can be used to achieve fast and accurate identification of faulty components. Therefore, the infrared camera and the visible light camera are fixed on the same pan/tilt of the channel robot, and the visible light camera and infrared camera rotate with the pan/tilt to make the image coordinates of the two coincide. When the infrared thermal imager detects that the temperature is too high, the visible light camera simultaneously rotates to the corresponding position to take an image. The device not only displays, stores, records, captures, and plays back the visible light and thermal imaging images, it displays the temperature value and coordinates of the target. By jointly monitoring visible light images and infrared thermal imaging, equipment failures can be discovered and captured in a timely and accurate manner. Methods for infrared thermal image analysis mainly include image segmentation methods for positioning and machine learning methods for image classification, shown in Fig. 7.6.

7.2.2.1 Bogie fault location and image segmentation

Image segmentation is an important part of the hot fault detection of the bogie. The purpose is to extract the fault location from the complex image background, which is convenient for in-depth analysis and identification of the fault location. The accuracy of

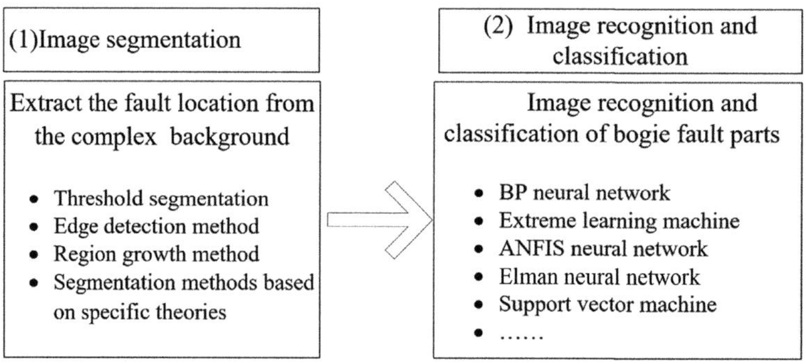

Figure 7.6
Intelligent analysis method of infrared thermal image. *ANFIS*, adaptive network-based fuzzy inference system; *BP*, backpropagation.

image segmentation is an important basis in the fault identification algorithm of train bogie equipment [40]. The higher the accuracy and integrity of the segmented image, the better the accuracy of the subsequent recognition algorithm. A large number of scholars at home and abroad have conducted a lot of research work in this area. However, because the infrared thermal image mainly reflects the thermal energy of the object, thermal radiation is converted into an electrical signal for display. Because objects often transfer and exchange heat, no matter how to the accuracy of the algorithm is improved, the segmented images will inevitably have errors, such as some important information being lost. This defect is determined by the characteristics of infrared thermography itself.

Image segmentation is based on the gray scale, color, texture, and other characteristics of the image to establish a suitable segmentation function, and set at least one threshold for the segmentation function; then, pixel points in the image are compared with the set threshold, and finally the image is segmented. Image segmentation algorithms can simplify or change the representation of the image, extract the region of interest from the original image, and eliminate the region of interest. After processing, the image is easier to analyze and understand. There are two types of image segmentation: a collection of subregions and a collection of contours on the image. The pixels of each subarea have similar characteristics, such as color, brightness, and texture under the measure of certain characteristics. Classical image segmentation methods include threshold-based segmentation, region growing, and edge detection [41].

7.2.2.1.1 Threshold segmentation

Threshold segmentation is the most commonly used method of image segmentation. The key is to determine the appropriate threshold to separate the background of the image from the region of interest. After the threshold is determined, the threshold is compared with the pixel points in the image one by one, the pixel larger than the threshold is set to 1 (or 0), and the pixel smaller than the threshold is set to 0 (or 1). The advantage of the threshold segmentation method is that calculation is simple, the speed is fast, and the calculation efficiency is high, but the accuracy is not high on some complicated occasions. Threshold segmentation methods developed in recent years have global, adaptive, and optimal thresholds.

7.2.2.1.2 Edge detection method

The edge detection method divides an image according to the characteristic that the edge of the object in the image has a gray-scale mutation [42]. The characteristic that the edge gray value of the image is discontinuous can be judged by detecting the first derivative extreme point or the second derivative. Commonly used first-order differential operators include the Canny operator, Roberts operator, Prewitt operator, and Sobel operator.

Commonly used second-order differential operators have the Laplace operator and Kirsh operator. Among them, the Roberts operator uses local difference operators to find edges with high edge precision.

However, if the image is not smoothed, it cannot suppress noise, and it works well for images with steep edges and less noise. The Sobel and Prewitt operators consider neighborhood information, which is similar to the weighted smoothing process of the image, and then perform the differential operation. Thus, it has a certain suppression effect on the noise. However, it cannot guarantee that there is no false edge in the detection result. In image edge detection, noise suppression and edge accurate positioning are difficult to achieve at the same time. Some edge detection algorithms such as the Sobel smooth noise and increase the inaccuracy of edge positioning. Some detection operators such as the Roberts increase sensitivity to edges but also to noise. The Canny operator is an operator that achieves a good balance between noise suppression and precise edge positioning. Therefore, the Canny operator is one of the most frequently used operators in edge detection algorithms.

7.2.2.1.3 Region growing method

The region growing method is a method of concentrating pixel points based on the similar properties of pixels in the same object region. Starting from the seed (single-pixel or small neighborhood point), the area is gradually enlarged by merging pixels or regions of the same nature until no points or areas can be merged. The region growing method has a good effect without prior knowledge. However, the method requires constant iteration, and the algorithm complexity and calculation amount are high. In general, there are three main issues to be aware of when using the region growing method for image segmentation. The first is the selection of seed points. The seed can select one or a group of pixels in the target area. This needs to be determined according to the characteristics of the specific segmentation problem.

Taking the infrared image as an example, the gray level of the image reflects the temperature information of the target. The pixel with the highest gray level can be selected as the seed pixel. Then, the growth criterion should be determined. It reflects how to include adjacent pixel points into the region during the region growing process. Finally, termination criteria should be selected to determine when to stop growing. The segmentation method based on region growing is a pixel-level processing method with good accuracy; it can achieve an ideal segmentation effect on complex occasions. The disadvantage is that the time and space complexity of the algorithm is high.

7.2.2.2 Image recognition and classification of bogie fault parts

Since *McCulloch* and *Pitts* proposed the first artificial neural network in 1943, it has grown tremendously [43]. Artificial neural networks are designed to mimic the mode of operation of biological brain neurons. The main idea is to abstract human brain neurons from the perspective of information processing, connect a large number of processing units to each other in a certain way, establish a simple model, and form different network neural networks according to different connection methods. The nonlinear model based on neural networks is to adjust the weight between neurons by inputting samples and output labels so that the output meets certain error requirements.

Because of its powerful data learning capabilities, it is widely used in classification problems. In the neural network, large-scale parallel computing processing can be performed, nonlinear information transformation and mapping can be realized, and there is self-organizing, self-adaptation, self-learning ability, good generalization, and easy hardware implementation. The support vector machine (SVM) is a machine learning method based on statistical learning theory and structural risk. It is also widely used in the field of infrared thermal imaging analysis. Deep learning has also emerged in recent years and can be regarded as a more complex and intelligent neural network. It has an increasingly important role in the field of image recognition. Among them, the deep learning theory and related applications will be described in detail in the next section. The following is a brief introduction to several commonly used neural networks and SVMs.

7.2.2.2.1 Backpropagation neural network

Rumelhart et al. first proposed a backpropagation (BP) neural network in 1986, which is a supervised algorithm with gradient descent [44]. The BP algorithm consists of the positive propagation of information and the BP of errors. In the BP process, the BP network continuously adjusts weights and thresholds through the gradient descent algorithm, minimizing network errors. The specific process of the training BP network is: (1) samples and tags are provided as input and output to the BP network, respectively. Input information is calculated layer by layer through the network hidden layer nodes and then transmitted to the network output layer to obtain the output; (2) the error between the actual output and the expected output is calculated, and the connection weight of each layer is returned in accordance with the error falling direction. Through continuous information forward calculation and error BP, the weights and biases of the network are continuously updated so that the actual output is continuously close to the expected output; (3) the preset accuracy of the network is achieved. The BP network is one of the most widely used neural networks. The structure of the simplest three-layer BP network includes an input layer, a hidden layer, and an output layer. Multiple neurons in each layer are not connected and affect each neuron in the next layer.

7.2.2.2.2 Extreme learning machine

In general, the final determination of the hidden layer node parameters of traditional neural networks requires a lot of iteration and optimization. This is one reasons why part of the network training process is too long and network training efficiency is low. At the same time, in some cases, the network does not converge after a large number of iterations. To solve the shortcomings of the BP algorithm, such as low learning efficiency and the cumbersome parameter setting, Huang et al. proposed the extreme learning machine [45]. Extreme learning is a novel solving method for the single hidden layer neural network. Compared with the traditional solving algorithm, it is simple and the solving speed is fast. Extreme learning machine network parameters (including weights and biases) are randomly selected and do not need to be adjusted. The network external right is determined by finding the least-squares solution. Therefore, in the process of determining the network parameters of the extreme learning machine, it is not necessary to perform a cumbersome BP adjustment process, which greatly simplifies the amount of calculation. At the same time, the extreme learning machine retains the basic structure of the traditional neural network and thus has the powerful learning ability of a traditional neural network. Therefore, the extreme learning machine is widely used for image classification.

7.2.2.2.3 Adaptive network-based fuzzy inference system neural network

Jang et al. proposed the adaptive network-based fuzzy inference system (ANFIS) in 1993 [46]. The traditional fuzzy inference system first fuzzifies the input data; then, it infers the data according to the inference mechanism set in advance, and finally it defuzzifies the data for output. Among them, reasoning mechanisms including fuzzy rules and membership function types are the key to the fuzzy inference system. However, selection of the membership function and formulation of the fuzzy rules require a large amount of professional knowledge and practical experience. This is the flaw of the traditional fuzzy inference system. The ANFIS neural network combines fuzzy logic and neural network to retain the fuzzy reasoning part, adjust the membership function of the fuzzy system by the self-learning ability of the neural network, and automatically generate the if–then rule. The mechanism of the ANIFS neural network is different from the traditional fuzzy inference system. The biggest feature of the ANFIS neural network is its self-learning ability, because the self-learning ability of neural networks is introduced into the fuzzy inference system. The fuzzy inference system of adaptive neural networks has strong generalization ability, fast convergence, and high reliability. Thus, this neural network has been widely used in various fields of image recognition.

7.2.2.2.4 Elman neural network

The Elman neural network is a typical local regression network with memory ability [47]. According to information about its operation, there are two kinds of neural networks: feed-forward neural and feedback neural. The so-called feed-forward neural network has a complex relationship between input and output through the hidden layer and nonlinear activation function. The output of the network is completely calculated with the input and the weight matrix. The feedback neural network is also called the recurrent neural network (RNN). Unlike the feed-forward neural network, the RNN such as the Elman neural network has a short-term memory function. The output of the neural network at time t is applied to generate output at time $t + 1$. In other words, the output of the Elman neural network is related to the output before the network. Because of this feature, it brings many advantages to the Elman neural network: (1) it is sensitive to the historical output of the network and meets the requirements of dynamic modeling; and (2) it has an associative memory function and high stability.

The Elman neural network is composed of four layers: input, hidden, output, and receiving. The receiving layer can carry out a one-step delay, which is the essential difference between BP neural networks and Elman neural networks. The receiving layer is also called the context layer because of its special location. The output of the hidden layer is forwardly fed to the output layer and returned to the receiving layer. The receiving layer receives the feedback signal from the hidden layer and stores it. After a certain delay, it is again input to the hidden layer. This short-term memory function adapts the Elman neural network to signals with time-varying features.

7.2.2.2.5 Support vector machine

The SVM is an important statistical learning model. It is essentially a classification model, and its principle is different from the traditional machine learning method based on the model training error minimization [48]. On the issue of learning, SVMs have obvious advantages over traditional models. At the same time, the solution of the SVM is related to only a few key support vectors. Therefore, it has low requirements on the number of samples, can obtain better learning effects in the case of small samples, and has better generalization and higher classification accuracy.

7.2.3 Location detection

The emergence of laser ranging and positioning technology has brought a huge revolution to the measurement industry. Laser ranging and positioning technology has been widely used in the field of train fault diagnosis and has an important role. The technology has unique performance and features and is essentially a noncontact measurement technology. The prospect of laser ranging and positioning technology is broad, and the market is also urgently demanding that laser ranging technology develop toward higher precision and smaller volume.

The purpose of fully automated intelligent analysis and positioning in the TEDS intelligent sensing system is to determine the location of faults further in the equipment using laser sensors mounted on the channel robot. Laser ranging methods include laser pulse method, laser phase method, laser interferometry, and laser triangulation. It is well-known that there is a direct relationship between the propagation distance of laser light in the atmosphere and the propagation time. Among them, the pulse method can use the laser beam emitted by the laser sensor in the air and the time difference to measure the distance. The specific working process is thus: The laser sensor emits laser light to the surface of the target to be measured, the laser is reflected back to the receiving system on the surface of the target, and the round-trip time T of the laser is calculated. The distance can be calculated according to the propagation speed of the light in the air. The distance of the measured target realizes the ranging function, which is calculated as [49]:

$$S = \frac{cT}{2} \tag{7.3}$$

where c indicates the speed of light, T the time difference, and S the measured distance.

The short-range laser triangulation positioning system is an important measurement technology in the field of noncontact measurement. Since the 1980s, the development of triangulation laser ranging technology has matured. It has been widely used in the fields of industry, national defense, and household life. Triangulation laser ranging positioning requires the laser sensor, image sensor, and target object to be measured at three points, forming a triangle in the geometric relationship. The principle of ranging is based on the triangular structure of the laser emitter, the image sensor, and the measured object. The laser emitted by the laser sensor hits the target to be measured, and then the light spot is imprinted on the complementary metal–oxide–semiconductor image sensor through the lens, and then the centroid coordinates of the laser spot are obtained according to digital image processing.

Because the shape of the optical path is a triangle, when the coordinate displacement of the spot on the photosensitive surface of the image sensor is known, the distance between the measured object and the laser can be calculated according to the relevant theory of triangle, to achieve positioning. The laser triangulation ranging system is simpler and more practical than the laser phase method and laser interferometry. Because of the limitation of the size of the sensor, the triangulation is mainly used for short-distance measurement. The farther the distance is, the lower the accuracy will be. The purpose of fully automated intelligent analysis and positioning is to calculate the distance between the laser sensor on the vehicle accurately and the faulty part of the train bogie. After calculation, the results are displayed on the display of the equipment, effectively realizing the requirements for positioning and control of the fault location in train operation and maintenance.

7.2.4 Big data analysis

The big data analysis module in the TEDS intelligent sensing system uses cloud computing to perform deep data mining on the train bogie equipment fault information, manages the life cycle of the equipment, and predicts the potential fault of the equipment according to the life-cycle curve of the equipment [50]. To some extent, accidents can be avoided and data support can be provided for the decision-making of technicians.

The module has the functions of automatically generating equipment defect reports, inspection task reports, and so on. At the same time, it can automatically generate analysis reports by comparing real-time data collection with historical data, classify and count different faults, and manage test data. Since the 20th century, high-speed trains have been in rapid use, and for trains in service, it is hard to avoid various types of failure. With continuous improvements in the level of informatization, the continuous emergence of various types of faults has produced massive real-time and historical data. Big data technology need to process large and diverse data effectively in a limited time and extract more valuable information from it. With the continuous development of big data technology, it has become increasingly mature. The following is a brief introduction to the relevant theories of big data technology.

Mainstream big data frameworks are Hadoop and Spark, which have well-functioning distributed storage, distributed offline computing, and distributed real-time computing, and the obvious advantages of large volume, large variety, and high-speed data collection, storage, and analysis [51]. The Hadoop Distributed File System (HDFS) of the Hadoop platform can form a big data distributed storage system on a cluster of multiple inexpensive machines. It divides large files into multiple data blocks according to actual requirements and distributes them to each node of the cluster to achieve high-throughput data storage and access. MapReduce is a programming model for big data batch computing. It divides batch tasks into two phases: map and reduce. The map phase is responsible for generating key-value pairs for data and provides a variety of sorting methods for sorting. The reduce phase performs certain processing and outputs the results to the HDFS. In addition, HBase and Hive are two important components based on Hadoop. Among them, HBase is a distributed nonrelational database that is suitable for real-time, random big data storage and access. Hive is a data warehouse tool that can perform data extraction, transformation, and loading operations through Hibernate Query Language statements. Because of its poor real-time performance, it is suitable for offline analysis.

Spark is a parallel big data framework based on memory computing, which is computationally efficient and fast [52]. Based on the Resilient Distributed Dataset, the Spark platform provides a number of operators that can easily perform a variety of transformation operations and action operations. Spark Streaming is a streaming computing framework built on top of Spark for real-time big data analytics. The Spark Structured Query Language (SQL) module is used to process structured data. In Spark programs, it can easily manipulate multiple data sources using SQL statements and the

DataFrame application programming interface. Because Spark is a memory-based framework, its performance in data computing is far superior to Hadoop's MapReduce framework. Therefore, in practical applications, considering the characteristics of these two big data frameworks, Hadoop's distributed file system is often used for data storage. Spark does real-time data processing and analysis. The functions of big data technology in the TEDS intelligent sensing system discussed next.

7.2.4.1 Data collection

Data sets are the basis of data analysis. Data sources include sensors (laser sensors, infrared camera sensors, and visible light cameras), a real-time acquisition system, visible light image analysis results, a sensor real-time acquisition system, and infrared thermal imaging diagnosis. Data collection and uploading use Flume to do massive log collection, which is standardized according to a unified specification format and uploaded to a distributed file system or a distributed database.

7.2.4.2 Data storage

Data storage is an important part of big data technology, which provides a persistent operation for data. There are three main forms of data storage: the HDFS, distributed database, and relational database. The distributed file system can store data in a specified folder at a specified time or size. The distributed database includes an HBase column database and Hive data warehouse. The relational database mainly refers to Redis and My SQL. Redis is a memory-based cached database. In the real-time computing framework, Redis caches the results and then transfers them to the My SQL database.

7.2.4.3 Data analysis

Data calculation and analysis are the core of the system. The module is divided into computational framework and problem modeling. The computing framework is a big data software framework for specific computing tasks, including offline analysis and real-time computing. As the key link of data analysis, problem modeling needs to embed visible light image analysis, infrared thermal imaging diagnosis, fault intelligent analysis, and other related algorithms into the big data framework. Specifically, it involves fault classification, fault cause analysis, feature extraction, data standardization, real-time algorithm selection, model testing, and model evaluation system establishment.

7.2.4.4 Model software

The intelligent model is programmed and systemized as a smart brain for train fault diagnosis, providing reliable diagnosis and early warning functions for train operation and maintenance safety.

7.3 Bogie fault diagnosis based on deep learning

With the rapid development of high-speed trains, the safe and stable operation of trains has become particularly important. Main factors in compromising the safety of high-speed train operations are the failure of key components and serious natural disasters. The health state of the bogie can affect the running performance of the train significantly. Its mechanical performance determines the reliability and safety of the whole high-speed train system. Therefore, the fault diagnosis method of high-speed train bogie is important in practical application. However, because the type of bogie failure is various and the characteristic signal is not obvious, it is difficult to extract fault information using conventional signal processing methods so that the bogie fault can be correctly identified. When considering high-speed railway safety, it is becoming increasingly urgent to develop a bogie fault identification method with high accuracy.

The fault diagnosis of mechanical equipment can be traced back to the 1950s at the earliest [53]. In 1967, the United States established the mechanical failures prevention group, which engaged in fault mechanism research, diagnosis technology research, reliability analysis, and so on [53]. Mechanical equipment fault diagnosis methods are model-driven and data-driven [54].

Model-driven fault diagnosis methods establish an accurate mathematical model of the mechanical component and combine the state of the mechanical component with the mathematical model to detect engine faults. The advantage of this type of method is that it uses deep information about the system for fault diagnosis. Disadvantages are that the system model is difficult to establish and the robustness of the system is not good owing to the existence of noise disturbances. With the complexity and nonlinearity of the mechanical components and working conditions, accurate mathematical models require higher model orders, which brings certain challenges to the accuracy of model parameter estimation and real-time performance.

With the application of signal processing technology and signal recognition technology, data-driven fault diagnosis technology has become an important means of diagnosis. Data-driven diagnosis methods can use signal analysis methods to obtain deeper feature vectors in the time and frequency domain of the system. The fault diagnosis is realized by establishing mapping between the feature vectors and faults. Commonly used signal processing methods include empirical mode decomposition (EMD), morphological filter, and wavelet packet transform [55]. With the development of artificial intelligence, machine learning technologies such as BP neural network, Elman neural network, and SVM have attracted increasing attention for data-driven fault diagnosis [56–58]. In this case, the fault diagnosis technology of mechanical equipment is developing toward intellectualization. Fault diagnosis technology has become an interdisciplinary subject that integrates mathematics, physics, computer, microelectronics, information processing technology, and artificial intelligence. The data-driven methods need no prior knowledge for the mechanical component. However, in the face of a large number of signals or complex structures, the learning ability of traditional machine learning algorithms is limited, and recognition results are often not ideal in many cases.

Deep learning is a breakthrough in machine learning research. A deep learning network has a better-fitting capacity than traditional neural networks [59]. Data are input from the visual layer and output learned features through nonlinear mapping of the hidden layer. As the number of hidden layers and the complexity of the network structure increase, the hidden layer can learn more abstract deep features to represent the data, which makes classification easier and improves the score. At the same time, because of the complex structure of multiple hidden layers, it can also be trained well in the face of a lot of samples and has broad application prospects. With the help of the deep neural network, the complex relationship between the features and faults can be learned, and high accuracy can be achieved [60]. A fault diagnosis framework based on deep learning is shown in Fig. 7.7.

7.3.1 Vibration signal acquisition and preprocessing

In general, the acquisition of original data is carried out by installing acceleration sensors in key parts of train bogies. To process vibration signals, commonly used traditional processing and analysis techniques are divided into time-domain, frequency domain, and

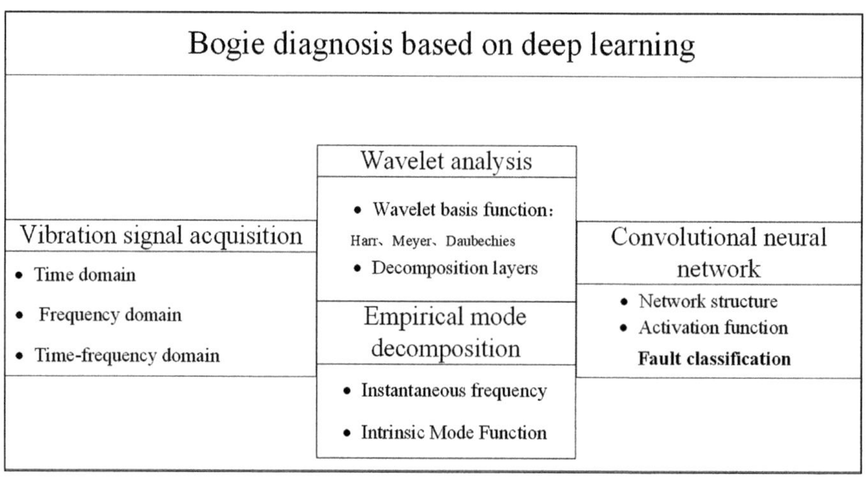

Figure 7.7
Bogie fault diagnosis structure diagram.

time-frequency domain analysis [61]. So-called signal time-domain describes the correspondence between signals and time. Any signal waveform in the time domain can be expressed as signal changes with time. The waveform contains all of the information of the signal. Moreover, characteristics of any signal first show its time characteristics. The waveform shows: (1) the corresponding time of signal amplitude; (2) whether the signal has a period and the length of the period; and (3) the rate of signal change. The image describing signal by time function is called the time-domain graph, and signal analysis in the time domain is called time-domain analysis. For vibration signals, common time-domain analysis indicators are variance, kurtosis, peak value, and impulse index, [62].

The signal graph in the frequency domain is generally called the spectrum, and signal analysis in the frequency spectrum is called the frequency domain analysis. The image of the signal in the time domain and frequency domain represents completely different information. The graph in the time domain shows the change of signal with time, whereas the graph in the frequency domain shows the distribution of frequency, the proportion of frequencies, and the phase of frequencies. The theoretical basis of frequency domain analysis is Fourier transform theory, whose essential function is to divide components according to the frequency from the whole signal. Commonly used frequency domain analysis methods are the transfer function analysis method, frequency inversion analysis method, and resonance demodulation technology.

In the fault diagnosis of train bogies, separate time and frequency domain information may not reflect the local characteristics of fault signals perfectly. The time-frequency domain analysis method not only reflects the local characteristics of the signal, it expresses the relationship between frequency and time. The research proves that it has a strong ability to deal with nonstationary signals produced by mechanical equipment. EMD, short-time Fourier transform, and wavelet transform are commonly used in time-frequency domain analysis. In the next section, several common time-frequency domain analysis algorithms are introduced.

7.3.2 Vibration signal feature extraction based on wavelet analysis

7.3.2.1 Fast Fourier transform

In 1965, Cooley et al. proposed the fast Fourier transform (FFT) method [63]. FFT is widely used in engineering because of its advantages of a fast operation speed and relatively stable spectrum analysis features in the analysis and processing of mechanical vibration signals. In fact, FFT is not new, but an efficient algorithm based on discrete Fourier transform (DFT), which is essentially the same. According to the physical meaning of Fourier transform, any continuous time-domain signal can be decomposed into numerous sinusoidal signals of different frequencies. DFT is a method to represent time-domain sampled signals by frequency-domain superposition. The frequency and amplitude of the sinusoidal wave constituting the original time-domain signal can be obtained by the DFT so that the signal can be converted from the time domain to the frequency domain. The operation of the DFT is [64]:

$$X(k) = \frac{1}{N}\sum_{n=0}^{N-1} x(n)e^{-j\pi nt/N} \tag{7.4}$$

The amplitude spectrum and energy spectrum of the Fourier transform are [64]:

$$|F(u,v)| = \left[R^2(u,v) + I^2(u,v)\right]^{1/2} \tag{7.5}$$

$$E(u,v) = R^2(u,v) + I^2(u,v) \tag{7.6}$$

where $R(u,v)$ and $I(u,v)$ represent the real and imaginary parts, respectively.

The FFT is based on the odd and even, virtual, and real characteristics of DFT, which simplifies the calculation process of the algorithm itself. FFT only improves the calculation speed of DFT. The algorithm is essentially the same; that is, if both inputs are the same, the output is the same. The only difference lies in the time complexity of calculation [65].

For example, for the DFT algorithm with 1000 sampling points, the computational complexity is about 2 million times, but the FFT needs only about 15,000 times. The FFT greatly improves the computational efficiency. The FFT algorithm includes two kinds of algorithms: time-based extraction and frequency-based extraction, which divide time-domain signal and frequency-domain signals into even–odd divisions, respectively. The frequency spectrum of the vibration signal is extracted by using FFT, which is essentially the decomposition and reconstruction of the signal on the orthogonal basis of linear space. In engineering applications, using the FFT algorithm to display the vibration data in the frequency domain can obtain the characteristics of the vibration data, to diagnose the fault of mechanical equipment.

7.3.2.2 Wavelet transform

The FFT is also imperfect. When dealing with some nonstationary signals, the FFT still has some limitations and can observe only the characteristics of the frequency domain. Wavelet transform can use the wavelet basis and window function to obtain the characteristics of the signal in the time domain and frequency domain [66]. Wavelet is a small waveform whose energy is concentrated in the time domain. On the one hand, the wavelet has attenuation, reflecting its small characteristics. On the other hand, the wavelet has fluctuation, showing the oscillation form of positive and negative amplitude alternation, reflecting its wave characteristics. Wavelet analysis is developed based on the Fourier transform.

The adaptive time-frequency analysis of signals is accomplished by scaling translation operation, multiscale refinement, and localization analysis of signals. Wavelet analysis is known as the microscope in the field of signal analysis because of its remarkable characteristics of frequency subdivision at low frequency and time subdivision at high frequency. As a time-frequency analysis method created after Fourier transform, wavelet analysis perfectly solves the defect that Fourier transform is insensitive to local information features. It can fully extract the time-frequency characteristics of signals, providing a powerful means for signal analysis. Wavelet analysis has been widely used in time series prediction, machine vision, image processing, data compression, and other fields [67–69].

Wavelet transform can be divided into continue wavelet transform and discrete wavelet transform [70]. Because of the wide application range of wavelet transform and the different purposes of wavelet transform, different kinds of wavelet basis functions have appeared. The selection of wavelet basis functions is an important part of wavelet transform. Several commonly used wavelet families include Haar, Morlet, Daubechies, Coiflets, Symlets, and Meyer. The applicability of wavelet-generating functions with different properties is different, which has a great influence on the effect of feature extraction. The wavelet should meet the requirement of the function of wavelet analysis.

To avoid distorting the filtering signal, the selected wavelet function should have the property of the linear phase, that is, the symmetry corresponding to the wavelet. If the wavelet is used as catching abrupt signal, the selected wavelet function should have a high vanishing moment. In addition, the compact support of the wavelet function is an important parameter. A wavelet with short support length has better computation performance and less truncation errors. However, the compact support has a trade-off with the vanishing moment. The shorter the support length, the smaller the vanishing moment.

In the practical application of production, the Morlet wavelet has a wide range of applications, which can realize feature extraction, edge detection, recognition, and classification of signals and images. The Daubechies basis function has the advantage of smoothing the process of signal reconstruction. The Daubechies is usually chosen as the wavelet basis function in digital signal processing. The Daubechies wavelet with N vanishing moments is abbreviated as DB N.

7.3.3 Feature extraction of vibration signal based on empirical mode decomposition

In the field of signal analysis, the methods of nonstationary signal analysis include short-time Fourier transform and wavelet transform, both of which are essentially Fourier transform. The EMD introduced in this section can adaptively decompose different scales or trends in complex time series into the sum of a limited number of intrinsic mode functions (IMFs), in which each IMF component represents different characteristic scales [71]. Moreover, the decomposed subsequences have instantaneous frequencies with clear physical meaning, which can well describe the oscillation of the original time series in each local region. The time series analysis of EMD includes two main processes. First, the time series is decomposed into empirical modes to obtain multiple IMFs. Then, the instantaneous frequencies of each IMF are obtained by Hilbert transform for each IMF. The two basic concepts (instantaneous frequency and IMF) are introduced next.

7.3.3.1 Instantaneous frequency

In actual production and life, most of the signals are nonstationary and nonlinear. For nonstationary processes, instantaneous frequency is a feature that cannot be ignored [72]. The concept of instantaneous frequency has been put forward for a long time; it exists widely in real life, such as in seismic signals, which are typical instantaneous frequency signals in life. However, the theory of instantaneous frequency is not so intuitive. It is generally believed that only the sine signal or cosine signal has a frequency, and it is a fixed value. Fourier cannot accurately express the time-variant signal, because the frequency of the signal is constantly changing. In practice, the frequency of the signal will not remain unchanged. Thus, the concept of instantaneous frequency is needed. Scholars have proposed the EMD as well as the definition of instantaneous frequency [73]. The original vibration is transformed into an analytic signal with instantaneous frequency by *Hilbert* transform. In practice, signals are usually complex signals composed of multiple frequency components, because the *Hilbert* transform can represent only simple signals with single-component frequency. Therefore, the EMD is needed to decompose the original vibration signal into simple signals with multiple-frequency components, which are IMF. Then, the instantaneous frequency is obtained by *Hilbert* transform, which makes the instantaneous frequency universal.

7.3.3.2 Intrinsic mode function

To decompose the coincidence signal into several single-component combinations, scholars have proposed the concept of the IMF: that is, a signal can be decomposed into multiple IMF components and a residual signal [74].

IMF needs to satisfy two conditions: (1) in the data set, the number of extreme points and zero-crossing points must be equal or at most one point different; (2) at any point, the mean value of the upper envelope determined by all local maximum points and the lower envelope determined by all local minimum points is zero.

7.3.3.3 Empirical mode decomposition

Generally speaking, most signals encountered in production and life are multicomponent. It is difficult to analyze the characteristics of these signals directly. Compared with multicomponent signals, the feature analysis of single-component signals is much easier, because their fluctuation mode is simple. The EMD is to decompose the original multicomponent signal into many single-component signals (i.e., IMF). Before EMD, it is necessary to assume that (1) there is at least one maximum and one minimum in the original time series; (2) the local time-domain characteristics of the data are uniquely determined by the time scales between the extreme points; and (3) if there is no extreme point in the original time series, it is necessary to have inflection points and find the extreme points by means of differentiation.

It is easy to see that these three conditions are relaxed, and most natural signals can be easily satisfied. Thus, EMD is widely used. For the EMD process of one-dimensional time series, it can be simply summarized as: (1) the local extremum points of the original time series can be found; (2) the difference of the extremum points can be processed to obtain the upper and lower envelopes of the time series, and then the mean envelopes of the upper and lower envelopes can be obtained; and (3) the sequence that conforms to the IMF can be screened sequentially (i.e., an original time). The sequence is decomposed into a set of IMF and a residual.

7.3.3.4 Advantages and defects of empirical mode decomposition

Through the introduction of wavelet analysis and the EMD algorithm, the EMD method has obvious advantages [75]:

(a) The EMD has few parameters to adjust and can directly decompose almost all nonstationary time series without choosing the appropriate basis function. Therefore, the EMD can analyze any kind of time series theoretically. However, wavelet decomposition based on the wavelet basis function is not so good in the range of signal processing;
(b) EMD can continuously decompose a multicomponent signal, and finally get a set of IMF and a residual term, whereas the original signal can be restored by multiple intrinsic modes. The modal functions and residual terms are obtained and realized. In other words, the original structure of sequence can be completely reconstructed by the IMFs and residual terms. The signal will not be distorted to any extent during the whole process of EMD processing, whereas the signal will be lost to a certain extent when the wavelet transform processes the signal. This is also an advantage of EMD.

Empirical mode decomposition also has some obvious shortcomings:

(a) Lack of a strict theoretical basis. Compared with Fourier transform and wavelet transform, EMD has no strict mathematical expression, which limits its development and application process in practical application to a certain extent;
(b) The existence of spectral mode aliasing will make the IMF lose its physical significance. The main reason for mode aliasing is the discontinuity of the signal, which

makes the EMD unable to decompose the components of the same frequency band correctly. In addition, the mode aliasing problem may also make the EMD algorithm unstable, and any disturbance may lead to the change in IMF;

(c) End point effect. In the EMD algorithm, cubic spline interpolation is needed to construct upper and lower envelopes for the extremum points of signals. However, in the actual signal, the end point value is not necessarily the extreme value. If the end point value is regarded as an extreme value, it will lead to the end point effect. Because the EMD has some obvious shortcomings, many scholars have made some improvements in the basis of the EMD. They put forward the ensemble EMD, complete EMD, and so on.

7.3.4 Bogie fault diagnosis based on convolutional neural network

The convolutional neural network, originating from the structure of the biological visual system, is a type of neural network. Compared with other neural networks, the biggest differences of convolution neural networks are the convolution layer that is added and that the sparse interaction and parameter sharing performance brought by the convolution layer greatly improve the learning ability of the network. Since its conception, the convolutional neural network has been active in image, speech recognition, and text processing, and has achieved good results [76]. In fact, the diagnosis of mechanical equipment is the problem of classification. With this, it is to classify the vibration signals of the train bogies. In this section, features extracted from vibration signals will be used as training samples for the deep learning model. The specific process is shown in Fig. 7.8.

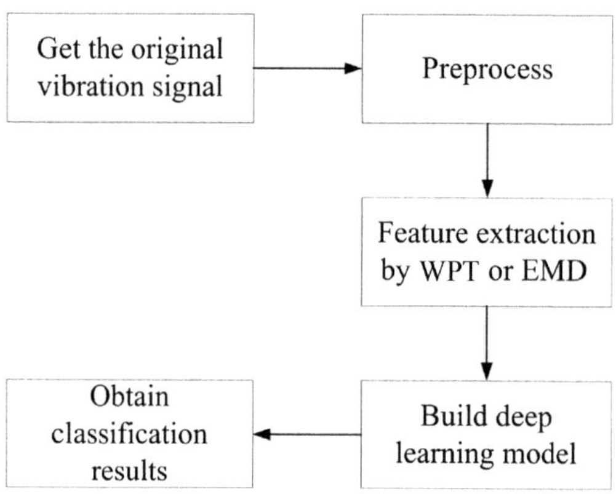

Figure 7.8
Flow of fault diagnosis. *EMD*, empirical mode decomposition; *WPT*, wavelet packet transform.

7.3.4.1 Convolutional neural network architecture

A complete convolution network is generally composed of the input, convolution, pooling, full connection, and output layers. However, by changing the number and order of each layer, convolutional neural networks with different performance can be achieved. The convolution layer is the key part of the convolution neural network. Convolution is the inner product of vectors in a concise sense. In signal classification, convolution operation has a role in filters. Convolution operations include continuous and discrete convolution. Specific equations are [77]:

$$y(t) = x(t) * h(t) = \int_{-\infty}^{+\infty} x(\xi)h(t-\xi)d\xi \tag{7.7}$$

$$y(n) = x(n) * h(n) = \sum_{t=0}^{N-1} x(i)h(n-i) \tag{7.8}$$

In the convolution neural network for vibration signal recognition, because the vibration signal is discrete points and not continuous, discrete convolution is needed for operation. Its calculation process is simply multiplication between matrices, using the selected convolution core and the input signal for convolution operation. In the neural network, the convolution core in the convolution layer has the advantage of parameter sharing and can speed the operation. The first item $x(t)$ in the previous equation is the input image, the second item $h(t)$ is the convolution core, and the left-most item $y(t)$ is the output item, that is, the feature output, which is extracted by the convolution operation.

In convolution operation, it is generally necessary to bias the result of feature extraction. The equation is [77]:

$$x_j^l = f\left(\sum_{i \in M_j} x_j^{l-1} * k_{ij}^l + b_j^l\right) \tag{7.9}$$

where x_j^l is the j, the feature graph output from layer l (currently the convolution layer), $f(x)$ is the activation function adopted by convolution layer, $*$ is the convolution operator, and k_{ij}^l is the convolution matrix of the convolution core. For each output graph, a bias term is added to the convolution operation. Through convolution operation, the desired features can be extracted and the features obtained by denoising are more obvious, which makes the recognition and classification of subsequent vibration signals more accurate.

Generally, the pooling layer is behind the convolution layer. The pooling operation can reduce the eigenvectors of the convolution layer output, improve the pooling function of the results, and avoid overfitting. There are many kinds of pooling methods, such as max, average, and random. The main idea is to use some statistical features of the adjacent output of a certain location to replace the output of the network at that location. The process of pooling is similar to that of convolution. It moves on the input image through a fixed size window according to the set step size. Among them, the maximum pooling algorithm is sensitive to the maximum value of the vector; the average pooling algorithm averages the element data of the vector; the random pooling algorithm randomly selects any value in the window according to a certain probability distribution. Whichever pooling function is used, pooling has the function of keeping input close to invariable when input disturbance occurs, which gives the network strong adaptability.

As the last layer of the convolutional neural network model, the full connection layer has the role of a classifier. After the input image is propagated through the convolution layers, the features extracted earlier are classified using the full connection layer network. The input above the full connection layer is obtained by lengthening the previous feature into a one-dimensional vector. The corresponding equations are [78]:

$$y^k = f\left(w^k x\right) \tag{7.10}$$

7.3.4.2 Activation function of convolutional neural network

Because the convolution, connection, and pooling layers are all linear computations without the activation function $f(x)$, the resulting model will be linear if the model only contains these, which cannot be applied to nonlinear scenarios. Therefore, a nonlinear activation function is added to make up for the deficiency of linear model expression and enhance the classification ability of the model. Correspondingly, selection of the activation function should meet certain requirements. First, the activation function should be continuous, which can avoid oscillation of the network in the training process and make the overall performance of the network more stable. Second, the activation function should be continuously differentiable (except for individual points) to facilitate training of the network. Third, the form of activation function should be simple and clear, which is greatly significant to reduce the complexity of the network. The activation function introduces nonlinearity into the network, which has a vital role in the learning generalization ability of the network. The activation functions commonly used in neural networks are the sigmoid function, rectified linear unit (ReLU) function, and tanh function. These activation functions are briefly introduced next.

7.3.4.2.1 Sigmoid activation function

The sigmoid function fully meets the three requirements mentioned earlier. It is continuously differentiable in the whole function domain and can map the input signal between 0 and 1 in a simple form. The sigmoid function has good properties as an activation function. From a mathematical point of view, it has a different effect on signal gain in the central and bilateral regions. Gain in the middle region is better than that in the bilateral region. From the perspective of neuroscience, there are excitatory states similar to neurons, such as the central area, and inhibitory states of neurons, such as bilateral areas. Expressions of sigmoid function and derivatives are [79]:

$$\text{sigmoid}(x) = \frac{1}{1 + e^{-x}} \tag{7.11}$$

$$\text{sigmoid}'(x) = \text{sigmoid}(x) \times (1 - \text{sigmoid}(x)) \tag{7.12}$$

The sigmoid function and its derivatives are shown in Fig. 7.9.

7.3.4.2.2 Tanh function

The tanh function and sigmoid function have many similarities. Generally speaking, the tanh function can be obtained by linear transformation of the sigmoid function, so they can be substituted for each other in many application scenarios. The biggest difference between them lies in the range of function value and the derivative value of the two functions. The tanh derivative value ranges between [0,1]. Compared with the sigmoid function, the phenomenon of gradient disappearance is not serious. Thus, it is widely used in deep neural networks. The tanh function maps the input signal between derivatives −1 and 1. It also satisfies the three requirements of the ideal activation function. Its function and derivative expressions are [80]:

$$\text{Tanh}(x) = \frac{e^x - e^{-x}}{e^x + e^{-x}} \tag{7.13}$$

$$\text{Tanh}'(x) = 1 - \text{Tanh}^2(x) \tag{7.14}$$

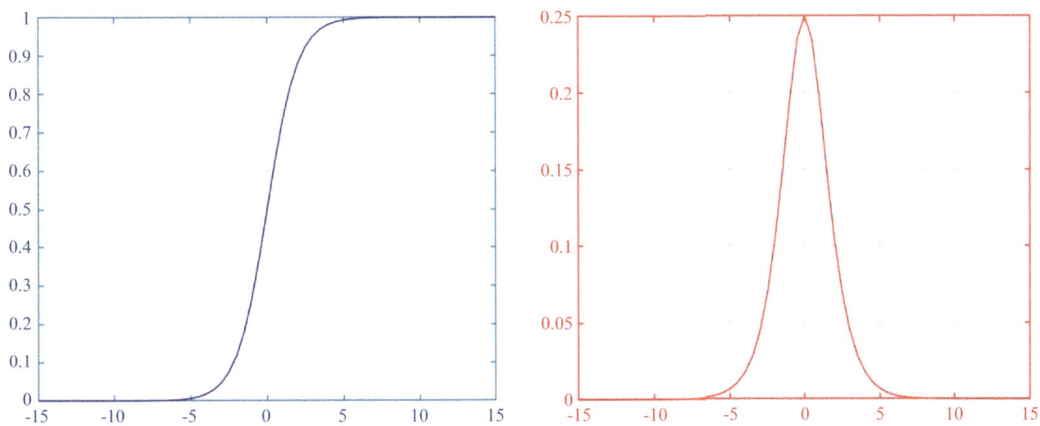

Figure 7.9
The sigmoid function.

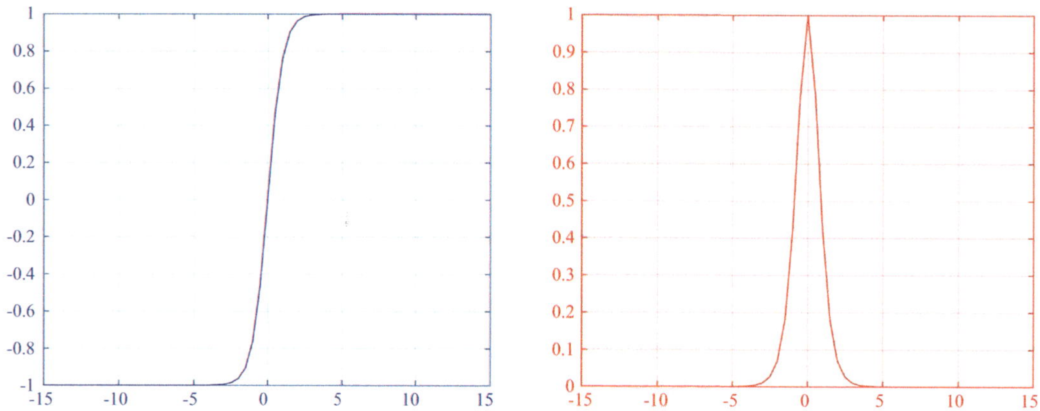

Figure 7.10
The tanh function.

The tanh function and its derivatives are shown in Fig. 7.10.

7.3.4.2.3 Rectified linear unit

As the activation function of deep neural network, the ReLU function has excellent performance and a simple structure. This function helps the deep neural network realize sparse activation. In the process of use, after initialization, the weight can make about half of the output of hidden units equate to 0. Sparse activation is not only more in line with the mechanism of human brain activity, it has a good advantage at the mathematical level. For a given input, the activated subset of neurons is single and the calculation of the subset is linear, which makes the ReLU function an activation function without the disappearance of the gradient. Application of the sigmoid function and tanh function in deep neural networks is limited because of the disappearance of the gradient. After the ReLU function was proposed as an activation function, many scholars improved it [81]. Expressions of ReLU function and its derivatives are [82]:

$$\mathrm{ReLU}(x) = \begin{cases} 0 & x \leq 0 \\ x & x > 0 \end{cases} \tag{7.15}$$

$$\mathrm{Re}LU'(x) = \begin{cases} 0 & x \leq 0 \\ 1 & x > 0 \end{cases} \tag{7.16}$$

The ReLU function and its derivatives are shown in Fig. 7.11.

7.3.4.2.4 Fault diagnosis result acquisition

A supervised learning algorithm is usually used in convolution neural network training. Its operation includes forward propagation and is error BP. Forward propagation is a feature of input or a predictive output obtained by transferring input to the neural network. Its

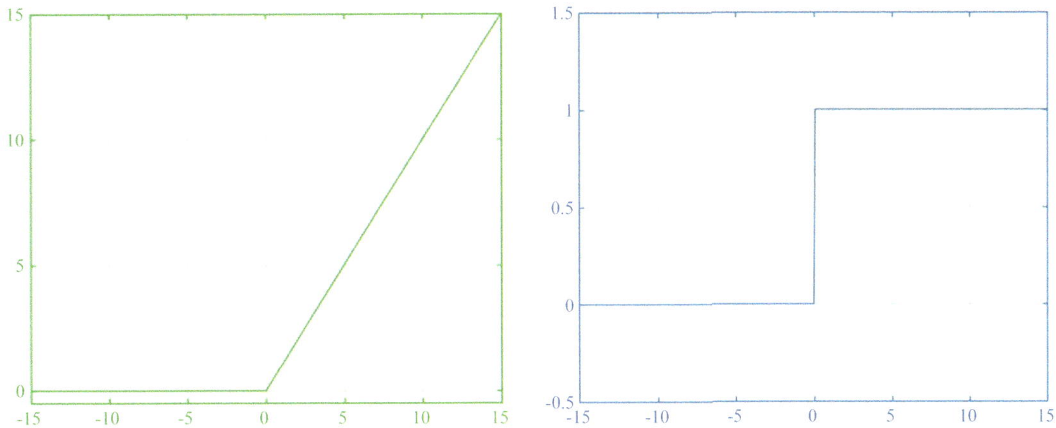

Figure 7.11
The rectified linear unit function.

main function is to forward transfer the feature information of input. In forward propagation, the information features are fed back to the description model and produce outputs. The loss function can calculate the classification error between the generated and target outputs. BP is when the output error propagates in some form from the hidden layer to the input layer and distributes the error to all units of each layer. The hidden layers can adjust the parameters to reduce the loss function. After the forward propagation and BP, the difference between the output value and the expected value of the network is small enough, and learning ends.

Of course, convolutional neural networks are not the only deep learning in the field of fault diagnosis. This chapter introduces only the basic theory of convolutional neural networks. After years of development, there are many types of convolutional neural networks, such as AlexNet, GoogLeNet, and ResNet. In general, fault diagnosis based on deep learning includes acquisition and preprocessing of vibration data, feature extraction of vibration data, and classification and identification using deep learning. After training with the deep learning model based on training samples, classification results can be obtained theoretically.

7.4 Conclusion and outlook

The rail transit channel robot combines high technology such as machinery, automation, computer, sensor, and artificial intelligence, and has a high application prospect. With the rapid development of rail transit in the world, the mileage of railway operations is increasing and more and more trains are on the line. The maintenance of rail transit faces enormous challenges owing to the large increase in operating trains. Moreover, it is difficult to accomplish such a heavy task as manual maintenance with high quality. To replace maintenance workers to complete the repeated and mechanical high-intensity work, the robot is slowly applied to maintain rail transit, in the form of the rail transit channel robot. Although the rail transit channel robot appeared later, it basically formed the structure of artificial intelligence technology as the core and integrated various sensing equipment, which can enable the automatic diagnosis of service train faults. Among them, artificial intelligence technology includes robot technology, infrared temperature measurement, image recognition, ground track technology, big data analysis, navigation, and machine learning. After many years of development, channel robots can meet different applications and objects. The channel robot described in this chapter is just one kind of these.

The rail transit channel robot mainly performs the inspection task of the train bottom; occasionally, it performs simple maintenance work. However, in the face of slightly complicated maintenance conditions, the channel robot can do nothing. It is expected that channel robots can perform complex maintenance tasks with a higher practical value. They must have excellent motion systems, reliable navigation systems, and accurate sensing capabilities.

The three important intelligent indicators of channel robots are autonomy, adaptability, and interactivity. Adaptability refers to the ability of a robot to adapt to a complex working environment. It not only recognizes and measures surrounding objects, it understands the surrounding environment and tasks to be performed. Thus, it can make correct judgments and operations. Autonomy means that the robot can determine the work steps and working methods according to the work tasks in a different environment. There are three main types of interaction: robots and environment, robots and people, and robots and robots, mainly involving the acquisition, processing and understanding of information. Channel robots have the ability to conduct practical maintenance only by continuously improving autonomy.

Multisensors and advanced perceptual algorithms are the keys to realizing the highly flexible and highly robust behavior of robots in real-world environments. They are also the effective means to improve the three intelligent indicators. The advantages of using multiple sensors are obvious: (1) multiple sensors can provide redundant information for the same environmental feature; (2) multiple features can be adopted to analyze the current scene quickly in parallel; and (3) once the sensor fails, the available sensor can take the place of the failed sensor quickly. Last but not least, while conducting robot algorithm research, increasingly complex and more realistic field tests needed to be performed, and more experience can be accumulated to develop a true rail transit channel robot with more comprehensive functions and more stable operation.

References

[1] C.-P. Day, Robotics in industry—their role in intelligent manufacturing, Engineering 4 (2018) 440–445.

[2] S. Zodey, K. Sharad, Pradhan. Matlab toolbox for kinematic analysis and simulation of dexterous robotic grippers, Procedia Eng. 97 (2014) 1886–1895.

[3] G. Xie, N. Zhang, S. Wang, Should upstream merger be regulated in a railway system with monopolistic operator, Transport. Res. Procedia 25 (2017) 407–415.

[4] X. Yang, Y. Xiang, The technological innovation efficiency of China's high-speed rail enterprises based on DEA approach, in: 2018 Portland International Conference on Management of Engineering and Technology, PICMET), 2018, pp. 1–9.

[5] M. Diao, Does growth follow the rail? The potential impact of high-speed rail on the economic geography of China, Transport. Res. A Pol. Pract. 113 (2018) 279–290.

[6] Z. Chen, J. Xue, A.Z. Rose, et al., The impact of high-speed rail investment on economic and environmental change in China: a dynamic CGE analysis, Transport. Res. A Pol. Pract. 92 (2016) 232–245.

[7] K. Friesen, The effects of the Madrid and London subway bombings on Europe's view of terrorism, Hum. Right Hum. Welfare 7 (2007) S10–S17.

[8] J. Gutiérrez, R. González, G. Gomez, The European high-speed train network: predicted effects on accessibility patterns, J. Transport Geogr. 4 (1996) 227–238.

[9] S.C. Chang, Railway development from the Japanese occupation period to the present–using Kaohsiung city as an example, Appl. Econ. 49 (2017) 4729–4741.

[10] P. Deng, L. Zhao, Q. Luo, et al., Study on the performance improvement of urban rail transit system, Energy 161 (2018) 1154–1171.

[11] A. Gasparetto, L. Scalera, From the unimate to the delta robot: the early decades of industrial robotics, in: Explorations in the History and Heritage of Machines and Mechanisms, Springer, 2019, pp. 284–295.

[12] N.J. Nilsson, A Mobile Automaton: An Application of Artificial Intelligence Techniques, Sri International Menlo Park Ca Artificial Intelligence Center, 1969.

[13] A. Lu, Y. Huo, J. Zhou, A multimedia stereo calibration algorithm based on rectangular pyramidal method used to aid visual navigation of ALVs under low illumination, Multimed. Tool. Appl. 78 (2019) 34673–34687.

[14] H. Brown Jr., M. Friedman, T. Kanade, Development of a 5-DOF walking robot for space station application: overview, in: 1990 IEEE International Conference on Systems Engineering, 1990, pp. 194–197.

[15] F. Rubio, F. Valero, C. Llopis-Albert, A review of mobile robots: concepts, methods, theoretical framework, and applications, Int. J. Adv. Rob. Syst. 16 (2019) 1–22.

[16] B. Dai, X.-M. Xiao, Z.-X. Cai, Current status and future development of mobile robot path planning technology, Contr. Eng. China 12 (2005) 198–202.

[17] J. Kim, A. Elizabeth, Croft. Online near time-optimal trajectory planning for industrial robots, Robot. Comput. Integrated Manuf. 58 (2019) 158–171.

[18] Z. Cao, B. Zhang, S. Wang, et al., Cooperative hunting of multiple mobile robots in an unknown environment, Acta Autom. Sin. 29 (2003) 536–543.

[19] M. Uchiyama, A. Konno, T. Uchiyama, et al., Development of a flexible dual-arm manipulator testbed for space robotics, in: EEE International Workshop on Intelligent Robots and Systems, towards a New Frontier of Applications, 1990, pp. 375–381.

[20] B. Paul, R. Wood, Tony Belpaeme, A touchscreen-based 'sandtray' to facilitate, mediate and contextualise human-robot social interaction, in: Proceedings of the Seventh Annual ACM/IEEE International Conference on Human-Robot Interaction, 2012, pp. 105–106.

[21] A. Wahrburg, E. Morara, G. Cesari, et al., Cartesian contact force estimation for robotic manipulators using Kalman filters and the generalized momentum, in: 2015 IEEE International Conference on Automation Science and Engineering, CASE, 2015, pp. 1230–1235.
[22] C.-Y. Weng, W.C. Tan, I.-M. Chen, A survey of dual-arm robotic issues on assembly tasks, in: ROMANSY 22–Robot Design, Dynamics and Control, Springer, 2019, pp. 474–480.
[23] C.H. Lee, S.H. Kim, S.C. Kang, et al., Double-track mobile robot for hazardous environment applications, Adv. Robot. 17 (2003) 447–459.
[24] L. Montano, J.R. Asensio, Real-time robot navigation in unstructured environments using a 3D laser rangefinder, in: Proceedings of the 1997 IEEE/RSJ International Conference on Intelligent Robot and Systems Innovative Robotics for Real-World Applications IROS'97. 2, 1997, pp. 526–532.
[25] F. Guy, M. Heymann, A. Bruckstein, Two-dimensional robot navigation among unknown stationary polygonal obstacles, IEEE Trans. Robot. Autom. 9 (1993) 96–102.
[26] Y. Fu, The Research on Dynamic Objectlocalization Based on Fusion of RFID and Laser Information, Southwest University of Science and Technology, 2019.
[27] G. Wang, Autonomous Navigation of Mobile Robot Based on Information Fusion in Indoor Environment, Shandong University of Technology, 2018.
[28] Z. Liu, Z. Li, A review on image process technique of thermal imager, Infrared Technol. 22 (2000) 27–32.
[29] J. R Speakman, S. Ward, Infrared thermography: principles and applications, Zoology 101 (1998) 224–232.
[30] A. Huda, S. Taib, Suitable features selection for monitoring thermal condition of electrical equipment using infrared thermography, Infrared Phys. Technol. 61 (2013) 184–191.
[31] K.Mo Koo, K.C. Shin, Automatic Charging Apparatus of Autonomous Mobile Robot and Automatic Charging Method Using the Same, Google Patents, 2008.
[32] J. Wang, M. Hu, C. Cai, et al., Optimization design of wireless charging system for autonomous robots based on magnetic resonance coupling, AIP Adv. 8 (2018) 055004.
[33] B.D. De Vos, F.F. Berendsen, M.A. Viergever, et al., A deep learning framework for unsupervised affine and deformable image registration, Med. Image Anal. 52 (2019) 128–143.
[34] D.G. Lowe, Distinctive image features from scale-invariant keypoints, Int. J. Comput. Vis. 60 (2004) 91–110.
[35] B. Ratna, V. Rajesh, Exudate detection and feature extraction using active contour model and sift in color fundus images, ARPN J. Eng. Appl. Sci. 10 (2015) 2362–2365.
[36] G. Ramu, S.T. Babu, Image forgery detection for high resolution images using SIFT and RANSAC algorithm, in: 2017 2nd International Conference on Communication and Electronics Systems (ICCES), 2017, pp. 850–854.
[37] H. Lahiani, M. Neji, Hand gesture recognition method based on HOG-LBP features for mobile devices, Procedia Comput. Sci. 126 (2018) 254–263.
[38] O. Gary, L. Petersson, Large scale sign detection using HOG feature variants, in: 2011 IEEE Intelligent Vehicles Symposium (IV), 2011, pp. 326–331.
[39] Y. Xu, X. Xu, C. Li, et al., Pedestrian detection combining with SVM classifier and HOG feature extraction, Comput. Eng. 42 (2016) 56–60.
[40] L. Wang, L. Zhang, X. Yang, et al., Level set based segmentation using local fitted images and inhomogeneity entropy, Signal Process. 167 (2020) 107297.
[41] B. Lei, J. Fan, Image thresholding segmentation method based on minimum square rough entropy, Appl. Soft Comput. 84 (2019) 105687.
[42] T.R. Fujimoto, T. Kawasaki, K. Kitamura, Canny-Edge-Detection/Rankine-Hugoniot-conditions unified shock sensor for inviscid and viscous flows, J. Comput. Phys. 396 (2019) 264–279.
[43] W.S. Mcculloch, W. Pitts, A logical calculus of the ideas immanent in nervous activity, Bull. Math. Biophys. 5 (1943) 115–133.

[44] D.E. Rumelhart, G.E. Hinton, R.J. Williams, Learning representations by back-propagating errors, Nature 323 (1986) 533–536.

[45] G.-B. Huang, Q.-Y. Zhu, C.-K. Siew, Extreme learning machine: theory and applications, Neurocomputing 70 (2006) 489–501.

[46] J.-S. Jang, ANFIS: adaptive-network-based fuzzy inference system, IEEE Trans. Syst. Man Cybern. 23 (1993) 665–685.

[47] J. Wang, A deep learning approach for atrial fibrillation signals classification based on convolutional and modified Elman neural network, Future Generat. Comput. Syst. 102 (2020) 670–679.

[48] M. Pal, G.M. Foody, Feature selection for classification of hyperspectral data by SVM, IEEE Trans. Geosci. Rem. Sens. 48 (2010) 2297–2307.

[49] Z. Shi, X. Pan, Q. Zhang, High-precision pulsed laser measuring distance by time delay method, Optic Precis. Eng. 22 (2014) 252–258.

[50] D. Fényes, B. Németh, P. Gáspár, A predictive control for autonomous vehicles using big data analysis, in: The International Federation of Automatic Control 2019, vol. 52, 2019, pp. 191–196.

[51] G. Singh Bhathal, A. Singh, Big data: Hadoop framework vulnerabilities, security issues and attacks, Array 1 (2019) 100002.

[52] S. Tsuboi, S. Miyokawa, M. Matsuda, et al., Influence of spark discharge characteristics on ignition and combustion process and the lean operation limit in a spark ignition engine, Appl. Energy 250 (2019) 617–632.

[53] S. Wang, M. Tomovic, H. Liu, New technology of aircraft hydraulic system, in: S. Wang, M. Tomovic, H. Liu (Eds.), Commercial Aircraft Hydraulic Systems, Academic Press, Oxford, 2016, pp. 171–251.

[54] M.S. Kan, A.C. Tan, J. Mathew, A review on prognostic techniques for non-stationary and non-linear rotating systems, Mech. Syst. Signal Process. 62 (2015) 1–20.

[55] Y. Shrivastava, B. Singh, A comparative study of EMD and EEMD approaches for identifying chatter frequency in CNC turning, Eur. J. Mech. Solid. 73 (2019) 381–393.

[56] D. Ma, Y. Liang, X. Zhao, et al., Multi-BP expert system for fault diagnosis of powersystem, Eng. Appl. Artif. Intell. 26 (2013) 937–944.

[57] L. Hongmei, W. Shaoping, O. Pingchao, fault diagnosis based on improved elman neural network for a hydraulic servo system, in: 2006 IEEE Conference on Robotics, Automation and Mechatronics, 2006, pp. 1–6.

[58] X. Zhang, Y. Liang, J. Zhou, et al., A novel bearing fault diagnosis model integrated permutation entropy, ensemble empirical mode decomposition and optimized SVM, Measurement 69 (2015) 164–179.

[59] W. Xu, S. Guo, X. Li, et al., A dynamic scheduling method for logistics tasks oriented to intelligent manufacturing workshop, Math. Probl Eng. (2019) 1–18, 2019.

[60] L. Guo, N. Li, F. Jia, et al., A recurrent neural network based health indicator for remaining useful life prediction of bearings, Neurocomputing 240 (2017) 98–109.

[61] Z. Peng, W.T. Peter, F. Chu, An improved Hilbert–Huang transform and its application in vibration signal analysis, J. Sound Vib. 286 (2005) 187–205.

[62] B. Sreejith, A. Verma, A. Srividya, Fault diagnosis of rolling element bearing using time-domain features and neural networks, in: 2008 IEEE Region 10 and the Third International Conference on Industrial and Information Systems, 2008, pp. 1–6.

[63] J.W. Cooley, J.W. Tukey, An algorithm for the machine calculation of complex Fourier series, Math. Comput. 19 (1965) 297–301.

[64] D. Shin, C. Kwak, G. Kim, An efficient algorithm for frequency estimation from cosine-sum windowed DFT coefficients, Signal Process. 166 (2020) 107245.

[65] M. Schmittfull, Z. Vlah, P. Mcdonald, Fast large scale structure perturbation theory using one-dimensional fast Fourier transforms, Phys. Rev. D 93 (2016) 103528.

[66] J. Chen, Z. Li, J. Pan, et al., Wavelet transform based on inner product in fault diagnosis of rotating machinery: a review, Mech. Syst. Signal Process. 70–71 (2016) 1–35.

[67] P. Wang, G. Zhang, F. Chen, et al., A hybrid-wavelet model applied for forecasting PM2.5 concentrations in Taiyuan city, China, Atmos. Pollut. Res. 10 (2019) 1884–1894.
[68] N.N. Bhat, K. Kumari, S. Dutta, et al., Friction stir weld classification by applying wavelet analysis and support vector machine on weld surface images, J. Manuf. Process. 20 (2015) 274–281.
[69] X. Hou, M. Han, C. Gong, et al., SAR complex image data compression based on quadtree and zerotree coding in discrete wavelet transform domain: a comparative study, Neurocomputing 148 (2015) 561–568.
[70] Z. Chang, Y. Zhang, W. Chen, Electricity price prediction based on hybrid model of adam optimized LSTM neural network and wavelet transform, Energy 187 (2019) 115804.
[71] D. Yu, J. Cheng, Y. Yang, Application of EMD method and Hilbert spectrum to the fault diagnosis of roller bearings, Mech. Syst. Signal Process. 19 (2005) 259–270.
[72] H. Li, X. Deng, H. Dai, Structural damage detection using the combination method of EMD and wavelet analysis, Mech. Syst. Signal Process. 21 (2007) 298–306.
[73] N.E. Huang, Z. Wu, S.R. Long, et al., On instantaneous frequency, Adv. Adapt. Data Anal. 1 (2009) 177–229.
[74] J. Cheng, D. Yu, Y. Yang, Research on the intrinsic mode function (IMF) criterion in EMD method, Mech. Syst. Signal Process. 20 (2006) 817–824.
[75] B. Zhu, W.-D. Jin, Feature extraction of radar emitter signal based on wavelet Packet and EMD, Inf. Eng. Appl. (2012) 1408–1415. Springer.
[76] L. Xiang, W. Zhang, Q. Ding, Understanding and improving deep learning-based rolling bearing fault diagnosis with attention mechanism, Signal Process. 161 (2019) 136–154.
[77] Y. Du, Bayesian deep convolutional neural network for SMS quality analysis, J. Vis. Commun. Image Represent. (2019) 102677.
[78] D. Jose, C. Desrosiers, L. Wang, et al., Deep CNN ensembles and suggestive annotations for infant brain MRI segmentation, Comput. Med. Imag. Graph. 79 (2020) 101660.
[79] X. Shao, H. Wang, Nonlinear tracking differentiator based on improved sigmoid function, Control Theory Appl. 31 (2014) 1116–1122.
[80] X. Li, A comparison between information transfer function sigmoid and tanh on neural, J. Wuhan Univ. Technol. 28 (2004) 312–314.
[81] A.L. Maas, A.Y. Hannun, A.Y. Ng, Rectifier nonlinearities improve neural network acoustic models, in: International Conference on Machine Learning, vol. 30, 2013 (3).
[82] W. Sun, M. Liu, Wind speed forecasting using FEEMD echo state networks with RELM in Hebei, China, Energy Convers. Manag. 114 (2016) 197–208.

CHAPTER 8

Rail transit inspection unmanned aerial vehicle (UAV) systems

8.1 Overview of inspection unmanned aerial vehicles

The rapid development of rail transit has greatly facilitated social life. Its coverage area has become increasingly wider and the environment of the area of protection has become more and more complex, which has brought great challenges to safety in operation. Traditional track inspection methods are manual inspection and fixed cameras, which are expensive, inefficient, and ineffective. Unmanned aerial vehicles (UAVs) based on the intelligent inspection method are suitable for the safety of rail transit, such as structural, environmental safety, and physical security monitoring [1]. The UAV collects images and videos in a designated area under autocruise or remote control, inspects rail transit in real time, and uses image recognition and feature detection methods to determine potential factors affecting safety in operation.

The widespread application and outstanding performance of UAVs have attracted extensive attention and have a huge market in the civilian field. Long-endurance UAVs are used by scientists to perform earth science missions such as water level monitoring of lakes and reservoirs, ecosystem health monitoring, and natural disaster monitoring [2]. Small and micro-UAVs are widely used in security, environmental pollution detection, traffic monitoring, and other fields because of their good versatility, portability, and maintainability.

The UAV is a multisystem integrated aircraft managed by a remote control station with no operator. It includes power electronics, aviation, reconnaissance, geographic information, image recognition, and other subsystems [3]. The UAV system consists of a UAV, a ground station (remote control, takeoff and landing equipment), and a payload (reconnaissance/surveillance load, weapon load, detection sensor, etc.). UAVs can be divided into different types according to their structure, power system, and so on. The overall structure of UAVs can be divided into fixed wings, helicopters, multirotors, umbrella wings, and airships. The power system of UAVs can be divided into electric and fuel.

There are three main applications in inspection UAVs: fixed-wing UAVs, rotor UAVs, and unmanned helicopters.

Robot Systems for Rail Transit Applications. https://doi.org/10.1016/B978-0-12-822968-2.00008-5

(a) Fixed-wing UAVs have faster speed and a longer flight distance. They have a larger payload than rotor UAVs and can carry more equipment and load. They have a low mechanical failure rate for maintenance, so they are more widely used for a range of mapping and inspection [4].
(b) Rotor UAVs are similar to manned helicopters and rely on lift force generated by rotation of the blades to move. They typically have four to eight blades. Compared with fixed-wing UAVs that require special takeoff and landing sites, rotor UAVs can take off and land vertically and can hover at specific locations to collect data, greatly facilitating mission deployment. Owing to the flexible operation and use of rotor UAVs, the applications of rotor UAVs is significantly wider than that of fixed-wing UAVs in the civil and general industrial fields [5].
(c) Unmanned helicopters combine the characteristics of helicopters and UAVs. They do not require takeoff and landing facilities. They have strong adaptability and are flexible in flight. Unmanned helicopters have larger loads and longer voyages than rotor UAVs. Missions that can be performed are more diverse.

The fixed-wing UAV consists of a wing, fuselage, empennage, landing gear, and power system. The rotor UAV includes the fuselage, power system, flight control system, and remote control system [6]. The unmanned helicopter consists of rotor, fuselage, tail rotor, landing gear, control system, transmission system, and power unit. The rail transit inspection UAV is loaded with high-resolution cameras and information processing platforms for real-time image acquisition, analysis, and processing of the track line environment.

The inspection UAV system consists of five subsystems: the UAV subsystem, launch and recycle subsystem (equipment related to launching and recycling, such as launchers, boosters, recycling umbrellas, and blocking nets), task equipment subsystem (monitoring equipment, reconnaissance equipment, communication relay equipment, etc.), remote control and information transmission subsystem (radio remote control equipment, information transmission equipment, relay forwarding equipment, etc.) and command and control subsystem (flight control and management equipment, integrated display equipment, etc.).

For UAV key technologies:

(a) Autopilot technology is the basis of the UAV inspection. The UAVs can be patrolled over long distances without manual control. The UAVs enable autonomous flight, takeoff, landing and target tracking through automatic matching of the geographical matching method, global positioning system (GPS) autonomous navigation and flight control.
(b) UAVs carry image transmission devices, which send the measured images to the ground for the operator to analyze and take action, so it is necessary to transmit clear video and images in real time.

(c) The target recognition technology provides support for the autonomous inspection of UAVs. When the UAVs execute the track inspection mission, they automatically recognize the position of the rail using the image features, extract the region of interest by referring to the railway limit area, and perform subsequent processing on the region of interest.

8.2 Main components of rail transit inspection unmanned aerial vehicles

The main components of the UAVs can be divided into the basic structure and auxiliary equipment, as shown in Fig. 8.1. The basic structure is the necessary condition for UAVs to fly, such as the fuselage, wing, flight control system, power system, and remote control system. The auxiliary equipment is needed for UAVs to perform a specific task, such as the reconnaissance, electronic countermeasure, and detection devices. For rail transit inspection UAVs, auxiliary equipment includes a vision sensor, ultrasonic sensor, infrared sensor, and so on.

8.2.1 Structures

The structural design of the aircraft consists of aircraft structural strength research and ground verification tests, including an aircraft structure antifatigue fracture reliability design, aircraft structural dynamic strength, composite material structural strength, aviation

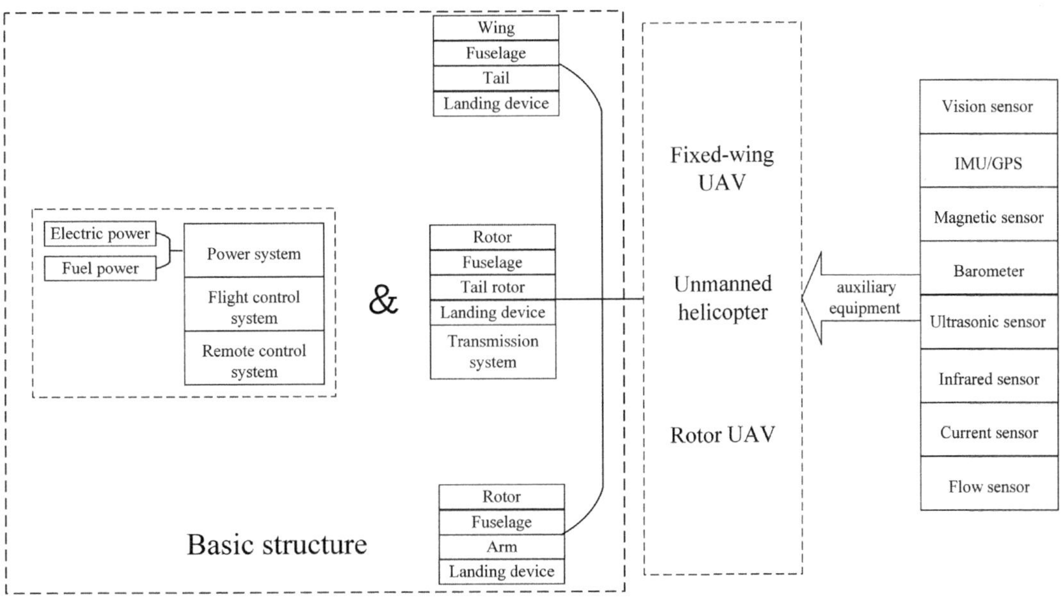

Figure 8.1
Main components of rail transit inspection unmanned aerial vehicles (UAVs). *GPS*, global positioning system; *IMU*, inertial measurement unit.

noise, aircraft structure comprehensive environmental strength, and aircraft structure test technology. The body material is made of high-strength lightweight materials (glass fiber, resin, and aluminum alloy) and the design of the body should be convenient for the installation and use of other devices.

8.2.1.1 Fixed-wing unmanned aerial vehicle

The basic structure of a fixed-wing UAV consists of the wing, fuselage, empennage, landing gear, and power unit.

(a) The wing of a fixed-wing UAV generates the power needed for flight, including the spar (the flange that withstands the bending moment and the web that withstands the shear force); false spar (forms a closed section with the skin to withstand torque, and on which the connection to the fuselage is hinged); stringer (connected to the rib and riveted in the skin to support the skin to increase its load-carrying capacity); rib (supports the skin as a lateral force skeleton and maintains the cross-sectional shape of the wing).
(b) The fuselage functions as the basis for loading and installation, including the skin (similar to wing skin), stringer (similar to wing stringer), longeron (similar to spar), and frame (similar to rib).
(c) The function of the empennage is to stabilize and control the pitch and yaw of a UAV. These include the horizontal empennage (consists of a fixed horizontal stabilizer and a rotatable elevator behind it) and the vertical empennage (consists of a fixed vertical stabilizer and a rotatable rudder behind it).
(d) The function of the landing gear is to support a UAV on the ground, including a pillar (for support and as the basis of the tire installation), shock absorber (absorbs landing impact energy), tire (contact with the ground to support the weight of the UAV, reduce the resistance of ground motion, and absorb part of the impact kinetic energy), and retracting mechanism (retract the landing gear and fix the pillar).
(e) The power unit functions to generate tension (propeller) and thrust (jet engine), allowing a UAV to move relative to the air.

8.2.1.2 Unmanned helicopter

The basic structure of an unmanned helicopter consists of a rotor, fuselage, tail rotor, landing gear, control system, transmission system, and power unit.

(a) The rotor is mainly composed of a blade and hub, which is the most critical part of the unmanned helicopter. It not only generates the lift force, it is also the source of the tension of horizontal movement of the unmanned helicopter. The plane of rotation of the rotor is the lift force surface and the control surface. The connection mode of the blade and hub is full-stranded, half-stranded, nonstranded and nonbearing.

(b) The fuselage of the unmanned helicopter is similar in structure and function to that of the fixed-wing UAV. It is used to load fuel, cargo, and equipment and as an installation base to integrate these parts. The fuselage functions as a bearing and transmitting force, which can bear all kinds of loads and dynamic loads.
(c) The main function of the tail rotor is to generate a lateral tension (or thrust) to form a deflection moment through the moment arm and balance the countertorque of the main rotor and control the course. It is equivalent to the vertical stabilizer of the fixed-wing UAV, which can improve course stability of the helicopter and provide a part of the lift force.
(d) The landing gear is designed to support weight on the ground and absorb impact upon landing. The landing gear for the helicopter includes the wheeled landing gear and skid landing gear. If the helicopter needs to have the ability to take off and land on the surface of the water or to have an emergency landing, it generally needs to have a water-sealed fuselage and a buoy to ensure lateral stability or an emergency landing buoy.
(e) The control system is used to control the flight of the unmanned helicopter, which is composed of the automatic tilting device, cockpit control mechanism, and control wire. Functions of the control system mainly include attitude stabilization and control, mission equipment management, and emergency control. The control system can partially or completely replace the pilot to control its attitude and path directly and can improve flight quality.
(f) The transmission system transmits power provided by the engine to the main rotor and tail rotor, which is composed of the main reducer, driveshaft, tail reducer, and intermediate reducer. There are three transmission methods for unmanned helicopters: single rotor, which has a simple structure and is most widely used; coaxial twin rotor, which has small space size but the structure is complex; and tilt rotor, combining vertical takeoff and landing of the helicopter and fast horizontal flight of fixed-wing aircraft [7].

8.2.1.3 Rotor unmanned aerial vehicle

The basic structure of rotor UAV is composed of a frame, power unit and flight control system.

(a) The frame consists of a fuselage, arm and landing gear. The fuselage is a platform used to carry all equipment. To some extent, the safety and stability of the rotor UAV are closely related to the fuselage. The size, material, shape, weight, strength, and other factors of the fuselage should be considered comprehensively when designing the fuselage of the rotor UAV.
(b) The arm determines the wheelbase of the rotor, which is an important parameter used to measure the size of the rotor UAV. It is usually defined as the diameter of the peripheral circle surrounded by the motor shaft. Normally, the wheelbase is the distance

between the two motors' axes on the diagonal. The size of the wheelbase limits the upper limit of propeller size, which limits the maximum tension that the propeller can produce and determines the carrying capacity of rotor UAV.

(c) The functions of landing gear include supporting the takeoff and landing of the rotary wing UAV, maintaining the horizontal balance of the fuselage, ensuring sufficient distance between the rotor UAV and the ground to avoid a collision between the propeller and the ground, and reducing turbulence caused by the propeller airflow striking the ground during takeoff and landing.

(d) Power units are divided into electric and fuel. Electric power is the main power of a rotor UAV. However, because of the current technological level of battery development, the limited energy density of batteries puts electric UAVs at a significant disadvantage in terms of load and holdover time. The electric power unit consists of a battery, electronic governor, motor, and propeller.

Batteries used by UAVs are generally lithium-polymer. The function of the electronic governor is to amplify the power of the control signal of the flight control board and send the drive signal to each switch tube that can make it saturate and turn off reliably, to control the speed of the motor. The rotor blade driven by motor rotation generates the lift and thrust of the UAV. By controlling the motor speed, the UAV can complete various flight states.

The fuel engines commonly used in UAVs are piston and gas turbine. The piston engine is a kind of machine that uses fuel combustion and expansion in the cylinder to generate pressure to push the piston to move and do work. It is mainly composed of a cylinder, piston, connecting rod, crankshaft, valve mechanism, propeller reducer, and casing. The gas turbine engine is mainly composed of an inlet, compressor, combustion chamber, turbine, and nozzle. When fresh air enters the inlet and flows through the compressor, gas is compressed into high-temperature, high-pressure gas, which enters the combustion chamber to be mixed with fuel oil and burned into gas. When the gas flows through the turbine, it drives the turbine to rotate, thus driving the compressor to work. Finally, the outflow gas continues to expand in the nozzle and discharge along the axial direction of the engine at a high speed.

(e) The flight control system is the control core of the UAV, which mainly consists of flight control, an accelerometer, barometer, gyroscope, geomagnetic instrument, positioning chip, main control chip and other components to achieve the functions of attitude stability, control of the UAV, equipment management, and emergency control. It is necessary to adjust the power system of the UAV in real time through the flight control system whether the UAV is in automatic flight state (such as hovering) or manual operation. In addition, to ensure the normal flight navigation function of the aircraft, the flight control system has functions such as safety redundancy, flight data recording, flight parameter adjustment, and automatic flight optimization [8].

(f) The remote control system is the manned equipment of the UAV. It is mainly composed of a remote control, receiver, decoder, and servo system. The remote control is the operating platform. The receiver receives the signal from the remote control for decoding, separates the action signal, and transmits it to the servo system. The servo system makes the corresponding action according to the signal.

8.2.2 Sensors

8.2.2.1 Sensors for pan/tilt camera

The pan/tilt camera is the key to the inspection UAV. The camera collects image information, the pan/tilt keeps the camera stable to improve the shooting effect, and the wireless transmission system transmits the picture to the ground equipment, which requires the characteristics of long-distance transmission, a clear and smooth image, stable transmission, antiinterference, and real-time performance.

The pan/tilt camera platform is the basis for the UAV to collect stable and clear image information. The camera platform system isolates the visual axis of the imager from the vibration of the transportation platform so that the visual axis of the imager is stable in a fixed space range. The camera platform system can complete the regular movement of the target real-time tracking function according to the instruction. The number of degrees of freedom (DOFs) of the camera platform is significant for the overall maneuvering performance of the system. To ensure stabile tracking of the inspection target, it is necessary to adopt a three-DOF mechanism to achieve a stable tracking function: yaw, roll, and pitch axis. The camera platform system is composed of an external mechanical structure, execution unit, sensor detection device, servo control system, and image algorithm system [9].

To enable the target stability tracking function of the camera platform system, various sensors are needed. One type of sensor is used to collect the surrounding environment and system targets, such as optical and infrared cameras. The other type is used to detect the speed and position of the servomotor, such as a rotary encoder and attitude sensor.

The optical camera is the most important equipment for an inspection UAV. It is used for visual navigation or monitoring. Under good light conditions, the optical camera takes pictures or videos and uploads them to the ground station for analysis by computer vision. Infrared cameras are complementary to optical cameras and are mainly used in poor visibility weather conditions. The infrared camera adopts thermal imaging technology to receive the infrared radiation energy distribution pattern of the target and reflect it to the photosensitive element of the infrared camera to obtain an infrared thermal image. Infrared cameras can penetrate smoke; they have clear subject targets and work both day and night.

As a commonly used position detection sensor, the rotary encoder is an important component for the platform, which can be classified into photoelectric and magnetoelectric types. Among these, the photoelectric type is the more widely used. The photoelectric encoder includes an incremental encoder and an absolute encoder. The photoelectric encoder converts the servomotor output shaft angular and other physical parameters into the corresponding pulse digital signal, with the advantages of high precision, strong antiinterference ability, and easy installation. Another kind of sensor used to detect pan/tilt platform attitude information is the attitude sensor, a three-dimensional (3D) motion attitude measurement system based on microelectromechanical systems (MEMS) technology. It includes a three-axis gyroscope, three-axis accelerometer, electronic compass, and other auxiliary three-axis motion sensors. In addition, the Advanced Reduced Instruction Set Computing Machine processor outputs angular velocity, acceleration, magnetic data, and other information through the sensor data algorithm based on quaternion attitude measurement.

8.2.2.2 Sensors for unmanned aerial vehicles

The UAV has a variety of sensors, including the inertial measurement unit (IMU), GPS, magnetic sensor, barometer, ultrasonic sensor, infrared sensor, current sensor, and flow sensor.

(a) The IMU is a device for measuring the attitude angle (or angular velocity) and acceleration of an object. The IMU is installed on the carrier directly in a strap-down inertial navigation system (INS); it obtains the physical value under the carrier coordinate system to complete navigation. Gyroscope and acceleration sensors are the core devices of the INS. With the help of gyroscope and acceleration sensors, the IMU can measure linear acceleration and the rotation angle rate of three directions and obtain the attitude, speed, and displacement of the carrier. IMUs on the market are mainly six- and nine-axis. The six-axis IMU contains a three-axis acceleration sensor and three-axis gyroscope [10]. The nine-axis IMU also has a three-axis magnetometer besides the acceleration sensor and gyroscope. In addition, for the IMU using MEMS technology, there is usually a built-in thermometer for real-time temperature calibration [11].

(b) The GPS provides location information of the UAV. The GPS system consists of a satellite, monitoring systems on the earth's surface, and user receivers. The GPS positioning principle is passive ranging positioning, by measuring the distance between the satellite and the receiver. Because the measured distance contains errors, the receiver must receive positioning signals from at least four positioning satellites; the current position coordinates and altitude of the receiver can be obtained by joint calculation.

(c) Magnetic sensors provide information for inertial navigation and orientation systems. They are a kind of sensor with an anisotropic magnetoresistive property. Compared with other sensors, they have a significant power consumption advantage, high precision, short response time, and other characteristics. They are suitable for use for the UAV.

(d) The absolute altitude of the UAV can be calculated from the value of atmospheric pressure measured by a barometer. When the UAV flies near the ground, atmospheric pressure is influenced by a ground effect, which makes it impossible to use a barometer to measure altitude. The solution is to use other sensors during takeoff or landing, such as ultrasonic sensors or laser range finders.

(e) An ultrasonic sensor is widely used in all kinds of ranging and proximity sensing applications. It generally composed of an ultrasonic transmitter and an ultrasonic receiver. The transmitter sends out an ultrasonic pulse; the pulse bounces off when it is in contact with the obstacle and is picked up by a receiver. The distance between the obstacle and the ultrasonic sensor is calculated by the time of flight and speed of the ultrasonic wave.

(f) The flow sensor is used to monitor the airflow inhaled by the gas engine of the UAV. This function can help the engine central processing unit determine the appropriate air–fuel ratio, which can improve power and reduce emissions. Most flow sensors use thermal diffusion, including a heating element and two temperature sensors. One temperature sensor is heated to a temperature above the environmental temperature and the other temperature sensor is used to sense the temperature of the medium being measured. The temperature difference between the two temperature sensors will change with the change in flow rate. According to the proportional relationship between the temperature difference and the flow rate, the flow rate of the gas can be calculated.

8.3 Key technologies in rail transit inspection unmanned aerial vehicles

The UAV system consists of the UAV, control station, communication link, and payload. The UAV includes a flight platform, power system, flight control system, and so on. The control station is the interface of the UAV to realize human–robot interaction. The control station sends instructions through the communication uplink to control the flight and the various payloads carried by the UAV. The data and images transmitted by the UAV are received through the communication downlink, including the load data, the status information of each subsystem on the aircraft, the position information of the UAV, and so forth.

8.3.1 Communication methods

The UAV needs to develop powerful, efficient, and secure communication technologies to ensure that it communicates with ground operators and air controllers without pilots. The UAV communication system consists of airborne and ground equipment. Airborne equipment includes the airborne antenna, remote control receiver, telemetry receiver, telemetry transmitter, video transmitter, and terminal processor. On the one hand, airborne equipment receives flight parameters from various sensors and sends these data to the ground station. On the other hand, it receives remote control commands from the ground station. Ground equipment consists of antennas, remote transmitters, telemetry receivers, telemetry station data terminals, and control and surveillance equipment.

The communication system can be divided into the analog and digital communication systems as well as the wired and wireless communication systems, according to signal forms or transmission medium. The UAV communication system is a digital wireless communication system.

Signals with messages are transmitted in the form of electromagnetic waves in wireless communication. Compared with wired communication, its mobility is good but its confidentiality is poor. Wireless channel parameters often change with time and environment. According to the different frequency bands, wireless channels can be divided into the ultralong wave, long wave, medium wave, shortwave, ultrashortwave, and microwave.

For the wireless transmission system, digital modulation is developed on three basic modulation modes: amplitude shift keying, frequency shift keying, and phase shift keying [12]. When choosing the modulation mode of the UAV communication system, main factors to consider include spectrum use, antiinterference ability, adaptability to transmission distortion, antifading ability, the complexity of equipment, and so forth.

The communication protocol defines the carrier and encoding of data transmission and enables data receiving and sending. According to the attributes of the data link, the protocol implementation can be divided into the flight control data link, task data link, and data link between UAVs.

(a) The flight control data link mainly consists of UAV controller parameter information, path planning information, and aircraft status information. The flight control data link is a necessary data link to guarantee the safety and stability of the UAV's flight. Controller parameter and path planning information is randomly sent with a small amount of data; aircraft status information is sent periodically with a large amount of data.
(b) The task data link transmits various task analysis data collected by the UAV, such as images and videos. The amount of data is large and requires a faster data transfer rate. There are two transfer methods of link task data: one is based on image transmission equipment and the other uses a wireless bridge.
(c) The data link between UAVs is mainly used to perform cooperative tasks and relay tasks with the UAV cluster. There are two main aspects of this communication. One is obstacle avoidance and path planning. Each UAV sends status information periodically to all members of the system. The other manages communication between UAVs according to the mission requirements of the ground station.

8.3.1.1 Communication key technologies

(a) Data link technology

The UAV carries out data transmission and information exchange through the data link. The transmission rate and processing capacity of the data link can enhance the intelligence of the UAV, which is the basis of achieving the full autonomy of the UAV [13].

With the development of technology, the improvement in airborne computer data processing capabilities allows drones to send processing results to a ground station, which reduces the performance requirements for the data link. At the same time, the use of power control and data compression technology also improves the efficiency of use of the data link.

When the UAV exceeds the wireless communication distance, it needs to communicate with the ground station through a relay device. Relay communication methods are divided into the ground relay and air relay. Ground relay equipment is placed high between the drone and the ground station, such as in high-rise buildings. Air relay equipment is placed on an aircraft or satellite. The space limitation of radio waves is small, the transmission channel is stable, and the transmission volume is large [13].

(b) Mobile ad hoc network

A mobile ad hoc network (MANET) is a dynamic self-organizing network system consisting of mobile nodes with wireless communication capabilities, which has arbitrary and temporary network extensions [14]. In a MANET network, routes between nodes are composed of multiple network segments, and two terminal nodes that cannot communicate directly can communicate by forwarding multiple intermediate nodes. After some communication networks are damaged, the remaining communication capabilities can be maintained and important communication commands are ensured, which is highly robust.

(c) Adaptive antenna array technology

In the complex electromagnetic environment of various communication devices, the communication link of the UAV will be affected by communication interference. If the antiinterference ability of the UAV communication system is weak, the UAV cannot accurately control the route and data return.

With the in-depth application of the UAV network, the adaptive antenna array began to be used for mobile communication and wireless access to complex signals. The basic principle of the adaptive antenna array is to adapt to the external electromagnetic environment automatically and adjust the gain and phase based on the difference of the amplitude, encoding, frequency domain, or space of the identification friend or foe signal. Forming the main beam in the direction of the useful signal and forming a zero-broadening in the direction of the interference can reduce the influence of the interfering signal on communication of the UAV [15].

8.3.1.2 Wireless communication frequency band

In the civil field, the UAV communication system cannot occupy the special communication channel. However, the Industrial Scientific Medical (ISM) band provides abundant frequency bands for communication and does not restrict user authorization. The ISM band is open to three main institutions: industry (902–928 MHz), science

(2.42–2.4835 GHz), and medicine (5.725–5.850 GHz) [16]. In the 900-MHz ISM band, digital radio stations can achieve long-distance wireless communication. Common bands used by UAV are 2.4 and 5.8 GHz, and this wireless communication method can realize a low-cost and universal UAV communication system. The transmission distance is usually tens or hundreds of meters [17].

The communication signals of the UAV have not only remote control signals with which the remote station interacts, but also data and image transmission signals. Most remote control signals use a wireless communication chip with a communication band of 2.4 GHz. This is short-range wireless transmission technology. Commonly used technical standards are ZigBee/Institute of Electrical and Electronics Engineers (IEEE) 802.15.4, Wireless fidelity (Wi-Fi)/IEEE 802.11b/g/n/, Bluetooth/IEEE 802.15.1, Wireless USB, and so on [18].

The data and image signal transmission frequency band is 5.8 GHz, which includes three 100-MHz frequency bands, 5.15–5.25 GHz for indoor wireless communication, 5.25–5.35 GHz for medium-range communication, and 5.725–8.825 GHz for broadband wireless access in the community [19]. The 5.8-GHz communication band complies with technical standards such as Wi-Fi/IEEE 802.11a/n/ac/ax, FCC Part 15, ETSI EN 301 489, ETSI EN 301 893, EN 50385, EN 60950, and so on [18].

The 5.8-GHz wireless product uses orthogonal frequency division multiplexing and point-to-multipoint, point-to-point networking, with a single sector rate of up to 54 megabits per second (Mbps) [20]. The channel has a higher frequency and relatively stronger antiinterference ability. At the same time, it can meet the needs of high-bandwidth applications to support a large number of users, and eight nonoverlapping channels make deployment more scalable and flexible. The 5.8-GHz band also uses Internet Protocol (IP) or circuit-based wireless transmission technology. The IP-based technology signaling protocol is simple and easy to implement. It has multiservice types and a uniform interface. It is easy to upgrade and is especially suitable for nonconnection.

8.3.1.3 Wireless transmission methods

Wireless fidelity (Wi-Fi) is a wireless local area network technology based on the IEEE 802.11b standard, which enables small-scale, high-speed signal transmission after the related electronic devices are connected to the local area network [21]. The mobile phone remote control system communicates the ground relay device through Wi-Fi, and the Wi-Fi module of the relay device communicates with the UAV. In this way, information interaction is realized and the control signal from the remote control system and the image data of the UAV are transmitted at the same time.

Conventional 4G/5G communication needs to set a relay station when processing over-the-horizon flight communication. Traditional point-to-point communication bandwidth is low, and it is impossible to enable the communication of a laser mapping device or a hyperspectral camera that has higher requirements for transmission bandwidth.

In the UAV communication mode based on 4G communication, UAV communication can be achieved as long as the location of the UAV is covered by 4G signals. The communication distance is not limited. At the same time, because the layout density of the 4G base station is high, the communication signal is not blocked by obstacles such as buildings and trees. Time division duplexing technology and frequency division duplexing technology applied to the 4G network can transmit bandwidths of 100 Mbpsand 150 Mbps, respectively, and can simultaneously implement data and image transmission [22].

5G mobile communication technology has the features of high speed, low latency, low power consumption, universal interconnection, ultrahigh bandwidth, high reliability, and wide coverage. Flight dynamics information can be transmitted between UAVs with zero delay. The UAV can also transmit large-capacity, high-definition pictures and videos to ground users.

Large-scale antenna array technology will provide a higher transmission rate for the UAV and effectively reduce interference through an advanced detection algorithm. Mobile edge computing technology provides the possibility of ultralow delay of UAV communication. By sinking the computing storage capacity and business service capacity to the edge of the network, it can bring lower delay and support more diverse and complex business requirements. Nonorthogonal multiple access is a key technology for 5G to improve spectrum efficiency [23]. Employing serial interference elimination technology and power reuse can bring higher spectrum efficiency and performance gain to the UAV and provide more possibilities for UAV development.

8.3.2 Data collection methods

The data collection system is an automatic test system integrating functions of UAV flight status monitoring, various ground test measurements, and fault diagnosis. Parameters of the airborne system of the UAV need to be collected, monitored, recorded, and fed back to the relevant system or personnel. Then, ground test personnel can understand the working status of the airborne system in real time.

The data collection system provides direct data for phenomenon analysis and to troubleshoot each system. It makes test flights more efficient and optimizes the test program. It can also make each test more targeted and speed the test flight process. During the test flight of the UAV, the real-time down transmission data of the system can provide a basis for decisions for command personnel.

The aircraft condition monitoring system (ACMS) is a modern onboard data collection and processing system. The system can collect the required airborne system data in real time and use the means of measurement and control or satellite communication to transmit the collected data to the ground station in real time. Then, the ground station analyzes data and distributes results to the corresponding personnel for judgment to enable status monitoring, fault diagnosis and positioning of the airborne system, which provides a basis for the ground personnel to make decisions.

The ACMS needs to classify and collect different types of parameters of different subsystems. The flight control, power, avionics, and measurement and control subsystems are all equipped with computers [24]. The ACMS mainly performs real-time monitoring functions through digital interfaces. The detection data of each subsystem can also be transmitted to the ACMS in real time through the digital interface.

Some parts of data of the electromechanical subsystem can be transmitted through the digital channel between the electromechanical management computer and the ACMS. Other parts of data are collected by ACMS directly.

8.3.2.1 Components of the aircraft condition monitoring system

The ACMS consists of a condition monitoring processor, parameter recorder and cache recorder.

(a) The condition monitor processor is the core of the ACMS. The condition monitoring processor is cross-linked with the avionics, electromechanical, flight control, measurement and control, and power systems. The condition monitor processor cross-links with the parameter recorder and cache recorder within the system. The condition monitoring processor completes condition monitoring of each airborne subsystem. The processor receives and executes downlink instructions, transmits downlink monitoring data in real time, forwards monitoring data, and provides the functions of the ground maintenance interface of the ACMS.
(b) The parameter recorder is a component of the ACMS. The condition monitoring processor transmits the collected data to the parameter recorder through the digital data interface. The parameter recorder feeds back the working state and recording state to the condition monitoring processor through the discrete data interface [25]. The condition data can also be downloaded from the parameter recorder by ground equipment.
(c) The cache recorder is a component of the ACMS. The condition monitoring processor transmits collected data to the cache recorder through the digital data interface [25]. The cache recorder feeds back the working state and recording state to the condition monitoring processor through the discrete data interface. The condition data can also be downloaded from the cache recorder by ground equipment.

8.3.2.2 Functions of aircraft condition monitoring system

The ACMS analyzes, encodes, and distributes collected data; these need to be transmitted in real time and are sent to the measurement and control system through the digital interface. The data that need to be forwarded to the electromechanical system are sent to the electromechanical system through the digital interface. All data are sent to the parameter recorder and cache recorder for real-time storage via the digital interface. The parameter recorder and the cache recorder need to receive data sent by the ACMS and record them in the storage medium [26].

Functional requirements of the ACMS are:

(a) Condition monitoring of each UAV subsystem. The ACMS monitors the data of avionics, electromechanical, flight control, power, measurement, and control subsystems of the UAV. For a subsystem with computers, the ACMS completes real-time monitoring through the digital interface. For a subsystem without a computer, the ACMS completes real-time monitoring by directly collecting discrete or analog data.
(b) Receive and execute downlink instructions. Because of the limitation of transmission bandwidth in different bands, the ACMS needs to respond to the ground station's instruction to adapt to the transmission bandwidth in different bands.
(c) Real-time transmission of monitoring data. The collected data need to be transmitted to the ground station through the measurement and control system. The operator can understand the working status of the UAV and make decisions in a timely and comprehensively manner. The ACMS transmits key data to the measurement and control system. Then, it transmits the data to the ground station through the measurement and control system.
(d) Forward monitoring data. The ACMS transmits flight control system data to the electromechanical system.
(e) Cache and record airborne data. Most data collected by the ACMS are not transmitted to the ground station. After the flight or test, the operators need to view all airborne subsystem data. The ACMS transmits all collected data to the cache recorder for recording.
(f) Antidamage recording airborne data. The recorded data are kept in the event of a UAV accident to facilitate subsequent analysis of the cause of the accident.
(g) Provide ground maintenance interface. When the ACMS is used on the ground, it transmits collected data to ground equipment through the ground maintenance interface. The operators can observe the status of each airborne subsystem in real time through the ground equipment.

8.3.2.3 Demand of aircraft condition monitoring system

According to the function, the demand analysis of the ACMS is:

(a) The ACMS should collect data from each subsystem. The ACMS needs to collect three kinds of signals: discrete data, analog data, and digital data. The ACMS needs to design a reasonable digital transmission bus to receive data from other airborne subsystem computers.

(b) The ACMS should regroup the collected data and send them to the cache recorder and parameter recorder for recording.
(c) The ACMS should regroup the collected information and send it to the measurement and control system for real-time transmission to the ground station. Then, it should send some parameters of the flight control system to the electromechanical system.

8.3.3 Scheduling methods

When the UAV performs a monitoring mission, it is necessary to schedule multiple UAVs from the mission center in the inspection area if there are several inspection points. The purpose of the scheduling system is to use many finite-range UAVs to complete as much data transmission as possible within a limited time. Characteristics of mission execution are:

(a) Only one UAV at each mission site performs monitoring missions.
(b) Not all mission points can complete the monitoring mission.
(c) It is necessary to dispatch a plurality of UAVs, but the flight path planning is performed on a single UAV.
(d) All UAVs depart from the same mission center and arrive at the assigned monitoring points, and then return to the starting point after the mission is completed.

The scheduling system of UAV can be divided into navigation, mission planning, and system management (Fig. 8.2).

8.3.3.1 Positioning

UAV positioning is the basis for the UAV to complete a navigation mission. UAV navigation requires that the UAV be correctly guided to the destination within a specified time along a predetermined path. To complete the mission, it is necessary to know navigation parameters such as the real-time position, navigation speed, and heading of the UAV [27].

A combined positioning and navigation system can output information in real time. The system can provide accurate direction reference and positioning coordinates for the UAV and predict the state of the aircraft based on the attitude information in real time. Integrated navigation combines multiple navigation technologies such as inertial, satellite, Doppler, and terrain-aided navigation.

8.3.3.1.1 Inertial navigation system/global positioning system integrated positioning

Taking the INS/GPS integrated navigation system as an example, the system consists of laser gyro strap-down inertial navigation, a satellite positioning system receiver, an

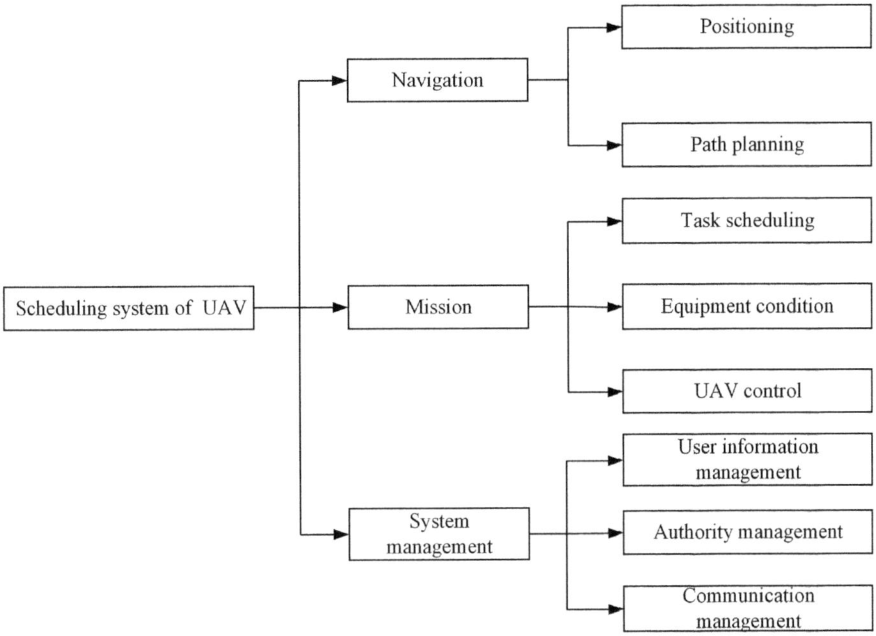

Figure 8.2
Scheduling system of an unmanned aerial vehicle (UAV).

integrated navigation computer, an odometer, an altimeter, and a base station radar system [28]. Using the long-term stability and moderate accuracy of GPS to compensate for the shortcomings of INS error propagation over time, the short-term high precision of the INS is used to compensate for the shortcomings of GPS receivers when errors are disturbed or signals are lost. The strap-down INS has the advantages of being simple, reliable, small, lightweight, and inexpensive.

At the same time, by using the attitude information and angular velocity information of the INS, the directional steering performance of the GPS receiver antenna can be improved. The global positioning satellite signal can be quickly captured or recaptured. With the high-precision position information and speed information continuously provided by the GPS, the position error, speed error, and other error parameters of the INS are estimated and corrected. These measures make the entire combined navigation system optimized and cost-effective.

Widely used INS/GPS integrated navigation algorithms mainly include the Kalman filter model and extended Kalman filter (EKF) model. The Kalman filter algorithm can estimate the system state from the estimated value at the previous time and the observed value at the current time.

The state estimation and prediction equation of the Kalman filter algorithm is [29]:

$$\begin{cases} X_k = \Phi_{k/k-1} X_{k-1} + \Gamma_{k-1} W_{k-1} \\ Z_k = H_k X_k + V_k \end{cases} \tag{8.1}$$

where X_k and Z_k are the state vector and observation vector at time k, $\Phi_{k/k-1}$ is the state transition matrix, W_{k-1} is the dynamic noise at time $k-1$, Γ_{k-1} is the system control matrix, H_k is the observation matrix at time k, and V_k is the observation noise at time k.

The process of the algorithm can be expressed as [30]:

(a) Preestimate :

$$\widehat{X}_{k/k-1} = \Phi_{k/k-1} \widehat{X}_{k-1} \tag{8.2}$$

(b) Calculate the preestimated covariance matrix:

$$P_{k/k-1} = \Phi_{k/k-1} P_{k-1} \Phi_{k/k-1}^T + \Gamma_{k-1} \widehat{Q}_{k-1} \Gamma_{k-1}^T \tag{8.3}$$

(c) Calculate the Kalman gain matrix:

$$K_k = \frac{P_{k/k-1} H_k^T}{H_k P_{k/k-1} H_k^T + \widehat{R}_k} \tag{8.4}$$

(d) Update estimates:

$$\widehat{X}_k = \widehat{X}_{k/k-1} + K_k \left(Z_k - H_k \widehat{X}_{k/k-1} \right) \tag{8.5}$$

(e) Calculate the updated covariance matrix:

$$P_k = (I - K_k H_k) P_{k/k-1} (I - K_k H_k)^T + K_k R_k K_k^T \tag{8.6}$$

Knowing state vector X_0 and covariance matrix P_0 at the initial time, state estimate $\widehat{X}_k$ at time k can be calculated: $k = 0, 1, 2, \cdots$.

Generally, the INS/GPS system sensor composition is a nonlinear system. The essence of the EKF algorithm is the process of transforming a nonlinear model into a linear system. The EKF algorithm is used to process path loss generated during the drone's movement, which can improve the positioning accuracy of the drone.

The system model of a nonlinear system can be expressed as [31]:

$$\begin{cases} X_k = f(X_{k-1}) + \Gamma_{k-1} W_{k-1} \\ Z_k = h(X_k) + V_k \end{cases} \tag{8.7}$$

where $f(X_{k-1})$ and $h(X_k)$ are state vectors of nonlinear systems.

The linearized state equation is presented as [32]:

$$\Phi_{k/k-1} = \frac{\delta f(X_{k-1})}{\delta X_{k-1}} |X_{k-1} = X^n_{k-1} \tag{8.8}$$

The state estimation bias of the nonlinear system is presented as [32]:

$$\Delta X_k = \Phi_{k/k-1} \Delta X_{k-1} + \Gamma_{k-1} W_{k-1} \tag{8.9}$$

The linearized measurement equation is presented as [32]:

$$\Delta Z_k = H_k \Delta X_k + V_k \tag{8.10}$$

$$H_k = \frac{\delta h(X_k)}{\delta X_k} \Big| X_k = X^n_{k/k-1} \tag{8.11}$$

The nonlinear system is transformed into a linear system after these calculations, and the state can be estimated by the Kalman filter algorithm.

8.3.3.1.2 Vision positioning

When the UAV is flying in the city or indoors, satellite signals will be unavailable owing to obstacles such as buildings, forests, and walls. Satellite navigation is not available. Therefore, UAV navigation based on the visual perception system has been widely studied and applied. These technologies include monocular/binocular visual odometry, visual simultaneous localization and mapping (SLAM), optical flow, and so on.

Visual odometry involves a device that relies on visual sensors (such as monocular/binocular cameras) to complete motion estimation. Odometry has an important characteristic: it focuses on motion only in local time and estimates motion in the time interval. Visual odometry uses the similarity between adjacent images to estimate camera motion. Methods of visual odometry are divided into one based on feature points and the direct method without using feature points. The one based on feature points is the mainstream method of current visual odometers [33].

(a) The method based on feature points selects some representative points in different images as feature points. The camera motion is estimated only for these feature points, whereas the spatial position of the feature points is estimated. No other information in the image will be used. The method converts the estimation of motion of an image into an estimation of motion between feature points. The problem lies in acquiring and matching feature points and calculating camera motion after obtaining the feature points [33].

The acquisition and matching of feature points is a widely studied computer vision problem. Commonly used feature points include Harris corner points, scale-invariant feature transform (SIFT) features, speeded up robust features (SURF), oriented features from accelerated segment test (FAST) and rotated binary robust independent elementary features (BRIEF) (ORB) features, and so on. To explain the differences among feature points, descriptors are used to describe them. According to the information of the feature points and descriptors, matching feature points in the image can be calculated [34].

After the feature point matching is completed, two one-to-one corresponding pixel point sets are obtained. In ordinary monocular images, only the pixel coordinates of the two-point sets are known. In binocular images and red–green–blue (RGB)-deep images, the distance from the feature point to the camera can also be known. Therefore, this problem has three different forms; 2D-2D: the camera motion is estimated by the pixel positions of the two images; 3D-2D: the camera motion is estimated by the 3D coordinates of one point set and the 2D coordinates of another point set; and 3D-3D: the camera motion is estimated by the 3D coordinates of the two-point sets [35].

(b) The direct method constructs an optimization problem and directly estimates camera motion based on the pixel information. This method saves the time of feature extraction, but because it uses all of the image information, the scale of the optimization problem is much larger than that of the method based on feature points. In addition, the direct method is effective only if the images are sufficiently similar [35].

Visual SLAM can obtain massive and redundant texture information from the environment and has strong scene recognition capabilities. The usable texture information has advantages in relocation and scene classification. At the same time, visual information can be easily used to track and predict dynamic targets in the scene, which is important for applications in complex dynamic scenes [36].

Vision SLAM is divided into three main modules: front-end visual odometry, back-end optimization, and loopback detection. The function of the visual odometry is to calculate the interframe motion of the camera and estimate the spatial position coordinates of the landmark points. The function of back-end optimization is to calculate maximum a posterior estimation based on the camera pose. The function of loop detection is to determine whether the camera has returned to the previous position to correct the error [36].

The optical flow method uses pixel changes in the time domain and the correlation between adjacent frames to calculate the motion information about objects between adjacent frames [37]. On an image plane, the motion of an object is often reflected by the different gray-scale distributions in the image sequence. Therefore, the motion field in space is transferred to the image and is represented as an optical flow field. The optical flow field is a 2D vector field, which reflects the gray-scale change trend of each point on the image. It can be regarded as the instantaneous velocity field generated by the pixels with gray scale moving on the image plane. The contained information is the instantaneous velocity vector information of each image point.

Optical flow methods can be divided into three types: gradient-based, matching-based, and energy-based [37]. (1) The gradient-based method uses the spatiotemporal gradient function of the time-varying image gray scale to calculate the velocity vector of the pixel. This method is easy to calculate and it can get better results, so it is widely used. However, the selection of parameters, preprocessing, noise, and so on will affect the accuracy of this method. (2) Matching-based optical flow methods include the feature-based and region-based methods. The feature-based method continuously locates and tracks the main features of the target and is robust to large motion and brightness changes of the target. The problem is that optical flow is usually sparse, and feature extraction and matching are difficult. The region-based method first locates similar target regions and then calculates optical flow based on the displacement of the regions. The same problem is that the optical flow calculated by this method is still not dense. (3) The energy-based method needs to perform spatiotemporal filtering on the input image to obtain an accurate velocity estimation of the optical flow field. However, it will reduce the temporal and spatial resolution of optical flow. The energy-based method often involves a large number of calculations. In addition, it is difficult to evaluate reliability.

In 2005, *Wu* et al. of the Georgia Institute of Technology designed and implemented a visual navigation system as an aid to inertial navigation [38]. The results of Kalman filter fusion inertial navigation and visual navigation were used to control the flight of the UAV. According to the imaging effect of the camera on the UAV on the ground scene of known size and position, combined with other sensors of the UAV, the height and position of the current UAV could be calculated.

In 2006, *Kong* et al. of Monash University in Australia designed and implemented a vision-based UAV navigation system to solve UAVs when GPS signals are inaccurate or GPS is unavailable [39]. The system adopts a feature-based method to match the image of the current area with real GPS image, and matched the image acquired by the UAV camera in real time to obtain the position of the current UAV.

In 2009, *Caballero* et al. used a homography matrix to calculate the rotation, translation, and so forth [40]. Between images acquired by the camera based on the natural scene, feature tracking algorithms do not need the aid of visual navigation beacons or other known location markers. The SLAM method was introduced to reduce the cumulative error of the tracking algorithm. The team used the HERO unmanned helicopter to carry out a large number of tests and verifications; navigation accuracy was good and could meet the application requirements.

In 2018, researchers at the University of Zurich developed an algorithm called DroNet, which could safely fly UAVs on city streets [41]. It was designed as a fast eight-layer residual network that generated two outputs for each input image: a steering angle that allowed the UAV to steer clear of obstacles while maintaining navigation, and collision probability that allowed UAVs to identify and react to dangerous situations.

According to different image feature spaces, the image matching methods can be divided into two types: gray-based and feature-based. Gray-based image matching is subject to large environmental interference with poor matching results. It is not robust to image rotation, scaling, and so on. It is less useful in real environments.

Feature-based image matching extracts feature points of the image to be matched and compares similarities between feature points to achieve the purpose of matching images. The accuracy of vision positioning obtained by image matching or image tracking alone is not high. Combining the results of image matching, image tracking can improve positioning accuracy and achieve the goal of visual navigation in low-altitude and small-range UAVs.

Image matching is the core algorithm of UAV visual positioning. The image matching algorithm running on UAV needs to consider the performance of algorithm execution speed, antirotation, and antiscaling.

The SIFT algorithm was first proposed by *Lowe* in 1999 and then summarized in 2004 [42,43]. It is used to detect and describe local features in the image. It looks for extreme points in the spatial scale and extracts their position, scale, and rotation invariants. SIFT features are based on points of interest in the partial appearance of the object, independent of image size and rotation, and have a high tolerance for light, noise, and slight changes. Based on these characteristics, they are highly significant and relatively easy to obtain, easily identifiable, and rarely misidentified in a large database of features.

The SIFT feature description also has a high detection rate for partially covered objects. According to the computer processing speed and a small feature database, the identification speed can be close to real-time operation. Also, the SIFT is suitable for fast and accurate matching in massive databases owing to a large amount of information on features.

The SIFT algorithm can be decomposed into four steps:

(a) Extreme value detection in scale space: Image positions are searched on all scales. The potential points of interest for scale and rotation invariant are identified by the Gaussian differential function.
(b) Key point positioning: the position and scale of each candidate position are determined through a fine-fitting model. Key points are chosen based on how stable they are.
(c) Direction determination: one or more directions are assigned to each key point position based on the local gradient direction of the image. All subsequent operations on the image data are transformed relative to the direction, scale, and position of the key points. Thus, the invariance to these transformations is ensured.

(d) Key point description: the local gradient of the image on the selected scale is measured in the neighborhood around each key point. These gradients are transformed into a representation that allows for larger local shape deformations and light changes.

The SIFT algorithm is not real-time. The SIFT algorithm cannot accurately extract feature points for targets with smooth edges. To improve the SIFT algorithm's shortcomings, some researchers put forward several improved algorithms. In 2006, *Bay* et al. presented a SURF algorithm [44]. The SURF algorithm uses the Hessian matrix to determine candidate points and then conducts nonmaximum suppression, which can reduce the computational complexity.

The SURF algorithm can be decomposed into five steps:

(a) Establishment of the Hessian matrix: This is the core of the SURF algorithm. It is a square matrix formed by the second partial derivative of a multivariate function and describes the local curvature of the function. The purpose of constructing the Hessian matrix is to generate image stable edge points and lay a foundation for feature extraction in the next step.

The Hessian matrix and its determinant are presented as [45]:

$$H(f(x,y)) = \begin{bmatrix} \frac{\partial^2 f}{\partial x^2} & \frac{\partial^2 f}{\partial x \partial y} \\ \frac{\partial^2 f}{\partial x \partial y} & \frac{\partial^2 f}{\partial y^2} \end{bmatrix} \tag{8.12}$$

$$\det = \frac{\partial^2}{\partial x^2}\frac{\partial^2 f}{\partial y^2} - \left(\frac{\partial^2 f}{\partial x \partial y}\right)^2 \tag{8.13}$$

When the determinant of the Hessian matrix obtains the local maximum, it determines that the current point is a brighter or darker point than other points in the surrounding neighborhood. Then, it positions the point as the key point. In the SURF algorithm, image pixel function $I(x,y)$is function $f(x,y)$.

(b) Establishment of a scale space: In the SIFT algorithm, images are the same size in the same octave layer of scale space, but the scale is different. Images in the different octave layer are also of a different size because it is obtained from the image of the previous layer. In Gaussian blur, the SIFT's Gaussian template size is always the same, just changing the image size between different octaves. The SIFT algorithm's Gaussian template size is always the same during Gaussian blur; only the image size is changed between different octave layers.

However, with the SURF algorithm, the size of the image remains unchanged. Images to be detected in different octave layers are obtained by changing the size of the Gaussian blur, and the size of the Gaussian template used in images in the same octave layer is also different. The SURF algorithm allows multilayer images in scale space to be processed simultaneously without the need for the secondary sampling of images, thus improving the performance of the algorithm.

(c) Feature point positioning: The localization process of feature points in the SURF algorithm is the same as the SIFT algorithm. Each pixel processed by the Hessian matrix is compared with 26 points in the neighborhood of 2D image space and descriptors to locate the key points preliminarily. After filtering out the key points with weak energy and mispositioning, the final stable feature points are selected.
(d) Direction assignment of feature point: the SURF algorithm adopts Haar wavelet features for direction assignment in the circular neighborhood of feature points. The sum of the horizontal and vertical Haar features of all points in a 60-degree sector in this circular neighborhood is computed. Then, the sector is rotated at a certain angle and the sum of the features is calculated again, until the sector with the largest Haar feature is found. Then, the direction of the sector is set as the feature point direction.
(e) Generation of feature descriptor: the SURF algorithm also takes a rectangular area block of 4×4 around feature points, and the direction of the rectangular area is along the direction of the feature point. The Haar wavelet's 25 pixels are counted in each subarea. Haar wavelets have four directions, which are the sum of the values in the horizontal direction, the sum of the values in the vertical direction, the sum of the absolute values in the horizontal direction, and the sum of the absolute values in the vertical direction. Both horizontal and vertical are relative to the feature point direction. The values in these four directions are used as the feature vectors of this subarea. There are 16 subareas in total, so it generates a 64-dimensional vector as the descriptor.

The ORB algorithm was proposed by *Rublee* et al. in 2011 [46]. It is a fast feature point extraction and description algorithm; feature extraction is developed from the FAST algorithm, and feature point description is improved according to the BRIEF feature description algorithm. The most important characteristic of the ORB algorithm is its fast computing speed, depending on the rapidity of the FAST algorithm and the representation of binary string unique to the BRIEF algorithm feature descriptors. It not only saves storage space, it can greatly reduce the matching time.

Feature point detection is based on the image gray value around the feature point. If there are enough pixels in the field around the candidate point that differ greatly from their gray value, it can be considered a feature point. Selecting a circular area with a radius of 3 pixels around the candidate points, 16 pixels in this area need to be compared.

If the gray value of n consecutive pixels on the circumference is larger or smaller than that of point P, P is considered the feature point and number n is generally set to 12. If N feature points need to be extracted from the image, the threshold of the FAST algorithm should be lowered so that feature points detected by the FAST algorithm are larger than number N. Then, Harris response value R of the feature points is calculated, and the first N points with large response values are taken as the FAST feature points.

The BRIEF algorithm calculates the feature descriptors of a binary string in the neighborhood of feature points. n pairs of pixel points Qi ($i = 1,2 \ldots, n$) are selected. Then, the gray value of each pair of pixels is compared. If the gray value of the point P is greater than Qi, it generates 1 in the binary string. Otherwise, it is 0 and all pairs of pixels are compared to generate a binary string of length n. The number n is usually 128, 256, or 512; mostly, it is 256.

Image tracking can select the target that needs to be tracked in the first initial image and then find the target in the subsequent image sequence. Light change, target scale change, target tracking, target deformation, motion blur, target fast movement, target rotation, target escaping parallax, background clutter, low resolution, and other phenomena in the tracking process are all challenges of target tracking.

In early years, continuously adaptive meanshift, optical flow, background difference, and other image tracking algorithms were popular and successfully applied in static background conditions. After 2008, these methods were gradually eliminated. More studies were conducted on image tracking in the dynamic background and complex scene environment. Current image tracking algorithms can be divided into tracking-by-detection, generative model, and deep learning methods. The tracking-by-detection algorithm applies the idea of target detection to target tracking. Samples are generated online, model is trained and tested online, the most likely location of the target is found, and the process to track the target location is repeated. The generative model algorithm can generate samples in the parameter space according to the tracking results during tracking, rather than directly sampling from the image space. Then, the generative model algorithm can find the optimal possible target in these samples as the tracking result. Deep learning methods are mainly divided into using the pretrained deep neural network and using the online trained neural network to output the target position.

Commonly used tracking by detection, generative, and deep learning models are:

(a) Multiple instance learning (MIL) algorithm

In 2009, *Babenko* et al. published the MIL algorithm, a typical tracking by detection method [47]. The MIL algorithm uses the online MIL boost algorithm for online learning, which is faster and can resist image occlusion.

(b) Struck tracking algorithm

In 2011, *Hare* et al. published the Struck algorithm, which proposed an adaptive visual target tracking framework based on structured output prediction [48]. By explicitly introducing output space to meet the tracking function, it avoids intermediate classification and directly outputs the tracking results. At the same time, to ensure real-time performance, the threshold mechanism was introduced to prevent the overgrowth of support vectors in the tracking process.

(c) Multiple experts using entropy minimization (MEEM) algorithm

In 2014, *Zhang* et al. published the MEEM algorithm [49]. In the process of video target tracking for a long time, because of shielding, deformation, rotation, and other reasons, samples are often polluted. The MEEM algorithm solved the problem of samples being polluted through multiple expert models. The MEEM algorithm applied multiple detectors to the current frame to determine confidence based on multiple predictions, which are selected from a previously stored classifier.

(d) Kernelized correlation filters (KCF) algorithm

In 2015, *Henriques* et al. published the KCF algorithm [50], which applies the correlation filtering method, whose core idea is to multiply the image by the cyclic matrix to produce a displacement of the image, thus obtaining a large number of samples. Putting samples of the displacement in a matrix will form a cyclic matrix. More specifically, the KCF algorithm produces a large number of samples by translation and assigns each sample a label, which is described by Gaussian distribution according to the distance from the center, which can be understood as confidence. In addition, the sample needs to be weighted through the cosine window before translation, to avoid the edge being too strong.

(e) Spatially regularized discriminative correlation filters (SRDCF) algorithm

In 2015, *Danelljan* et al. published the SRDCF algorithm [51], an improvement on the KCF algorithm, which increases the quality of the tracking model through space regularization. By using a spatial penalty factor with Gaussian distribution, the output at the boundary can be significantly suppressed by adding penalties with different weights to different positions.

(f) Deep learning tracker (DLT) algorithm

In 2013, *Wang* et al. proposed the DLT algorithm [52]. As the first tracking algorithm to apply the deep network to single-target tracking, the DLT algorithm first proposed the idea of offline pretraining and online fine-tuning. It used stacked denoising autoencoders to conduct offline training on a lot of data. The purpose of offline training is to extract features for online tracking, which was accomplished by particle filtering.

(g) Hedged deep tracking (HDT) algorithm

In 2016, *Qi* et al. published the HDT algorithm [53]. In the tracking process, six depth features were extracted using the VGG16 convolutional neural network based on the target position clip-out of the previous frame of an image. These depth features were used to train independent correlation filters to calculate their respective responses. Each weak tracker was used to estimate the preliminary target location. Through the adaptive weighting algorithm, all weak trackers were used to integrate a strong tracker. In the process of online tracking, through the calculation of weight, the loss for all weak tracker was averaged by adaptive algorithm and the weight of each weak track was counted, which could minimize weak tracker accumulated error.

(h) Continuous convolution operators tracker (CCOT) algorithm

In 2016, *Danelljan* et al. proposed the CCOT algorithm [54]. On the basis of feature fusion improvement, a continuous correlation filter in the time domain is created by learning from the continuous resolution sequence. It uses feature maps with different resolutions as the input of the filter so that the traditional feature and depth feature can be deeply combined. Finally, the estimated position of the target is obtained from the output responses.

(i) Efficient convolution operators (ECO) for tracking algorithm

The CCOT algorithm needs to train more filters to process high-dimensional features. To improve time efficiency and space efficiency, *Danelljan* et al. published the ECO algorithm in 2017 [55]. The ECO algorithm constructs a smaller set of filters that can be effectively used for matrix decomposition; it reduces the size of the model and prevents too-high a dimension leading to inefficiency and fitting problems. At the same time, the ECO algorithm uses the Gaussian mixture model to represent different target appearances, which make the training set diversified and prevented the overfitting of continuous several frames of samples. Also, the ECO algorithm changes the strategy of template updating to reduce the updating frequency and improve efficiency.

8.3.3.2 Unmanned aerial vehicle path planning

The path planning problem is that in a completely known or unknown environment, the robot plans one or more continuous collision-free safety paths to reach the end point based on the optimization criteria, starting from the current position. The optimization indicators of the path planning algorithm include execution time, path smoothness, overall path safety, and path length. Path planning algorithms generally include static path planning and dynamic path planning algorithms according to replanning efficiency [56]. Static path planning refers to all obstacle and threat area information in the map and its constraints as known information. Using this information as the input of the path planning algorithm, the obstacles in the operating environment of the UAV will not change with time or change with the position of the UAV. The reasonably feasible path is obtained by the path planning algorithm in one operation. After the path planning algorithm ends, the UAV performs the mission along the path derived by the path planning algorithm.

The input information of the dynamic path planning algorithm is part of the known information in the map environment: that is, the location and size information of some obstacles that the UAV knows through various channels before the flight, and the location and area of some threat areas. During the flight, the path planning algorithm still works. The obstacle or threat information in the map environment is captured in real time by the visual sensor installed on the UAV. When the visual sensor of the UAV captures the same information as the local map environment, the dynamic path planning algorithm is waiting. If the visual sensor detects that the predetermined optimal route of the UAV is blocked in the local map information, the dynamic path planning algorithm immediately uses the known data saved by the algorithm to obtain a new path quickly in the current map environment.

It can be seen that the dynamic path planning algorithm is capable of quickly replanning a new path. The dynamic path planning algorithm does not clear the data when replanning, but it can quickly replant using data generated by the algorithm before execution.

8.3.3.2.1 Classical path planning algorithms

According to the sequence of publication and the basic principles of the algorithm, the path planning algorithm can be roughly divided into a traditional algorithm, graphics method, intelligent bionics algorithm, and other algorithms.

(a) Traditional algorithm

Traditional path planning algorithms include the artificial potential field method, simulated annealing algorithm, tabu search algorithm, and fuzzy logic algorithm.

The basic idea of the artificial potential field method is derived from the concept of field in physics [57]. The movement of the robot in the environment is designed as a movement in an abstract artificial gravitational field. The target point will generate gravitation to the robot, thereby forming a gravitational field, and obstacles will generate repulsive force on the robot, thereby forming a repulsive field.

The combined force of gravity and repulsion determines the path of the mobile robot. The advantage of this method is that the planning path is smooth, safe, and easy to describe, but the performance of the algorithm will be affected by the local optimization problem and how to design the gravitational field.

The simulated annealing algorithm is a universal global optimization algorithm. The idea comes from the annealing process of solid materials. It searches randomly in the solution space with a certain probability combined with the probabilistic hopping characteristics [58]. The method is simple to describe, flexible to use, and low in computational cost, and has less of an initial condition limitation. However, owing to its randomness, convergence speed is relatively slow and is affected by parameter setting.

The tabu search algorithm is a subheuristic global optimization intelligent algorithm that imitates the human intelligent process through a stepwise optimization [59]. Starting from an initial feasible solution, the solution moves to the direction that optimizes the objective function. To jump out of the local optimum, the tabu search can break the tabu level and release some tabu states, thereby achieving global optimization.

The fuzzy logic algorithm imitates the logical thinking method of the human brain; it uses fuzzy sets and fuzzy rules for logical reasoning and makes fuzzy comprehensive judgments to obtain the results of global path planning [60]. The fuzzy logic algorithm combines physiological perception and motion principles and performs path planning through a lookup table based on real-time information collected by the sensor. Fuzzy logic algorithms mimic the thinking logic of the human brain and can avoid mathematical modeling, easily converting expert knowledge into control signals. It is robust and fault-tolerant. However, it is difficult to summarize fuzzy rules and adjust them online. The most difficult problems are optimal membership functions, control rules, and online tuning methods.

(b) Graphics method

When traditional algorithms solve practical problems; environment modeling is the weak link. The graphics method provides the basic method of environment description, but graphics methods lack a searching ability and most of them need to be combined with other methods. Common graphic methods include the grid, C space, free space, and Voronoi diagram methods.

The grid method uses coded grids to represent the map, and grids containing obstacles are labeled as obstacle grids, otherwise as free grids, to conduct a path search on this basis [61].

The C-space method expands obstacles into polygons in motion space and searches for the shortest path by using the feasible straight lines among the start, end, and vertices of all polygons as the path range. The C-space method is intuitive and easy to find the shortest path. However, if the mobile robot mission goal changes, it needs to be remodeled [62].

Aiming at the defect of the C-space method, the free space method constructs free space with the predefined basic shape and the free space is represented as a connected graph. Then, path planning is carried out by searching the connected graph. When the starting point and the end point change, they change only their position in the free space and only need to be repositioned instead of redrawing the whole graph [63]. The disadvantage is that the complexity of the algorithm will be increased if there are too many obstacles.

A Voronoi diagram divides the space with some basic graphics called elements and determines the edges of elements with the midline between every two points. Finally, the whole space is divided into the Voronoi diagram, and then the algorithm is used to search the optimal path formed by the edges of polygons optimally [64]. The advantage is that the obstacle is surrounded by the element, which can effectively avoid the obstacle. The disadvantage is that redrawing the graph is time-consuming, so it is unsuitable for a large dynamic environment.

(c) Intelligent bionics algorithm

When the UAV performs flight path planning in a complex environment, it faces dynamic environmental changes, and the bionic algorithm has a great role. Algorithms that imitate biological habits in nature have been the object of many scholars' research, such as the ant colony optimization (ACO) algorithm, neural network algorithm, particle swarm optimization (PSO) algorithm, and genetic algorithm (GA).

The ACO algorithm is a heuristic algorithm that imitates the foraging behavior of ant colonies. Each ant will mark a certain amount of biological information on the route it travels. In unit time, the scatters visited by the ants the most frequently and which have the highest density of biological information are the shortest path [65]. In addition, the marking information can also be recognized by subsequent ants to guide the direction, which will have a positive feedback role to speed the solution. This algorithm has strong global capability and parallelism and is easy to calculate. However, the calculation will be large, and there is also the problem of the local optimal solution, which can be improved by introducing elite ants.

The neural network algorithm is an excellent method in the field of path planning, especially for the deep neural network that is more widely used. The basic idea is to simulate the information transmission process of neurons in animals by training the artificial defined network, which has a strong generalization ability and fitting ability [66]. It is increasingly are used in combination with other intelligent algorithms.

The GA simulates the genetic selection and evolution model. It is an iterative search algorithm implemented in accordance with the principles of genetics [67]. The algorithm is easy to combine with other algorithms, but its operational efficiency is insufficient.

The PSO, which simulates the behavior of birds when they fly to hunt for food, is simpler in rules than the GA; it eliminates the need for crossover and mutation operations and has a memory function [68]. Advantages of the algorithm are that it is simple, robust, and insensitive to population size, and it has good convergence, but it is also difficult to jump out of the local optimal solution.

(d) Other algorithms

Some artificial algorithms have been widely used because of their excellent characteristics. These algorithms generally have strong path searchability and can have a good role in the discrete path topological network. These algorithms include the A* algorithm, Dijkstra algorithm, fallback algorithm, and Floyd algorithm.

The Dijkstra algorithm is the most typical optimal path search algorithm. It starts from the starting point and gradually expands outward to find the optimal path. Owing to its ergodic property, the success rate of searching the shortest path is high and the robustness is good [69]. However, because of its ergodic property, the disadvantage of low computational efficiency in large and complex maps will be further amplified.

The A* algorithm is the most widely used heuristic intelligent algorithm. By setting an appropriate heuristic function, the distance from the search node to the beginning and the end point is considered comprehensively [70]. Nodes with optimal heuristic function values are selected for the extension. It has the advantages of few extension nodes, a small computation cost, good robustness, and a quick response. The disadvantage is that the node limitation caused by the robot's volume is ignored in practical application.

The fallback algorithm is an improved algorithm of the Dijkstra algorithm, which is mainly applied to routing with multiple quality of service requirements. The different qualities of service requirements are used as the objective function. The Dijkstra algorithm is used for path search. According to whether the multiple quality of service requirements is satisfied, the Dijkstra algorithm desides to repeat or terminate [71].

The Floyd algorithm is a global traversal search algorithm based on the known point-to-point distance to get the shortest path between points [72]. The establishment of a weight matrix and three for-loop statements are the key to it. Compared with the Dijkstra algorithm, it is a great improvement. The dense graph has the best effect, and the edge weight can be positive or negative. The change in starting point and end point has little influence on the algorithm, and its efficiency is higher than that of the Dijkstra algorithm. However, it also has the disadvantage of high time complexity and is unsuitable for computing large amounts of data.

8.3.3.2.2 Reinforcement learning algorithms

The process of reinforcement learning, which is similar to biological learning behavior in adapting to environmental changes, has two characteristics: (1) the algorithms have active interaction with the surrounding environment; and (2) the environment responds to the test and evaluates it, and the organism adjusts its behavior accordingly.

Reinforcement learning is a mapping from environment to action. The training data for reinforcement learning come from the experience of agents interacting with the environment. On this basis, the strategy of selecting actions is improved to achieve the purpose.

In the framework of reinforcement learning, learning is a tryout method. The agent obtains the current state s and reward value r from the environment, updates the internal state value, and outputs action a. Under action a, the environment switches to a new state s' and generates r, which is immediately fed back to the agent [73]. The agent then selects the next action according to the new state s'. The action a affects not only the immediate reward, but also the subsequent and final cumulative reward.

The components of a reinforcement learning system, in addition to agents and environment, need to have four basic elements:

(a) The mechanism of agent–environment interaction selection is called policy, which is expressed as $\pi : S \times A \rightarrow [0, 1]$, where S is a state set and A is an action set. The policy is a mapping from the state set to the action set, describing the set of actions taken in each possible state. This policy is the core of reinforcement learning. It determines the quality of the entire method.
(b) The reward function is a mapping from the state to the enhanced signal, which is an evaluation of the previously selected action, expressed as $r : s \rightarrow r$. The reward function reflects the current task of reinforcement learning and can be used as a reference for agent improvement strategies. The reward function is usually a scalar, with positive numbers representing rewards and negative numbers representing punishments. The goal of reinforcement learning is to maximize the cumulative reward.
(c) The value function is the cumulative reward that an agent can get from this state and the evaluation of medium and long-term benefits in the learning process. Different from the immediate use of the reward function r, the value function pays more attention to long-term rewards to obtain the maximum benefits. The value function is the mathematical expectation of the cumulative weighted reward of the agent from the initial state to the terminal point under the guidance of policy π, which is often expressed as V.
(d) The main function of the environment model is to simulate the behavior of the environment, which can help the agent consider possible situations that may occur in the interaction process, but which have not been experienced. With current state s and action a, and using the environment model, the agent can comprehensively consider the decision made after interacting with the environment and the result predicted by the model, to carry out planning and predict new state s' and the reward.

The relationship among the four elements of reinforcement learning theory can be described as a pyramid structure. The environment model is unknown in most cases; the data and experience needed for learning can be obtained only from interaction between the agent and the environment. In the process of selecting actions, the theory not only relies on the immediate report of the selected actions, it considers the uncertainty of the environment and the long range of the target. Therefore, a value function between the policy and the reward function is constructed to selects the optimal action.

Most studies proposed on reinforcement learning are based on Markov decision processes (MDPs) [74].

Definition 8.1. Markov property means that the next state s_{t+1} of the system is only related to the current state and is independent of the previous state.

Definition 8.2. If and only if $P[s_{t+1}|s_t] = P[[s_{t+1}|s_1, \cdots, s_t]]$, the state can be considered a Markov state.

Definition 8.3. Markov process is a binary group (S, P), S is a state set, $P$and is the state transition probability [75]:

$$P = \begin{bmatrix} p_{11} & \cdots & p_{1n} \\ \vdots & \vdots & \vdots \\ p_{n1} & \cdots & p_{nn} \end{bmatrix} \tag{8.14}$$

The Markov decision process is described by the tuple (S, A, P, r, γ), in which r is the reward function and γ is the discount factor used to calculate the cumulative reward, $0 \leq \gamma \leq 1$ [76].

The objective of reinforcement learning is to find the optimal policy by solving a Markov decision process. The policy π refers to a distribution on the action set for state s [77]:

$$\pi(a|s) = p[A_t = a|S_t = s] \tag{8.15}$$

For policy π, the cumulative reward is defined as [74]:

$$G_t = \gamma r_{t+1} + \gamma r_{t+2} + \cdots = \sum_{k=0}^{\infty} \gamma^k r_{t+k+1} \tag{8.16}$$

Meanwhile, for policy π, it is assumed that starting from state s_1, the state sequence has various possibilities. The updates of states can be discontinuous, such as $s_1 \rightarrow s_2 \rightarrow s_4 \rightarrow s_5$.

Definition 8.4. When an agent adopts a policy, π, the cumulative reward obeys a distribution and the expected value of the cumulative reward is defined as a state-valued function [74]:

$$V^{\pi}(s) = E_{\pi}\left[\sum_{k=0}^{\infty} \gamma^k r_{t+k+1}|s_t = s\right]$$

$$= E_{\pi}[r_{t+1} + V^{\pi}(s_{t+1})|s_t = s]$$

$$= \sum_{a} \pi(s, a) \sum_{s'} P(s, a, s')(r(s, a, s') + \gamma V^{*}(s'))|s_t = s, s_{t+1} = s' \tag{8.17}$$

in which $r(s, a, s') = E[r_{t+1} + \gamma r_{t+2} \cdots |s_t = s, a = a_t = \pi(s_t), s_{t+1} = s']$ represents the expected reward of selecting the action a and transitioning to the state s' in state s.

Eq. (8.17) is the Bellman equation. The action value function $Q^{\pi}(s,a)$ represents the expected reward obtained by the learning system after selecting the action according to policy π [74]:

$$Q^{\pi}(s,a)=E_{\pi}\left[\sum_{k=0}^{\infty}\gamma^{k}r_{t+k+1}|s_t=s,a_t=a\right] \tag{8.18}$$

The optimal policy π^* can be obtained by solving the optimal value function, and the corresponding optimal state value function and action value function are respectively expressed as V^* and π.

The optimal state value V^* also satisfies Bellman's optimal equation and can be expressed as [74]:

$$V^*(s)=\max_{\pi}V^{\pi}(s)=\max_{a\in A}\sum_{s'\in S}P(s,a,s')(r(s,a,s')+\gamma V^*(s')) \tag{8.19}$$

For $\forall s\in S,\forall a\in A$ and any optimal policy π', the optimal action value function can be expressed as [74]:

$$Q^*(s)=\max_{\pi}Q^{\pi}(s,a)=\sum_{s'\in S}P(s,a,s')\left[\left(r(s,a,s')+\gamma\max_{a\in A}Q^*(s',a')\right)\right] \tag{8.20}$$

Thus, it can be concluded that the optimal policy can be expressed as [74]:

$$\begin{aligned}\pi^*(s)&=\operatorname*{argmax}_{a\in A}\sum_{s'\in S}P(s,a,s')(r(s,a,s')+\gamma V^*(s'))\\ \pi^*(s)&=\operatorname*{argmax}_{a\in A}Q^*(s,a)\end{aligned} \tag{8.21}$$

According to the path planning mission, it can be divided into a model-based reinforcement learning algorithm and a model-free reinforcement learning algorithm.

When the state transition probability and reward function model of the environment is known, the method for solving the optimal policy is dynamic programming; it is also called the model-based method [78]. Four different dynamic programming methods are:

(a) Linear programming, which transforms the solution of value function into a linear programming problem according to the Bellman equation;
(b) Policy iteration. Based on Behrman's optimal equation, the optimal policy is found by alternating policy evaluation and policy iteration;
(c) Iteration method of value function, an extension of the dynamic programming algorithm of a finite time period in an infinite time period; a successive approximation algorithm;
(d) Generalized policy iteration integrating the characteristics of policy iteration and value iteration methods.

However, for path planning tasks, in most cases, the relevant model is unknown and difficult to construct. Therefore, the model-free reinforcement learning method is more suitable for solving path planning problems.

In many practical applications, the model of state transition probability or reward function is unknown, and the agent and environment need to interact to acquire training samples to learn the optimal policy.

The Monte Carlo method estimates the whole in part and solves the problem using random numbers, which can obtain the approximate solution of the problem through statistical simulation or sampling [73]. This method is used only for tasks that have a termination state. Policy evaluation mainly uses the law of large numbers to estimate the value function with the sample average of the reward value of each state, and finally finds the optimal policy.

The temporal difference algorithm can observe the difference value of a variable at two continuous time points. Supposing the system is in time t, the current state is s_t, under the guidance of the policy π, the state s_t changes to s_{t+1}, and obtains reward r_t immediately, the estimated value of the new value function is expressed as [79]:

$$V'(s_t) = r_t + \gamma V(s_{t+1}) \tag{8.22}$$

The temporal difference of the system at time t can be expressed as [79]:

$$\delta_t = r_t + \gamma V(s_{t+1}) - V(s_t) \tag{8.23}$$

The temporal difference method observes the long-term impact of each step action through Eq. (8.23) to evaluate the past action and update its current value function depending on the subsequent value function.

The state action reward state action (SARSA) algorithm is an on-policy learning method. The policy that generates the actions and the policy that needs to be evaluated in the SARSA algorithm are the same, also known as online learning. Because the decision of behavior and iteration of value function are carried out at the same time, it is necessary to evaluate and improve the policy based on actions generated by the policy. The current actual state action value function is used to generate the optimal policy. Its update function is shown as [80]:

$$Q(s_t, a_t) \leftarrow Q(s_t, a_t) + \alpha(r_t + \gamma Q(s_{t+1}, a_t) - Q(s_t, a_t)) \tag{8.24}$$

The Q learning algorithm is an off-policy learning method. The policy that generates the actions and the policy that needs to be evaluated in the Q learning algorithm are different. Q function values are estimated rather than actual values, and the maximum Q function is selected from the estimated values of different actions in each step for updating, which predicts the optimal policy based on previous experience. Its updating function is shown as [80]:

$$Q(s_t, a_t) \leftarrow Q(s_t, a_t) + \alpha\left(r_t + \gamma \max_a Q(s_{t+1}, a_t) - Q(s_t, a_t)\right) \tag{8.25}$$

The path planning of the UAV should consider the flight kinematic constraints of the UAV itself: that is, the constraint of continuous path curvature. The conventional path smoothing method is to smooth straight lines constituting the shortest path. Another method is to apply the curves suitable for UAV flight to path planning directly, including the Dubins, Clothoid, B-spline, and Pythagorean hodograph curves [81].

The Dubins curve satisfies the conditions of curvature constraints and the specified tangent direction of the beginning and end, but the curvature is discontinuous. The curvature of the Clothoid curve varies linearly with the arc length and is satisfied with the condition of continuous curvature. The premise of the calculation of the Clothoid curve is that the curvature at the connection origin is zero, but in actual path planning, it may be necessary to connect the arc curve. The path generation process of the Clothoid curve is complicated and the flexibility is poor. The B-spline curve has continuous curvature, but there is no rational expression for the curvature. It is difficult to determine whether each point satisfies the maximum curvature constraint. Curvature of the Pythagorean Hodograph curve is continuous, the curve is smooth, and the curve length and curvature have rational expressions.

Faced with complex mission requirements, multiple UAVs are often required to perform missions collaboratively. Compared with single UAV path planning, the optimal path obtained by collaborative path planning needs to satisfy the coordination constraints and may not be the optimal path for a single UAV itself. Collaborative path planning can be divided into the task coordination of a single target by multiple UAVs, and the task coordination of multiple targets by multiple UAVs [82]. Modeling collaborative path planning problems is an optimization problem with constraints. There are constraints on the UAV flight path:

(a) Restrictions in the environment and no-fly zones.
(b) Flight performance constraints such as energy limitation, maximum range limitation, flight altitude, minimum turning radius, and maximum climb angle/dive angle.
(c) Cooperative restrictions. Coordinated space limitation is the needed space between UAVs required for formation flight. Coordinated time limitation requires UAVs to arrive at an end point within a time period, at no order in a time period, or simultaneously.

8.3.3.3 Scheduling of unmanned aerial vehicles

8.3.3.3.1 Problem statement

For a given mission set with N_t subtasks and N_u assignable UAVs, the goal of mission planning and scheduling is to find optimal mapping from the mission set to the UAVs to maximize the sum of the benefits obtained from the mission, on the premise of satisfying the mission's own and external constraints.

The global benefit of mission execution is the sum of the local benefits obtained by each UAV during the corresponding mission execution, so the global benefit is determined by the decision and planning variables of the mission. The decision variable contains only pairing information about mission and UAVs. The planning variable is the relevant variable that affects the benefit of the mission; for UAVs, it generally includes fuel consumption, income obtained from the mission, the UAV position, the target position corresponding to the mission, and so on.

Mission planning and scheduling can be described as [83]:

$$\begin{aligned} &\max_{\mathbf{d}} \sum_{i=1}^{N_u} \sum_{j=1}^{N_t} \mathbf{R}_{ij}(\mathbf{d}, \boldsymbol{\theta}) \\ &s.t.\mathbf{C}(\mathbf{d}, \boldsymbol{\theta}) \leq \mathbf{b} \\ &\mathbf{d} \in \boldsymbol{\chi} \end{aligned} \tag{8.26}$$

where $\mathbf{d}$ is decision variable matrix, if $d_{ij} = 1$, the mission j is assigned to the UAV i, $i \in \mathbf{I} = \{1, \cdots, N_u\}, j \in \mathbf{J} = \{1, \cdots, N_t\}, \boldsymbol{\chi} = \{0, 1\}^{N_u \times N_t}$; $\boldsymbol{\theta}$ is a set of variables that affect the benefit of mission; $\mathbf{R}_{ij}$ is the benefit obtained by the UAV i executing mission j under the condition of d_{ij} and θ_i; and $\mathbf{C}(\mathbf{d}, \boldsymbol{\theta}) \leq \mathbf{b}$ is the constraints that the mission planning needs to meet, mainly including dynamics, load, and no-fly zone constraints for the UAV.

Eq. (8.26) shows that mission benefit $\mathbf{R}$ and constraint conditions $\mathbf{C}$ are related to the decision variable matrix $\mathbf{d}$. Real mission scenarios are often dynamic and time-varying, so the benefit is also closely related to time. Moreover, the relevant constraints are closely related to time. Introduce the time factor into Eq. (8.26) as [83]:

$$\begin{aligned} &\max_{\mathbf{d}, \boldsymbol{\tau}} \sum_{i=1}^{N_u} \sum_{j=1}^{N_t} \mathbf{R}_{ij}(\mathbf{d}, \boldsymbol{\tau}, \boldsymbol{\theta}) \\ &s.t.\mathbf{C}(\mathbf{d}, \boldsymbol{\tau}, \boldsymbol{\theta}) \leq \mathbf{b} \\ &\mathbf{d} \in \boldsymbol{\chi} \end{aligned} \tag{8.27}$$

where τ is the mission execution time matrix corresponding to decision variable matrix $\mathbf{d}$ and τ_{ij} is the time when the UAV i executes mission j.

The difficulty of solving the mission planning problem is the collaboration between the UAVs and the dependence between missions. The solution of the model is in the solution space, and the size of the solution space is related to the number and dimension of decision variables. As the scale of the mission planning problem increases, the alternative solution space also increases. Therefore, it is necessary to simplify the general model given by Eq. (8.27) properly and clarify the constraint conditions.

Simplified condition 1. The uniqueness of mission execution means that a certain mission can be assigned to only one UAV at most. The simplified conditions can be written as [84]:

$$\begin{aligned} &\sum_{i=1}^{N_u} d_{ij} \leq 1, \forall j = \mathbf{J} \\ &d_{ij} \in \{0, 1\} \end{aligned} \tag{8.28}$$

The simplified condition can significantly reduce the dimension of decision space matrix χ and ensure cooperation between UAVs.

Simplified condition 2. The mission benefit obtained by UAV i performing the assigned mission is only related to the UAV's mission decision matrix $\mathbf{d}_i$ and execution time $\mathbf{\upsilon}_i$. Under this assumption, the mission benefit is determined only by the capability of the corresponding UAV and the time to execute the mission. Eq. (8.27) can be simplified as [84]:

$$\max_{\mathbf{d},\tau} \sum_{i=1}^{N_u} \left(\sum_{j=1}^{N_t} \mathbf{R}_{ij}(\mathbf{d}_i, \boldsymbol{\tau}_i, \boldsymbol{\theta}) \right) \tag{8.29}$$

Based on simplified conditions 2, and considering that elements in the mission decision matrix can only be 0 or 1, the objective function can be written as [84]:

$$\max_{\mathbf{d},\tau} \sum_{i=1}^{N_u} \left(\sum_{j=1}^{N_t} \mathbf{R}_{ij}(\boldsymbol{\tau}_{ij}, \boldsymbol{\theta}) d_{ij} \right) \tag{8.30}$$

To refine the objective function further, set $\mathbf{P}_i = \left\{p_{i1}, \cdots, p_{i|p_i|}\right\}$; $\mathbf{P}_i$ contains the following two types of information: (1) the UAV i will perform which numbered missions; and (2) the sequence of mission numbers in $\mathbf{P}_i$ indicates the sequence in which missions are executed.

For $\forall j = \{1, \cdots, |P_i|\}$, p_{ij} represents the mission j of UAV i, and Eq. (8.30) can be rewritten as [84]:

$$\max_{\mathbf{d},\tau} \sum_{i=1}^{N_u} \left(\sum_{j=1}^{N_t} \mathbf{R}_{ij}(\tau_{ij}(\mathbf{P}_i(\mathbf{d}_i)), \boldsymbol{\theta}) d_{ij} \right) \tag{8.31}$$

Mission planning mainly completes two steps: (1) assignment mapping between the mission and UAV; and (b) the order of execution of assigned missions. Therefore, for $\forall i \in \mathbf{I}$, the objective function of mission planning is presented as [84]:

$$\max_{\mathbf{P}_i}\left(\max_{\tau_i}\sum_{j=1}^{N_t}\mathbf{R}_{ij}(\tau_{ij}(\mathbf{P}_i(\mathbf{d}_i)),\boldsymbol{\theta})d_{ij}\right) \tag{8.32}$$

In Eq. (8.32), inner ring optimization means seeking the optimal execution order for any $\mathbf{P}_i$; outer ring optimization means searching for the best mission set.

Simplified condition 3. The mission planning problem in Eq. (8.27) is simplified into two subproblems: mission assignment and path planning. To reduce the difficulty of solving the problem, it is assumed that path set $\mathbf{G}$ can be used to replace the accurate path as the path cost estimation in the mission assignment process, and $\mathbf{G} \in G(\mathbf{d}, \boldsymbol{\upsilon})$. $G(\mathbf{d}, \boldsymbol{\upsilon})$ is all path sets that satisfy UAV dynamic and environmental constraints: that is, the mission assignment and path planning problems are solved by decoupling. For $\forall i \in \mathbf{I}$, the input of path planning is $\mathbf{P}_i$ (mission decision-making information and task execution order) and $\boldsymbol{\upsilon}_i$ (mission execution time). Eq. (8.32) can be written as [84]:

$$\begin{aligned}&\max_{\mathbf{d},\boldsymbol{\tau}}\sum_{i=1}^{N_u}\left(\sum_{j=1}^{N_t}\mathbf{R}_{ij}(\tau_{ij}(\mathbf{P}_i(\mathbf{d}_i)),\boldsymbol{\theta},\mathbf{G})d_{ij}\right)\\&s.t.\mathbf{G}\in G(\mathbf{d},\boldsymbol{\tau})\\&\mathbf{C}(\mathbf{d},\boldsymbol{\theta})\le\mathbf{b}\\&\sum_{i=1}^{N_u}d_{ij}\le 1, d_{ij}\in\{0,1\}\end{aligned} \tag{8.33}$$

Simplified condition 4. The UAV capability limitation means that a UAV can perform only a limited number of missions owing to its own capability and payload limitations. The simplified condition can be regarded as a constraint condition in mission planning as [84]:

$$\begin{aligned}&\sum_{j=1}^{N_t}d_{ij}\le L_i, \forall i\in\mathbf{I}\\&d_{ij}\in\{0,1\}\end{aligned} \tag{8.34}$$

Eq. (8.27) can be written as [84]:

$$
\begin{aligned}
&\max_{\mathbf{d},\tau} \sum_{i=1}^{N_u} \left(\sum_{j=1}^{N_t} \mathbf{R}_{ij}(\tau_{ij}(\mathbf{P}_i(\mathbf{d}_i)), \boldsymbol{\theta}, \mathbf{G}) d_{ij} \right) \\
&s.t. \mathbf{G} \in G(\mathbf{d}, \boldsymbol{\tau}) \\
&\mathbf{C}(\mathbf{d}, \boldsymbol{\theta}) \leq \mathbf{b} \\
&\sum_{i=1}^{N_u} d_{ij} \leq 1, d_{ij} \in \{0, 1\} \\
&\sum_{j=1}^{N_t} d_{ij} \leq L_i, i \in \mathbf{I}, j \in \mathbf{J}
\end{aligned}
\tag{8.35}
$$

Constraints in mission planning problems include noncoupled and coupled constraints.

The uncoupled constraint can be defined thus: for $\forall i = \mathbf{I}$, if the constraint condition affects only the planning result of the UAV i but does not affect the planning result of UAV $j(\forall j \in \mathbf{I}/i)$, the constraint condition is called an uncoupled constraint. Common uncoupled constraints include the functional/resource constraints of missions; the time constraint for mission execution; dynamics constraints of the UAV; the holdover time constraint of the UAV; communication capability constraints between UAVs; detection capability constraints of the UAV; and external environmental constraint. For complex real mission scenarios, coupling constraints include two categories: (1) coupling of mission assignment, include unilateral dependence, interdependence, and mutual exclusivity; and (2) coupling between mission timings.

8.3.3.3.2 Control methods

Mission scheduling of the UAV assigns missions to individuals in the system reasonably according to the inspection area; it achieves the purposes of efficiently performing missions and optimizing the UAV system. There are three types of control methods for the UAV mission assignment: centralized, distributed, and hierarchical distributed control.

(a) Centralized control system

Signal transmission and control in a centralized control system is performed by a single control center. The most commonly used models for mission assignments based on centralized control systems are the mixed-integer linear programming (MILP) model, multiple traveling salesman problem model, vehicle routing problem model, and dynamic network flow optimization model. Centralized solution methods can be divided into optimization and heuristic methods [85]. Optimization methods include an exhaustive method (breadth-first or depth-first), integer programming, constraint planning, graph theory method, and the like. Optimization can ensure the optimal solution under some simplified assumption. However, because of nondeterministic polynomial characteristics of mission assignment and collaborative planning problems, the larger the problem size, the more difficult it is to solve the optimization method and the longer the time. The heuristic method does not necessarily obtain the optimal solution to the problem; it is the most satisfactory solution obtained in a certain time. Heuristic methods can be further divided into intelligent optimization algorithms and traditional heuristic algorithms.

Optimization algorithms are explained thus.

The exhaustive method enumerates all solutions in the solution space to obtain the optimal solution. This method is suitable only for discrete and small-scale problems. When the size of the solution space is too large, the method takes too long and cannot solve the problem.

The MILP model solves small-scale mission scheduling problems by establishing objective functions and constraints according to established goals [86].

The constraint programming method is composed of a variable set and constraint set. All variables in the variable set have a corresponding range, and the value of the variable can only be selected from its range. It is a general method for solving combinatorial optimization problems.

The graph theory method expresses the characteristics of the mission and the members receiving the mission through the graph method, establishes a match between the mission and the members of the system with the graph theory method, and designs a reasonable and feasible mission assignment scheme.

Heuristic methods are explained thus:

The step of list scheduling method establishes the priority of the mission and obtains the processing order of the mission first; then, it assigns the mission to the system members according to the processing order of the mission.

The clustering method clusters each task into one cluster until the task cluster and the numbers of system members reach the same.

Intelligent optimization algorithm. The intelligent optimization algorithm aims to obtain a satisfactory solution within the acceptable time range and adjust the solution time and quality. Its advantages are that it is easy to implement, the calculation is not complicated, and the quality of the solution is relatively high. Commonly used intelligent optimization algorithms are evolutionary algorithms and swarm intelligence algorithm (SIA) [87]. Evolutionary algorithms include the GA, genetic programming, evolution strategy, and evolutionary programming. SIAs include the PSO, ACO, artificial immune algorithm, tabu search algorithm, and simulated annealing algorithm.

(b) Distributed control system

In terms of signal transmission, the distributed control system is not only between the UAV and the ground station; the UAVs can also communicate with each other. Distributed control systems have higher requirements for UAV, which require the UAV to have independent capabilities of computing, analysis, and decision-making. However, this control system has better flexibility.

Mission assignment methods based on distributed control system structure include ContractNet, auction, distributed model predictive control, decentralized Markov decision process, and dynamic distributed constraint optimization problem.

ContractNet is one of the most widely used distributed mission assignment methods. ContractNet solves each problem with communication to prevent conflicts. The ContractNet method has the roles of publisher and bidder, composed of four interactive stages of invite bids—bid—win the bidding—confirmation [88]. In the ContractNet collaboration method, the role of system members does not need to be specified in advance; any system members can be managers, as can workers. The difference is that only the member issues a mission notice or replies to the mission notice. Therefore, the mission can be hierarchically decomposed and assigned.

Missions that the UAV needs to perform can be regarded as auction items. The mission distributor and the mission receiver of the UAV jointly constitute participants, and both sides have their respective revenue functions and bidding strategies [89]. The auction method uses explicit rules to guide interaction between buyers and sellers, which is highly operable and can reasonably allocate resources within a short time to obtain an optimal or better solution for the problem.

(c) Hierarchical distributed control system

A hierarchical distributed control system has the advantages of both a centralized and distributed control system, which is always a hybrid control system structure. All UAVs are stratified and classified according to certain rules. First, they are grouped according to categories. A centralized control system structure is selected for UAVs at the same level and UAVs in the same group, whereas a distributed control system structure is selected for different groups and control centers.

This control system structure has great flexibility when UAVs perform combat missions. It can independently adjust the amount of information exchange with the control center according to the real situation, reduce the calculation time, and adjust the mission allocation strategy in a timely manner according to actual needs.

8.4 Rail transit intruding detection based on inspection unmanned aerial vehicles

The intrusion of foreign matter onto the track is a serious threat to the safety of operations. The detection method of an intrusion based on UAV adopts a real-time environment sensing system with a pan/tilt camera to inspect the track. The method of rapid extraction and matching of track visual features enables the automatic identification of intrusions, so as to take measures in advance to ensure the safety of the train's operation.

The UAV is unaffected by the terrain and can cruise at high speed with a limited load. The UAV-based track intrusion detection system must have the characteristics of strong system stability, system compatibility and scalability, a small system hardware load, and a real-time sensing system.

The track intrusion object identified by the UAV refers to obstacles that affect safe running of the train within the track's limit, including: (1) tracks buried and blocked because of a natural disaster; (2) tracks blocked because of infrastructure failure; and (3) the illegal intrusion of humans and animals.

There are problems to be solved in the detection of a track line environment based on UAVs:

(a) Reasonable feature descriptions should be extracted for track sections. The traditional background difference method or interframe difference method cannot meet the requirements of UAV detection. The UAV cannot collect all of the original images of tracks in the section as a standard template, which requires a large storage capacity and cannot be updated.
(b) The detection algorithm can deal with multiple foreign matter invasion limits better. Owing to the complexity of the track environment and intrusion situation, it is difficult to detect and identify intruding foreign matter. The detection method should be universal and adaptable to changes in the environment.
(c) The real-time performance of the algorithm should meet the requirements of the UAV inspection speed. The payload of the UAV is limited and the computing capacity of the onboard central processing computer is limited. Therefore, the detection algorithm should be as simple and efficient as possible and avoid complex data processing to meet the requirements of the UAV inspection speed.

A stable processing algorithm is important for allowing the system to function. Although the geometric structure of the track is significant, it is difficult to extract track features

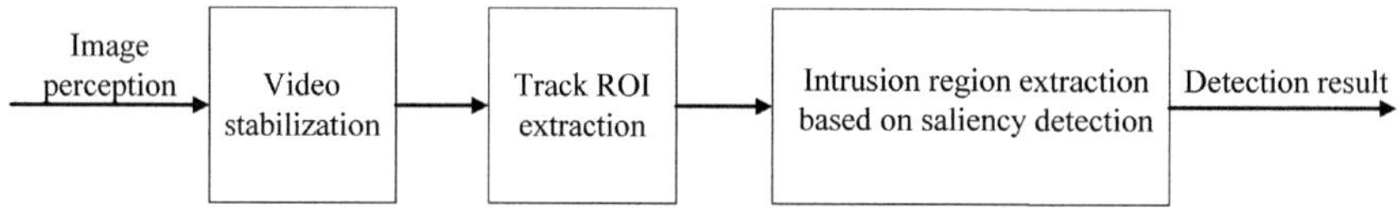

Figure 8.3
Flowchart of perception algorithm. *ROI*, region of interest.

(A) The real-time image stabilization of the collected video is finalized.
(B) The track region of interest is extracted from the perception image according to railway boundary regulations, to reduce unnecessary operations and speed the subsequent image processing rate.
(C) The intrusion is identified based on a track map.

accurately owing to the complex and changeable track environment and numerous interference factors. The process of perception algorithm is shown in Fig. 8.3.

8.4.1 Image stabilization

Image stabilization includes mechanical image and electronic image stabilization technology. Mechanical image stabilization technology uses gyro and other inertial sensors to measure the motion of the system and compensates for motion according to the motion parameter control system so that the camera can keep the visual axis stable and achieve image stabilization. Electronic image stabilization applies image processing methods to compensate for jitter. It can remodify, arrange, reduce, or eliminate distortions such as irregular translation, rotation, and scaling between image sequences. This makes the video more stable and suitable for processing operations such as target detection, target tracking, and recognition [90].

Electronic image stabilization algorithms can be divided into motion-based and nonmoving image stabilization algorithms. The motion-based image stabilization algorithm of the video is divided into global motion estimation, motion compensation, and image generation.

8.4.1.1 Global motion estimation

The methods of global motion estimation include the optical flow, feature point, and block matching methods, structure from motion, and so on. Motion compensation includes the filtering, curve fitting/interpolation, and optimization-based methods. Image generation involves graphics-related knowledge, such as image fusion, image mosaic, and image repair. The methods include 2D full-frame transformation and content-preserving warping methods.

There are two methods for global motion estimation: one considers only the motion of the camera in the 2D image plane and uses the motion model between adjacent frames to describe global motion and calculate the path of the camera in a 2D plane. The other considers the actual motion of the camera in space and calculates the motion trajectory of the camera to represent global motion. The image stabilization algorithm that considers only motion in the image plane is also called a 2D image stabilization algorithm, whereas the one that considers the actual motion trajectory of the camera is a 3D image stabilization algorithm. The global motion estimation of the 2D image stabilization algorithm can be divided into an optical flow method, feature point method, and block matching method.

The optical flow method first estimates the optical flow field according to the luminance gradient of the time-domain spatial image; then, it estimates the geometric transformation relation between adjacent frames according to optical flow using the least intercept square method [91]. When estimating optical flow, the current frame is divided into some nonoverlapping blocks to improve the efficiency of calculation. When the block is not large, the motion vector of all pixels in each block can be assumed to be the same, so that the motion of the central pixel of each block can be represented by the motion of the block. The optical flow method is not affected by the complex environment; its calculation is simple and effective. Its disadvantage is that it is sensitive to optical flow calculation error and it is difficult to select reliability evaluation factors.

The feature point method estimates the motion model matrix according to the corresponding relationship of feature points between adjacent frames. Commonly used image features include shape, texture, and color. It needs to extract a certain number of feature points in two consecutive frames and carry out feature point matching or tracking to find matching feature point pairs and obtain an interframe motion vector [92].

The feature point method can estimate motion with large displacement and has a good effect on video with violent shaking. Its main disadvantage is that accuracy of feature point selection and tracking cannot be guaranteed in some cases. The accuracy of feature point selection and matching largely determines the accuracy and robustness of global motion estimation. When a single feature cannot get a good effect, more color features or texture features can be considered to assist.

The block matching method divides the current frame into many blocks and assumes that all pixels within the block have the same motion vector. Then, it searches for the best matching block and obtains the current block of the corresponding position [93]. In this way, the global motion vector between adjacent frames can be estimated based on a matching criterion for the current frame. There are two commonly used matching standards: mean square error and mean absolute error. The block matching method is simple to implement, but it is easily affected by shadow, occlusion, and other effects in the matching process, which leads to a high mismatching rate and poor robustness.

Global motion estimation of the 3D image stabilization algorithm mainly depends on structure from motion (SfM), which refers to the estimation of 3D structure information from a 2D image sequence [94]. It needs to match the feature points of every two images first, and then make 3D reconstruction with the matched feature points. The computational complexity of the SfM is high, and video is required to contain enough motion information to reconstruct a 3D structure. When there is no parallax information in the video, or there is a zoom of the lens, it is difficult for the SfM to obtain an effective 3D structure. In addition, nonlinear effects in the video have a significant impact on SfM results.

8.4.1.2 Motion compensation

Motion compensation is the core of image stabilization of video. Motion compensation is an inverse problem with multiple solutions; it can be dealt with in various ways, including filtering, curve fitting/interpolation, and optimization-based methods.

The filtering method is a low-pass filter that filters the parameters of the global motion model or the global motion trajectory and retains only the low-frequency part. The principle is that subjective motion is usually slow and smooth, which corresponds to the low-frequency signal [95]. The vibration is sudden and violent, which is equivalent to high-frequency noise. In this way, subjective motion and vibration can be separated through the low-pass filter to achieve the effect of image stabilization. Common low-pass filters include the Gaussian low-pass filter and Kalman filter.

Curve fitting/interpolation obtains a smooth curve close to the original trajectory using fitting or interpolation through the original trajectory information, to remove high-frequency vibration [96].

The optimization-based method constructs the optimization problem about the smooth trajectory with certain criteria and then solves the problem to obtain a smooth trajectory or corresponding parameters [97].

8.4.1.3 Image generation

Image generation processes the original video according to subjective motion obtained in the motion compensation process to obtain a stable output video. Image generation methods include 2D full-frame transformation and content-preserving warping.

2D full-frame transformation is a relatively simple processing method. The main idea is to transform the original frame into a stable video frame through the transformation matrix obtained from motion compensation. Content-preserving warping is also a one-to-one transform, with the difference that it divides a single image into many small chunks and computes a transformation matrix for each chunk [98].

8.4.2 Extraction of region of interest

The UAV can automatically identify the rail position during the track line patrol and then automatically extract the region of interest (ROI) by referring to railway boundary regulations for subsequent processing of the ROI. On the one hand, it can provide support for the autonomous patrol control of the UAV. On the other hand, it can effectively reduce the amount of calculation and speed the identification and detection rate.

The steps of the algorithm are: (1) the input image is preprocessed to remove the influence of light and noise; (2) edge detection and extraction are carried out to obtain a track edge feature map. Based on the edge feature graph, probability Hough transform is used for straight-line fitting, and the straight-line length and slope are screened. According to the end points of the screened straight line segment, the rail position can be obtained by further fitting; (c) the Kalman filter is designed to track the slope and intercept of rail line in real time; and (d) the ROI in the perceptive image is extracted by referring to the railway boundary [99].

The track environment collected by the visual sensor carried by the UAV contains a lot of background information about the track, especially the ballast gravel area around the rail. Such unnecessary disturbance information will interfere with data processing and affect the efficiency of detection. Image preprocessing includes image graying, histogram equalization, and image smoothing.

Most color images adopt the RGB color mode. When images are processed, three components of RGB should be processed separately. However, in fact, the RGB mode cannot reflect the morphological characteristics of the image well, so it is necessary to conduct gray-scale processing on the image.

The essence of histogram equalization is to homogenize the part concentrated in the gray range of the original image: that is, to make the number of pixels in a certain gray range roughly the same through nonlinear transformation of the image. Histogram equalization can effectively increase the image contrast and eliminate the influence of illumination change to some extent.

Image smoothing is a practical digital image processing technology used to reduce and suppress image noise. Mean filtering is a typical linear filtering algorithm. Given a template of target pixels on an image, the template includes adjacent pixels around it. The average value of all pixels contained in the template is used to replace the original pixel value. Gaussian smoothing uses the Gaussian function filter to carry out the convolution operation on the image, to eliminate image noise [100].

According to the basic characteristics of the target edge detection algorithm, the boundary between the target and the background is extracted. The image characteristics in the area are discontinuous; in addition to the gray value changes, there can also be mutations such as color and texture structure, and there can also be boundaries between different areas. In actual image segmentation, only the first and second derivatives are used, and various operators are the instantiation calculation of this derivative segmentation principle. There are commonly used edge detection operators [101]:

(a) The Sobel operator is mainly used for edge detection. It is a discrete difference operator used to compute the image brightness function gradient approximation, which is typical of the Sobel operator edge detection operator based on the first derivative. The operator is introduced in a similar local average operation, so the noise has a smooth effect and can eliminate the influence of noise well.
(b) The Roberts operator is one of the simplest operators. It uses the local difference operator to find the edge. It uses the difference between two adjacent pixels in the diagonal direction to approximate the gradient amplitude to detect the edge. The effect of vertical edge detection is better than that of oblique edge, with high positioning accuracy and sensitivity to noise.
(c) The Prewitt operator is a kind of edge detection of the first-order differential operator. It uses the gray difference among the upper, lower, and left and right adjacent points of the pixel point to reach the extreme value at the edge to detect the edge and removes part of the false edge to smooth the noise. Its principle is to complete neighborhood convolution with the image using two direction templates in the image space, one of which detects the horizontal edge; the other detects the vertical edge.
(d) The Laplace operator is a kind of isotropic operator and second-order differential operator, which is more suitable when it cares only about the position of the edge without considering the difference in the gray value of pixels around it. The Laplace operator is more responsive to isolated pixels than to edges or lines, so it is applicable only to noiseless images. In the presence of noise, it is necessary to perform low-pass filtering before using a Laplace operator to detect edges.
(e) The Canny operator is complicated to implement. Before processing, the Canny operator uses a Gaussian smoothing filter to smooth the image to remove noise. The Canny segmentation algorithm uses first-order partial derivatives of finite difference to calculate the gradient amplitude and direction. In the process, the Canny operator will also go through a process of the maximum inhibition. Finally, the Canny operator is also used to connect two threshold edges.

After the edge image is obtained, the next step is line fitting. A standard Hough transform is a commonly used method of line fitting that is achieved by using the duality of point and line, essentially mapping the image to its parameter space. By using the probability Hough transform, line segments are detected in the edge graph, and then the end point sets of line segments at the rail position are obtained by further screening. Finally, the rail position is obtained by linear fitting according to these end point sets.

8.4.3 Saliency detection

The saliency detection method is draws on biological neuroscience and the cognitive psychology of the visual selective attention mechanism system [102], determining which can express the content of the image of human interest or can most express the area of image content or lead the attention to significant attributes of the target, such as an area with a different structure and color.

Early traditional saliency detection methods are usually based on the basic features of images, such as color, shape, and texture. These methods based on the human visual mechanism can extract artificially designed saliency features and use them to calculate the saliency of the input image and obtain a saliency map. In addition, some algorithms combined with a machine learning algorithm can design a top-down saliency detection algorithm to pursue a better detection effect [103]. Common algorithms include support vector machine, Bayesian network, and conditional random fields. With the continuous development of deep learning, various methods related to deep neural networks have achieved remarkable results in different fields. Owing to the strong learning ability of the deep network structure, the model has achieved certain success in the field of saliency detection. The typical saliency model can be divided into three layers: feature extraction, feature saliency graph representation, and global saliency graph representation [104].

After computation of the saliency, the extraction steps of the intrusion area are:

Step 1: the threshold value of the significant figure is calculated and the binary image is obtained;

Step 2: the connected domain and area is calculated and the small connected domain is eliminated;

Step 3: the morphological corrosion and expansion operations are conducted on the binary image;

Step 4: the area of each connected domain is sorted in the binary image;

Step 5: the connected domain is filtered from large to small and the small connected domain is removed. Each connected domain is fitted with a rectangle, and whether it is a rail or sleeper is determined according to the length−width ratio of the rectangle. If not, the region is regarded as an intrusion region.

8.4.4 Unmanned aerial vehicle intruding detection application in rail transit

The rail transit intruding detection system includes a monitoring center, inspection UAV, and image processing big data center. The monitoring center calculates the distance between the UAV and the train in real time according to the model of train, track line slope, and real-time speed, and sends it to the UAV. At the same time, the monitoring center issues warning information according to the information about intrusion from the image processing big data center. The UAV system includes a flight device and vision sensor array, ultrasonic anemometer, train speed measurement device, distance sensor, wireless communication module, and IMU/GPS module. The image processing big data center is used to process image information collected by the UAV, including image acquisition, image processing, and wireless communication modules.

Steps for the system to perform an intrusion detection mission are [105]:

Step 1: The rail is divided into equal intervals. Each track section is equipped with monitoring units, and each monitoring unit is equipped with a group of UAVs.

Step 2: When it is detected that a train is passing in the line area under the jurisdiction of a certain monitoring unit, the inspection mission should be performed. The monitoring center judges the coordinate of the mission position and calculates the power consumption coefficient ε_i of each UAV under the jurisdiction of two monitoring units adjacent to the mission position. Then it selects the monitoring unit to which the UAV belongs with the lowest power consumption coefficient. The power of UAV involved in calculating the power consumption coefficient should be greater than the power required to complete the inspection mission. The power consumption coefficient is calculated as [106]:

$$\begin{cases} \varepsilon_i = ln\left(\dfrac{\mathrm{I_i}}{\mathrm{I_{ti}}}\right) \\ \mathrm{I}_i = \phi_i \bullet S_i \end{cases} \tag{8.36}$$

in which ε_i is the power consumption coefficient of the UAV i; I_i and I_{ti} are, respectively, the power consumption and current remaining power of the UAV i to complete the mission; φ_i is the wind direction influence factor of the UAV i along the flight path direction; and S_i is the linear distance between the workstation of the UAV i and the mission position coordinates.

According to the real-time wind speed of the monitoring unit where the UAV is located, it is decomposed into the wind speed of the direction vector between the UAV and the task coordinate, and the wind speed influence factor φ is obtained from the prestored corresponding table of wind speed and wind direction influence factor. The predetermined wind speed impact factor measures the flying distance of the UAV under certain power consumption E at each wind speed. When the flying direction of UAV and the wind speed direction are on the same side, the wind direction influence factor is the downwind factor φ_{dw}. At different sides, the wind direction influence factor is the headwind factor φ_{hw}. These two factors are calculated as [106]:

$$\begin{cases} \phi_{dw} = ln\left(\dfrac{E}{\mathrm{S}_{dw}}\right) \\ \phi_{hw} = ln\left(\dfrac{E}{\mathrm{S}_{hw}}\right) \end{cases} \tag{8.37}$$

where E is power consumption during flight; S_{dw} and S_{hw} are, respectively, the distance traveled by the UAV under downwind and headwind; and φ_{dw} and φ_{hw} are, respectively, downwind factor and headwind factor, which means the power consumption of the UAV when flying per unit distance.

Furthermore, if the remaining power of the UAV under the jurisdiction of the two adjacent monitoring units cannot complete the task, the monitoring center divides the original task into multiple subtasks and sends subtask instructions to the monitoring unit. The dividing principle is to make the best use of the endurance ability of the UAV under the jurisdiction of the monitoring unit within the interval. After the task assignment is completed, the monitoring center will issue takeoff instructions to the UAV and send the real-time train position and speed to the UAV.

Step 3: According to the real-time position and speed of the train sent by the monitoring unit, the UAV flies to the front of the train and maintains a distance from the train, which is the braking distance of the train under emergency braking. When the UAV detects an intrusion, it will give an early warning and the train can stop in time to ensure its safety.

Step 4: After the UAV reaches the mission position, the image sensor is used to collect image information about the location and send it to the big data center for image processing through wireless transmission. Meanwhile, the flight speed is adjusted according to the train speed information sent by the monitoring center to keep pace with the real-time speed of the train.

Step 5: The image processing big data center analyzes the collected image information and determines whether there is a suspicious object in the current image through the intrusion object recognition algorithm described earlier. If a suspicious object is found, the process proceeds to Step 6; otherwise, Step 4 is repeated.

Step 6: The image processing big data center finds the intrusions and sends GPS information to the monitoring center, which issues the corresponding level of early warning information.

Step 7: When the measured train moves to the next rail section, the UAV carries out the task handover. The monitoring center issues instructions to the latest rail area where the train to be measured enters for task assignment, making the UAV in the rail area take off to track the train to be measured simultaneously. The original UAV enters the latest rail area to recharge and sends the real-time speed and position of the train to be measured to the monitoring center. According to the early warning information, the monitoring center reads the train information and sends an alarm, and provides the speed limit information according to the train operation control rules.

8.5 Conclusion and outlook

This chapter introduces the application of the UAV in the field of rail transit inspection. Compared with traditional manual inspection and fixed monitoring equipment at the crossing, the inspection UAV has huge advantages, such as a wide detection range, diversified information collection, flexibility, low risk, and high efficiency.

The whole UAV system includes a UAV subsystem, mission equipment subsystem, a measurement and control and information transmission subsystem, a command and control subsystem, and a launch and recovery subsystem. The basic structures of fixed-wing UAV and unmanned helicopter and rotary-wing UAV, as well as the auxiliary equipment of UAV for rail transit inspection are introduced in detail.

For the key technologies of the UAV system, the communication, data collection, and task scheduling systems are introduced. For the communication system, this chapter introduces the digital communication system based on the wireless channel used in the UAV communication system. The communication frequency band of the wireless communication channel and the digital modulation mode of the system data transmission and the communication protocol of the UAV communication system are introduced. Then, it introduces the key technologies of UAV communication, such as data link technology, mobile ad hoc network, and adaptive antenna array technology, as well as the widely used wireless communication standards such as Wi-Fi and 4G/5G networks.

The data collection system is an automatic test system that integrates UAV flight status monitoring, various ground test measurements, and fault diagnosis functions. The ACMS needs to satisfy the functions of status monitoring for each airborne subsystem, receiving and executing downlink instructions, real-time downlink monitoring data, forwarding monitoring data, cache recording airborne data, and antidamage recording airborne data, and provide ground maintenance interface.

The scheduling system includes location and navigation, task planning, and system management. The integrated navigation method of IMU/GPS and the visual navigation method of the UAV are introduced. For UAV path planning, several classical path planning algorithms and path planning methods based on reinforcement learning are introduced. The UAV scheduling solves the problem of UAV task allocation and optimizes the efficiency of UAV task execution. This chapter introduces three control methods of UAV task allocation: centralized, distributed, and hierarchical distributed.

Finally, the application of UAV in rail transit patrol inspection and the detection method of track foreign body intrusion based on the image are introduced. After collecting images of the rail area, the UAV completes real-time image stabilization of the images. Then, it preprocesses the images, extracts the ROIs, and finally discriminates intruding according to the saliency of the target.

The UAV can use sophisticated technologies such as artificial intelligence, signal processing, and autopilot, and it has been widely applied owing to its advantages of small size, unmanned piloting, and long range. It has been applied in many aspects such as natural environment investigation, scientific research, agriculture, national sovereignty, and public security. This kind of diversified application makes all countries accelerate the pace of exploration and development of the UAV.

In the field of UAV manufacturing, the hardware of the UAV will continue to improve. Better battery and energy management techniques should be developed to extend the hang time, and more efficient power systems should be developed to allow for larger payloads. For example, solar panels that can recharge batteries during flight and hydrogen fuel cells that allow for longer flight times are two important development directions.

Materials such as carbon fiber—reinforced polymers, combined with advanced manufacturing technology, will make drones lighter, cheaper, and potentially easier to manufacture. The improved sensor of UAV reduces the load demand and improves the capability of UAV.

As an advanced autonomous unmanned system, the UAV is developing toward the direction of low manual intervention, high autonomy, and high intelligence. The dynamic uncertainty of the environment and the complexity of tasks also require the development of the UAV cluster.

Autonomous control of the UAV is an important basis for the formation and cooperation of multiple UAVs. The main challenge of autonomous control of UAVs is to design an appropriate decision mechanism and corresponding algorithms. The autonomous control system needs timely information about its internal state and external world, accurate estimation of its available resources, and prompt decision-making. For autonomous path planning, it is important to obtain an effective algorithm for online generation and execution of a motion plan.

The UAV cluster system can show more outstanding coordination, intelligence, and autonomous ability than the artificial system. Autonomous cluster of UAV has: (1) the capacity to solve the conflict between multiple UAVs in a limited space; (2) a low-cost and highly decentralized form to meet functional requirements; (3) a dynamic self-healing network; (4) distributed cluster intelligence; and (5) a distributed detection method. The UAV autonomous cluster system is being developed for manned aircraft, new variant UAVs, and an intelligent cluster system.

References

[1] I.-H. Kim, H. Jeon, S.-C. Baek, et al., Application of crack identification techniques for an aging concrete bridge inspection using an unmanned aerial vehicle, Sensors 18 (2018) 1881–1895.

[2] C. Zhao, C. Zhang, S. Yang, et al., Calculating e-flow using UAV and ground monitoring, J. Hydrol. 552 (2017) 351–365.

[3] L. Li, The UAV intelligent inspection of transmission lines, in: International Conference on Advances in Mechanical Engineering and Industrial Informatics, 2015, pp. 1542–1545.

[4] C. Deng, S. Wang, Z. Huang, et al., Unmanned aerial vehicles for power line inspection: a cooperative way in platforms and communications, J. Commun. 9 (2014) 687–692.

[5] S. D'oleire-Oltmanns, I. Marzolff, K. Peter, et al., Unmanned aerial vehicle (UAV) for monitoring soil erosion in Morocco, Rem. Sens. 4 (2012) 3390–3416.

[6] G. Pajares, Overview and current status of remote sensing applications based on unmanned aerial vehicles (UAVs), Photogramm. Eng. Rem. Sens. 81 (2015) 281–330.

[7] Z.Y. Yin, B.B. Fu, T.B. Xue, et al., Development of helicopter power transmission system technology, Appl. Mech. Mater. 86 (2011) 1–17.

[8] Y. Kang, B.-J. Park, A. Cho, et al., Development of flight control system and troubleshooting on flight test of a tilt-rotor unmanned aerial vehicle, Int. J. Aeronaut. Space Sc. 17 (2016) 120–131.

[9] J. Huang, W. Cai, UAV low altitude marine monitoring system, in: International Conference on Wireless Communication and Sensor Network, 2014, pp. 61–64, 2014.

[10] J.-H. Kim, S. Sukkarieh, S. Wishart, Real-time navigation, guidance, and control of a UAV using low-cost sensors, in: The 4th Conference on Field and Service Robotics, 2003, pp. 299–309.

[11] K.H. G Hasan, R. Ahsan, et al., Evaluation of a low-cost MEMS IMU for indoor positioning system, Inert. J. Emerg. Sci. Eng. 1 (2013) 70–77.

[12] S. Jawla, R. Singh, Different modulation formats used in optical communication system, IOSR J. Electron. Commun. Eng. 8 (2013) 15–18.

[13] J. Chen, Q. Fei, Q. Geng, The design of data link for multi-UAVs, in: 4th International Conference on Intelligent Human-Machine Systems and Cybernetics, vol. 1, 2012, pp. 106–109, 2012.

[14] P. Goyal, V. Parmar, R. Rishi, Manet: vulnerabilities, challenges, attacks, application, Int. J. Comput. Eng. Manag. 11 (2011) 32–37.

[15] S. Chandran, Adaptive Antenna Arrays: Trends and Applications, Springer Science & Business Media, 2013.

[16] J.J. Deva Gifty, K. Sumathi, IEEE 802 standard network's comparison under grid and random node arrangement in 2.4 GHz ISM band for single and multiple CBR traffic, Int. J. Adv. Netw. Appl. 8 (2017) 3118–3123.

[17] J. Romeu, A. Albert, J. Alonso, et al., Small UAV radiocommunication channel characterization, in: The Fourth European Conference on Antennas and Propagation, 2010, pp. 1–5.

[18] N. Neji, T. Mostfa, Communication technology for unmanned aerial vehicles: a qualitative assessment and application to precision agriculture, in: International Conference on Unmanned Aircraft Systems, 2019, pp. 848–855, 2019.

[19] T. Schwengler, M. Glbert, Propagation models at 5.8 GHz-path loss and building penetration, in: IEEE Radio and Wireless Conference, 2000, pp. 119–124, 2000.

[20] A.B. Ibrahim, A. Abas, Data transmission analysis using MW-5000 at 5.8 GHz frequency, Int. J. Electr. Comput. Eng. 8 (2018) 254–258.

[21] J. Lansford, A. Stephens, R. Nevo, Wi-Fi (802.11 b) and Bluetooth: enabling coexistence, IEEE network 15 (2001) 20–27.

[22] P. Gautam, S. Kaur, R. Kaur, et al., Review paper on 4G wireless technology, Int. J. Adv. Sci. Technol. 2 (2014) 15–19.

[23] B. Li, Z. Fei, Y. Zhang, UAV communications for 5G and beyond: recent advances and future trends, IEEE Internet Things J. 6 (2018) 2241–2263.

[24] D. Chen, X. Wang, J. Zhao, Aircraft maintenance decision system based on real-time condition monitoring, Proc. Eng. 29 (2012) 765–769.
[25] E. Ortiz, A. Babbar, V.L. Syrmos, Extreme value theory for engine health monitoring and diagnosis, in: IEEE Control Applications,(CCA) & Intelligent Control,(ISIC), 2009, pp. 1069–1074, 2009.
[26] M.C. Dani, C. Freixo, F.-X. Jollois, et al., Unsupervised anomaly detection for aircraft condition monitoring system, IEEE Aerosp. Conf. (2015) 1–7, 2015.
[27] L.R. García Carrillo, A.E. Dzul López, R. Lozano, et al., Combining stereo vision and inertial navigation system for a quad-rotor UAV, J. Intell. Rob. Syst. 65 (2012) 373–387.
[28] J. Wang, M. Garratt, A. Lambert, et al., Integration of GPS/INS/vision sensors to navigate unmanned aerial vehicles, Int. Arch. Photogram. Rem. Sens. Spatial Inf. Sci. 37 (2008) 963–969.
[29] H. Simon, Kalman Filtering and Neural Networks, John Wiley & Sons, 2004.
[30] L. Zhao, W.Y. Ochieng, M.A. Quddus, et al., An extended Kalman filter algorithm for integrating GPS and low cost dead reckoning system data for vehicle performance and emissions monitoring, J. Navig. 56 (2003) 257–275.
[31] E.A. Wan, Rudolph Van Der Merwe, The unscented Kalman filter for nonlinear estimation, in: IEEE 2000 Adaptive Systems for Signal Processing, Communications, and Control Symposium, 2000, pp. 153–158.
[32] C. Fritsche, A. Klein, D. Wurtz, Hybrid GPS/GSM localization of mobile terminals using the extended Kalman filter, in: 6th Workshop on Positioning, Navigation and Communication, 2009, pp. 189–194, 2009.
[33] M.O. Aqel, M.H. Marhaban, M. Iqbal Saripan, et al., Review of visual odometry: types, approaches, challenges, and applications, SpringerPlus 5 (1897), 2016.
[34] Y. Li, X. Zhu, H. Lu, et al., Review on visual odometry technology, Appl. Res. Comput. 29 (2012) 2801–2805.
[35] W. Ding, De Xu, X. Liu, et al., Review on visual odometry for mobile robots, Acta Autom. Sin. 44 (2018) 385–400.
[36] H. Lategahn, A. Geiger, B. Kitt, Visual SLAM for autonomous ground vehicles, in: IEEE International Conference on Robotics and Automation, 2011, pp. 1732–1737, 2011.
[37] P. Gao, W. Zhao, J. Shen, Detection of crowd state mutation based onbackground difference algorithm and optical flow algorithm, J. Zhejiang Univ. 52 (2018) 649–656.
[38] A.D. Wu, E.N. Johnson, A.A. Proctor, Vision-aided inertial navigation for flight control, J. Aero. Comput. Inf. Commun. 2 (2005) 348–360.
[39] W. Kong, G. Egan, T. Cornall, Feature based navigation for UAVs, in: IEEE/RSJ International Conference on Intelligent Robots and Systems, 2006, pp. 3539–3543.
[40] F. Caballero, L. Merino, J. Ferruz, et al., Vision-based odometry and SLAM for medium and high altitude flying UAVs, J. Intell. Rob. Syst. 54 (2009) 137–161.
[41] A. Loquercio, A. I Maqueda, C.R. Del-Blanco, et al., Dronet: learning to fly by driving, IEEE Robot. Autom. Lett. 3 (2018) 1088–1095.
[42] D.G. Lowe, Object recognition from local scale-invariant features, in: IEEE International Conference on Computer Vision, vol. 99, 1999, pp. 1150–1157.
[43] D.G. Lowe, Distinctive image features from scale-invariant keypoints, Int. J. Comput. Vis. 60 (2004) 91–110.
[44] B. Herbert, T. Tuytelaars, L. Van Gool, Surf: Speeded up robust features, in: European Conference on Computer Vision, 2006, pp. 404–417.
[45] P. Panchal, S. Panchal, S. Shah, A comparison of SIFT and SURF, Int. J. Innov. Res. Comput. Commun. Eng. 1 (2013) 323–327.
[46] E. Rublee, R. Vincent, K. Kurt, et al., ORB: an efficient alternative to SIFT or SURF, in: IEEE International Conference on Computer Vision, vol. 11, 2011, pp. 2564–2571.
[47] B. Babenko, M.-H. Yang, S. Belongie, Visual tracking with online multiple instance learning, in: IEEE Conference on Computer Vision and Pattern Recognition, 2009, pp. 983–990.

[48] S. Hare, S. Golodetz, A. Saffari, et al., Struck: structured output tracking with kernels, IEEE Trans. Pattern Anal. Mach. Intell. 38 (2015) 2096–2109.

[49] J. Zhang, S. Ma, S. Sclaroff, MEEM: robust tracking via multiple experts using entropy minimization, in: European Conference on Computer Vision, 2014, pp. 188–203.

[50] J.F. Henriques, C. Rui, P. Martins, et al., High-speed tracking with kernelized correlation filters, IEEE Trans. Pattern Anal. Mach. Intell. 37 (2014) 583–596.

[51] M. Danelljan, G. Hager, F.S. Khan, et al., Learning spatially regularized correlation filters for visual tracking, in: IEEE International Conference on Computer Vision, 2015, pp. 4310–4318.

[52] N. Wang, D.-Y. Yeung, Learning a deep compact image representation for visual tracking, in: Advances in Neural Information Processing Systems 26 (NIPS 2013), 2013, pp. 809–817.

[53] Y. Qi, S. Zhang, L. Qin, et al., Hedged deep tracking, in: IEEE Conference on Computer Vision and Pattern Recognition, 2016, pp. 4303–4311.

[54] M. Danelljan, A. Robinson, F.S. Khan, et al., Beyond correlation filters: learning continuous convolution operators for visual tracking, in: European Conference on Computer Vision, 2016, pp. 472–488.

[55] M. Danelljan, G. Bhat, F.S. Khan, et al., Eco: efficient convolution operators for tracking, in: IEEE Conference on Computer Vision and Pattern Recognition, 2017, pp. 6638–6646.

[56] D. Zhu, M. Yan, Survey on technology of mobile robot path planning, Control Decis. 25 (2010) 961–967.

[57] O. Khatib, Real-time Obstacle Avoidance for Manipulators and Mobile Robots. Autonomous Robot Vehicles, Springer, 1986, pp. 396–404.

[58] H.E. Romeijn, R.L. Smith, Simulated annealing for constrained global optimization, J. Global Optim. 5 (1994) 101–126.

[59] I.H. Osman, Metastrategy simulated annealing and tabu search algorithms for the vehicle routing problem, Ann. Oper. Res. 41 (1993) 421–451.

[60] F. Khosravi Purian, E. Sadeghian, Mobile robots path planning using ant colony optimization and Fuzzy Logic algorithms in unknown dynamic environments, in: International Conference on Control, Automation, Robotics and Embedded Systems, 2013, pp. 1–6, 2013.

[61] M. Ma, A.S. Morris, Path planning in unknown environment with obstacles using virtual window, J. Intell. Rob. Syst. 24 (1999) 235–251.

[62] P. Raja, S. Pugazhenthi, Optimal path planning of mobile robots: a review, Int. J. Phys. Sci. 7 (2012) 1314–1320.

[63] R.A. Brooks, Solving the find-path problem by good representation of free space, IEEE Trans. Syst., Man, Cybern. 13 (1983) 190–197.

[64] S. Garrido, L. Moreno, A. Mohamed, et al., Path planning for mobile robot navigation using voronoi diagram and fast marching, in: IEEE/RSJ International Conference on Intelligent Robots and Systems, 2006, pp. 2376–2381.

[65] S. Mazzeo, I. Loiseau, An ant colony algorithm for the capacitated vehicle routing, Electron. Notes Discrete Math. 18 (2004) 181–186.

[66] Le Zarate, M. Becker, B. Garrido, et al., An artificial neural network structure able to obstacle avoidance behavior used in mobile robots, in: IEEE Annual Conference of the Industrial Electronics Society, vol. 3, 2002, pp. 2457–2461.

[67] J. Tu, S.X. Yang, Genetic algorithm based path planning for a mobile robot, in: IEEE International Conference on Robotics and Automation, vol. 1, 2003, pp. 1221–1226.

[68] Y. Shi, Particle swarm optimization: developments, applications and resources, in: IEEE Congress on Evolutionary Computation, vol. 1, 2001, pp. 81–86.

[69] H. Wang, Y. Yuan, Q. Yuan, Application of Dijkstra algorithm in robot path-planning, in: Second International Conference on Mechanic Automation and Control Engineering, 2011, pp. 1067–1069, 2011.

[70] G. Nannicini, D. Delling, L. Liberti, et al., Bidirectional A* search for time-dependent fast paths, in: International Workshop on Experimental and Efficient Algorithms, 2008, pp. 334–346.

[71] Y. Shen, FallBack routing algorithm based on hierarchical shortest path, Comput. Digital Eng. 4 (2007) 23–26.
[72] W. Shi, K. Wang, Floyd algorithm for the shortest path planning of mobile robot, Chin. J. Sci. Instrum. 10 (2009) 2088–2092.
[73] R.S. Sutton, A.G. Barto, Reinforcement Learning: An Introduction, MIT press, 2018.
[74] M. Van Otterlo, W. Marco, Reinforcement Learning and Markov Decision Processes. Reinforcement Learning, Springer, 2012, pp. 3–42.
[75] A.T. Bharucha-Reid, Elements of the Theory of Markov Processes and Their Applications, Dover Publications, 2012.
[76] R. Givan, S. Leach, T. Dean, Bounded-parameter Markov decision processes, Artif. Intell. 122 (2000) 71–109.
[77] M.L. Littman, Value-function reinforcement learning in Markov games, Cognit. Syst. Res. 2 (2001) 55–66.
[78] D.P. Bertsekas, Dynamic Programming and Optimal Control. Athena Scientific, 1995.
[79] H.R. Maei, Gradient Temporal-Difference Learning Algorithms, University of Alberta, 2011.
[80] Q. Wang, Z. Zhan, Reinforcement learning model, algorithms and its application, in: International Conference on Mechatronic Science, Electric Engineering and Computer, 2011, pp. 1143–1146, 2011.
[81] X. Yang, W. Zhou, Y. Zhang, On collaborative path planning for multiple UAVs based on pythagorean hodograph curve, in: IEEE Chinese Guidance, Navigation and Control Conference, 2016, pp. 971–975, 2016.
[82] H.-B. Duan, X.-Y. Zhang, J. Wu, et al., Max-min adaptive ant colony optimization approach to multi-UAVs coordinated trajectory replanning in dynamic and uncertain environments, J. Bionic Eng. 6 (2009) 161–173.
[83] W. Wu, N. Cui, J. Guo, Distributed task assignment method based on local information consensus and target estimation, Control Theory Appl. 35 (2018) 566–576.
[84] W. Wu, N. Cui, W. Shan, et al., Distributed task allocation for multiple heterogeneous UAVs based on consensus algorithm and online cooperative strategy, Aircraft Eng. Aero. Technol. 90 (2018) 1464–1473.
[85] S. Rasmussen, P. Chandler, J. Mitchell, et al., Optimal vs. heuristic assignment of cooperative autonomous unmanned air vehicles, in: AIAA Guidance, Navigation, and Control Conference and Exhibit vol. 5586, 2003.
[86] M. Alighanbari, Task Assignment Algorithms for Teams of UAVs in Dynamic Environments, Massachusetts Institute of Technology, 2004.
[87] X. Chen, Y. Qiao, Summary of unmanned aerial vehicle task allocation, J. Shenyang Aerosp. Univ. 33 (2016) 1–7.
[88] R.G. Smith, The Contract Net Protocol: High-Level Communication and Control in a Distributed Problem Solver. Readings in Distributed Artificial Intelligence, Elsevier, 1988, pp. 357–366.
[89] J. Zong, M. Liao, Dynamic target assignment method based on multi-agent decentralized cooperative auction, J. Beijing Univ. Aeronaut. Astronaut. 33 (2007) 180–184.
[90] P. Rawat, J. Singhai, Review of motion estimation and video stabilization techniques for hand held mobile video, Signal Image Process.: Int. J. 2 (2011) 159–168.
[91] Michael J. Smith, Boxerbaum Alexander, Gilbert L. Peterson, et al., Electronic image stabilization using optical flow with inertial fusion, in: IEEE/RSJ International Conference on Intelligent Robots and Systems, 2010, pp. 1146–1153, 2010.
[92] S.-K. Kim, S.-J. Kang, T.-S. Wang, et al., Feature point classification based global motion estimation for video stabilization, IEEE Trans. Consum. Electron. 59 (2013) 267–272.
[93] A. Barjatya, Block matching algorithms for motion estimation, IEEE Trans. Evol. Comput. 8 (2004) 225–239.
[94] Z. Cui, P. Tan, Global structure-from-motion by similarity averaging, in: IEEE International Conference on Computer Vision, 2015, pp. 864–872.

[95] A. Litvin, J. Konrad, W. Clement Karl, Probabilistic video stabilization using Kalman filtering and mosaicing, in: Image and Video Communications and Processing, vol. 5022, 2003, pp. 663–674, 2003.
[96] M.L. Gleicher, F. Liu, Re-cinematography: improving the camera dynamics of casual video, in: The 15th ACM International Conference on Multimedia, 2007, pp. 27–36.
[97] S. Liu, Y. Lu, P. Tan, et al., Bundled camera paths for video stabilization, ACM Trans. Graph. 32 (2013) 1–10.
[98] F. Liu, M. Gleicher, H. Jin, et al., Content-preserving warps for 3D video stabilization, ACM Trans. Graph. 28 (2009) 1–9.
[99] K. Vu, A.H. Kien, W. Tavanapong, Image retrieval based on regions of interest, IEEE Trans. Knowl. Data Eng. 15 (2003) 1045–1049.
[100] J.-S. Lee, Digital image smoothing and the sigma filter, Comput. Vis. Graph Image Process 24 (1983) 255–269.
[101] G. Shrivakshan, C. Chandrasekar, A comparison of various edge detection techniques used in image processing, Int. J. Comput. Sci. Issues 9 (2012) 269–276.
[102] Q. Yan, L. Xu, J. Shi, et al., Hierarchical saliency detection, in: The IEEE Conference on Computer Vision and Pattern Recognition, 2013, pp. 1155–1162.
[103] R. Zhao, W. Ouyang, H. Li, et al., Saliency detection by multi-context deep learning, in: The IEEE Conference on Computer Vision and Pattern Recognition, 2015, pp. 1265–1274.
[104] J. Harel, C. Koch, P. Perona, Graph-based visual saliency, in: Advances in Neural Information Processing Systems 20 (NIPS 2007), 2007, pp. 545–552.
[105] H. Liu, Y. Li, R. Xiong, Unmanned Aerial Vehicle Intelligent Recognition and Early Warning Method and System for Foreign Body Beyond-Limit Detection along Railway, National Intellectual Property Administration, China, 2018.
[106] H. Liu, Y. Li, C. Li, Method and Control System for Measuring Wind Speed along Railway Based on Unmanned Aerial Vehicle Fleet Intelligent Endurance Control, National Intellectual Property Administration, China, 2017.

Index

Note: 'Page numbers followed by "f" indicate figures and "t" indicates tables'.

C

H

I

K

L

Q

R

S

T

U

V

W

Z

9780128229682